Social and Affective Neuroscience of Gaming

Lucas Murrins Marques
Ricardo Riyoiti Uchida • Pedro Makoto Uchida
Victor Henrique Oyamada Otani
Joao Ricardo Nickenig Vissoci
Editors

Social and Affective Neuroscience of Gaming

Insights from Fundamentals to Public Mental Health Applications

Springer

Editors
Lucas Murrins Marques
Mental Health Department
Santa Casa de São Paulo School
of Medical Sciences
Sao Paulo, Brazil

Artificial Intelligence Research Division
Infinity Doctor's Inc.
Dover, DE, USA

Pedro Makoto Uchida
Pontifícia Universidade Católica de
Campinas Campinas
Sao Paulo, Brazil

Joao Ricardo Nickenig Vissoci
Division of Emergency Medicine
Duke Global Health Institute
Durham, NC, USA

Ricardo Riyoiti Uchida
Mental Health Department
Santa Casa de São Paulo School
of Medical Sciences
Sao Paulo, Brazil

Victor Henrique Oyamada Otani
Mental Health Department
Santa Casa de São Paulo School
of Medical Sciences
Sao Paulo, Brazil

ISBN 978-3-032-00324-9 ISBN 978-3-032-00325-6 (eBook)
https://doi.org/10.1007/978-3-032-00325-6

This Springer imprint is published by the registered company Springer Nature Switzerland AG
The registered company address is: Gewerbestrasse 11, 6330 Cham, Switzerland

Foreword: "Gaming, Neuroscience, and the Future of Mental Health Research"

Over the past two decades, I have witnessed firsthand how our understanding of the human brain has been transformed by interdisciplinary efforts bridging neuroscience, psychology, and public health. Yet, even amidst these advances, there remain emerging frontiers—new cultural and behavioral landscapes that challenge our existing paradigms and invite bold, integrative exploration. The world of gaming is one such frontier. Once regarded as a marginal or purely recreational activity, digital gaming is now deeply woven into the fabric of contemporary life, shaping how people engage with themselves, with others, and with the broader world. This book, *Social and Affective Neuroscience of Gaming*, enters this space with the depth and clarity it demands.

What makes this volume particularly noteworthy is its commitment to scientific rigor without losing sight of cultural relevance and translational potential. The chapters compiled here examine gaming not as an isolated behavior, but as a dynamic, multifaceted experience—one that engages core brain systems, modulates affect, fosters identity exploration, and catalyzes social connection. In doing so, the contributors offer a compelling framework for understanding gaming as both a psychological phenomenon and a potential tool for intervention, especially in the context of mental health promotion and global well-being.

From a public health perspective, this volume is timely and necessary. As societies grapple with rising rates of mental health disorders, loneliness, and stress-related conditions, we must expand our toolkit for prevention, resilience-building, and emotional regulation. Games—particularly those designed with empathy, interactivity, and immersion in mind—can serve not only as mirrors of human cognition and emotion but as potential modulators of them. The evidence presented here makes a persuasive case: when thoughtfully designed and responsibly implemented, games may offer low-cost, scalable, and culturally adaptable means to support mental health across the lifespan.

What also impressed me in this work is its methodological diversity. The integration of neuroimaging, behavioral analysis, clinical trials, and educational strategies reflects a mature field in motion—one that values complexity, avoids reductionism, and embraces interdisciplinary synthesis. The use of multimodal approaches mirrors what we strive for in translational neuroscience: research that is not only scientifically sound but socially meaningful and clinically applicable.

Several chapters in this book also touch on themes central to my own work: the need to understand brain plasticity in real-world settings, the importance of emotional regulation as a therapeutic target, and the role of innovation in expanding access to care. I was particularly drawn to the sections that examine gaming among youth, neurodiverse populations, and communities historically underserved by conventional health systems. These are precisely the groups that require fresh, adaptive models of engagement—models that recognize their lived experiences and speak their cultural languages. Digital games, when used wisely, can offer just that.

This book is not merely descriptive—it is a call to action. It challenges researchers, clinicians, educators, and developers to think beyond disciplinary boundaries and reimagine what tools like games can mean for health, education, and community. As someone committed to advancing the science of healing through innovation, I find this vision both inspiring and essential.

I commend the editors and contributors for their thoughtful, evidence-based contributions to this rapidly evolving field. Their work will undoubtedly serve as a reference point for future research and application. Most importantly, it will spark necessary conversations about how we can harness human creativity—not only to entertain but to understand, to connect, and to heal.

Felipe Fregni
e-mail: ffregni@post.harvard.edu

Professor of Physical Medicine & Rehabilitation,
Harvard Medical School,
Boston, MA, USA

Professor of Epidemiology, Harvard T.H. Chan
School of Public Health,
Boston, MA, USA

Research Center Director, Spaulding
Neuromodulation Center, Department
of Physical Medicine & Rehabilitation,
Spaulding Rehabilitation Hospital
and Massachusetts General Hospital,
Boston, MA, USA

Program Director, Clinical Research
Collaborative Learning Training
Program – Principles and Practice
of Clinical Research (PPCR), ECPE,
Harvard T.H. Chan School of Public Health,
Boston, MA, USA

Preface

The idea for this book emerged from a growing recognition of the need to bridge neuroscience, psychology, and public health with the immersive and expanding universe of gaming. In recent decades, digital games have become deeply integrated into contemporary life—not merely as sources of entertainment but as environments where people learn, connect, cope with adversity, explore identities, and even seek emotional refuge. This transformation demands a more nuanced, interdisciplinary understanding of how gaming interfaces with brain function, affective experience, social behavior, and mental health outcomes.

Social and Affective Neuroscience of Gaming: Insights from Fundamentals to Public Mental Health Applications brings together a diverse group of scholars, clinicians, students, and game developers, all united by a shared interest in the interplay between gaming and the human mind. The book is structured into five interconnected parts: from the neurobiological fundamentals of gaming to its implications in development, mental health, social connectedness, and public health strategies. Across these sections, we explore a wide range of themes—from dopaminergic circuits and multisensory integration to online social dynamics, empathy development, therapeutic potentials, and ethical responsibilities within the industry.

Many chapters were born from mentorship-based research, graduate training programs, or collaborative projects in Brazilian institutions and international partnerships. Several contributors are themselves game developers or work closely with the gaming industry, bringing practical expertise to complement theoretical and empirical perspectives. This dual perspective enriches the book with insights that are both academically rigorous and grounded in real-world design and gameplay experience.

Importantly, this volume is not only about understanding games, it is also about guiding their thoughtful use and design. For researchers and educators, the book offers a conceptual and methodological framework to explore games as tools for cognitive development, emotional regulation, and social interaction. For clinicians and mental health professionals, it provides new perspectives on the therapeutic potential of gaming. And for game developers, the chapters offer scientific grounding to inspire designs that are engaging, inclusive, emotionally intelligent, and mindful of users' well-being.

We hope this book contributes to a growing dialogue between science, society, and creative industries. Games are no longer peripheral to human experience—they are at its center, shaping the ways we play, relate, express, and grow. Understanding this influence is essential for advancing research, promoting health, and designing experiences that resonate not only cognitively but affectively and socially.

We are deeply grateful to the authors who contributed their time, vision, and expertise to this project, and to the institutions and research programs that made their work possible. We also extend our appreciation to the professionals in the gaming industry who collaborated or inspired this work with their design sensibilities and commitment to building better, more meaningful digital worlds. May this book serve as a bridge—between disciplines, between communities, and between ideas and action.

Sao Paulo, Brazil — Lucas Murrins Marques
Sao Paulo, Brazil — Ricardo Riyoiti Uchida
Sao Paulo, Brazil — Pedro Makoto Uchida
Sao Paulo, Brazil — Victor Henrique Oyamada Otani
Durham, NC, USA — Joao Ricardo Nickenig Vissoci

Acknowledgements

Chapter 1 EDP is supported by a scholarship from CEBRASPE/CARREFOUR. AYAO is supported by São Paulo Research Foundation (FAPESP, #24/20442-2). KL is support by São Paulo Research Foundation (FAPESP, #22/03206-8). JRS is supported by São Paulo Research Foundation (FAPESP, #21/05332-8; 24/09675-5; and 23/16997-6) and Allan Faconi-Souto for the drawings that served as illustrations.

Chapter 10 AA is supported by São Paulo Research Foundation (FAPESP, #24/09587-9). JV is supported by São Paulo Research Foundation (FAPESP, #24/09578-0). MH is supported by São Paulo Research Foundation (FAPESP, #24/22303-0).

Chapter 12 GGR is supported by CNPq—INCT (National Institute of Science and Technology on Social and Affective Neuroscience, grant n. 406463/2022-0).

About the Book

This book seeks to address key aspects of the relationship between gaming and social and affective neuroscience, bringing together interdisciplinary contributions that examine how games influence emotional, cognitive, behavioral, and social processes. The chapters are organized to progressively guide the reader through neurobiological foundations, emotional design, community dynamics, therapeutic applications, and broader cultural and ethical considerations.

The first part of the book explores how games activate and interact with core brain systems, particularly those involved in reward, motivation, and emotional processing. Topics include dopaminergic and limbic engagement, the modulation of stress and mood, and the mechanisms through which games promote emotional regulation and resilience. By delving into how different types of gameplay impact brain function and emotional states, this section sets the foundation for understanding the immersive power of games.

Subsequent chapters examine the role of emotional design and game mechanics in shaping player experiences. Rather than focusing solely on narrative and gameplay structure, these chapters emphasize multisensory stimulation, aesthetic immersion, and interface responsiveness. The authors discuss how colors, music, character development, and narrative arcs influence emotional responses and facilitate identification, empathy, and narrative engagement, contributing to both short-term arousal and long-term emotional memory.

The book then turns to the social domain, analyzing how games foster connection, collaboration, and community formation. It examines the dynamics of online gaming environments, the psychological and emotional bonds formed between players and non-playable characters (NPCs), and the significance of virtual communities as spaces of belonging, exploration, and mutual support. This section also considers how gaming enables identity construction, social learning, and emotional co-regulation.

Another important dimension addressed in this volume is the clinical and therapeutic application of gaming. Specific chapters explore how games can be harnessed as tools for emotional support, social skills training, and mental health intervention. Case examples include the use of video games for individuals on the autism spectrum, adolescents with depression, people recovering from trauma, and transgender youth exploring their identities in virtual environments. These discussions highlight the potential of games to transcend entertainment, offering meaningful experiences that support psychological and emotional development.

Finally, the book addresses ethical, public health, and cultural issues linked to gaming. It explores the responsibilities of the gaming industry in designing emotionally engaging yet psychologically safe experiences, the risks associated with exploitative mechanics like loot boxes and excessive screen time, and the need for inclusive narratives and representation. It also investigates how games can function as instruments of cultural exchange and symbolic meaning-making, shaping not only entertainment markets but also identities, communities, and values.

This book is intended for researchers, graduate students, clinicians, educators, digital health professionals, public health policymakers, and game developers. By combining empirical research, theoretical insight, and applied perspectives, it provides a comprehensive, evidence-based framework for understanding the emotional, cognitive, and social dimensions of gaming and its relevance for mental health, cultural expression, and human development.

Contents

Editors and Contributors

About the Editors

Lucas Murrins Marques is an Instructor Professor at the Mental Health Department of the Santa Casa de São Paulo School of Medical Sciences (DSM-FCMSCSP), where he also supervises graduate students in the Stricto Sensu Postgraduate Program in Health Sciences. He is also an advisor in the Graduate Program in Medical Sciences at the University of São Paulo School of Medicine (FMUSP). In addition to his academic roles, he serves as Head of Digital Health Research at Infinity Doctors Inc., an international digital health and telemedicine. He holds a bachelor's degree in Psychology from Mackenzie Presbyterian University (UPM, 2014), including a sandwich program at the University of Minho in Portugal through the Science Without Borders program (CNPq). He earned both his master's and Ph.D. in Developmental Disorders from UPM (2016 and 2020), supported by FAPESP scholarships. From 2019 to 2021, he worked as a Senior Researcher in Wellbeing Sciences at Natura & Co. Between 2021 and 2024, he was a postdoctoral fellow in Neurophysiology at the Institute of Physical Medicine and Rehabilitation (IMREA), University of São Paulo School of Medicine (HCFMUSP), also funded by FAPESP. In 2023, he completed a certification in Principles and Practices of Clinical Research at the Harvard T.H. Chan School of Public Health. He is also a member of the editorial boards of the *International Journal of Clinical and Health Psychology* (IJCHP), *Frontiers in Cognition*, and *Frontiers in Psychology*. His main areas of interest include Well-being, Mental Health, Neurophysiology, Public and Global Health, Biomarkers, Epidemiology, Preventive Medicine, Complexity Science, Data Science, and Machine Learning.

Ricardo Riyoiti Uchida is an Associate Professor and Head in the Mental Health Department at the Santa Casa de São Paulo School of Medical Sciences (DSM-FCSMSCSP). He is a mentor in the Stricto Sensu Postgraduate Program in Health Sciences at FCMSCSP. He coordinates the Lato Sensu Postgraduate Courses in

Psychiatry, Cognitive Behavioral Therapy, and Mental Health at FCMSCSP. Additionally, he supervises the Psychiatry and Psychopathology courses in the fourth year and Psychiatry Internship in the sixth year of the medical undergraduate program. Certified by the Brazilian Association of Psychiatry (ABP). He serves as a preceptor for residents in the Psychiatry Residency Program at ISCMSP. He is a founding member of Scio Education and serves as the Head of Science and AI at Infinity Doctors Inc.

Pedro Makoto Uchida is currently studying psychology at Pontifícia Universidade Católica de Campinas (PUCC), developing various areas of knowledge. In addition, Pedro has a great interest in the area of games, behavior, and mental health, thus participating in national and international research groups to contribute to the creation of articles in the field.

Victor Henrique Oyamada Otani is a Researcher and Instructor Professor at the Mental Health Department—Santa Casa de São Paulo School of Medical Sciences (DSM-FCMSCSP), Coordinator of Lato Sensu Postgraduate Courses in Psychiatry, Coordinator of the Innovation Unit at DSM-FCMSCSP, and former Supervisor of the Psychiatry Residency Program at the Santa Casa de Misericórdia de São Paulo (ISCMSP). He graduated in Medicine from FCMSCSP; trained as a psychiatrist by ISCMSP and certified by the Brazilian Association of Psychiatry (ABP); and obtained his master's and doctorate degrees from FCMSCSP. He serves as a preceptor for residents in the Psychiatry Residency Program at ISCMSP. He is a founding member and CEO of Scio Education and board member at Infinity Doctors Inc. Coordinator of the Critical Skills Initiative Project.

Joao Ricardo Nickenig Vissoci with a background in psychology and data science, earned his PhD in Social Psychology from Pontificia Universidade Catolica de Sao Paulo/Brazil. Currently an Assistant Professor at Duke University in Emergency Medicine, Global Health, and Neurosurgery, he previously served as an Assistant Professor of Public Health in Brazil. He completed postdoctoral fellowships in data science at the University of Sao Paulo and Duke Global Health Institute. He co-directs the Global Emergency Medicine Innovation and Implementation (GEMINI) Research Center and is a faculty member of the Research Design and Analysis Core (RDAC) at DGHI. As chief of the Division of Translational Health Sciences in Duke's Department of Emergency Medicine, he also leads the Geoprocessing and Technology research group (GETS) at the State University of Maringá/Brazil. His research focuses on leveraging data and technology to address access and equity gaps in low-resource settings, translating evidence into impactful practice or policy. His interests extend to global mental health, emphasizing the role of daily activities in mental health reintegration. His work spans population health, health systems, and quality of care in Brazil, as well as traumatic brain injury, mental health, and technology in Tanzania and Uganda.

Contributors

Ana Abutara Mental Health Department, Santa Casa de São Paulo School of Medical Sciences, São Paulo, Brazil

Thales Marcon Almeida Mental Health Department, Santa Casa de São Paulo School of Medical Sciences, São Paulo, Brazil

Douglas Amaral Mackenzie Presbyterian University, São Paulo, Brazil

Rafael Ângulo Condoretti Barros Mental Health Department, Santa Casa de São Paulo School of Medical Sciences, São Paulo, Brazil

Amanda Douat Cardoso Computational Intelligence in Psychometrics and Epidemiology Research – Mackenzie Presbyterian University, São Paulo, Brazil

Cecilia Cheng The University of Hong Kong, Pokfulam, Hong Kong

Raimundo da Silva Soares Jr. Instituto D'Or de Pesquisa e Ensino, Rio de Janeiro, Brazil

Diogo de Lacerda Veiga Mental Health Department, Santa Casa de São Paulo School of Medical Sciences, São Paulo, Brazil

Carlos Bandeira de Mello Monteiro Faculty of Medicine, University of São Paulo – FMUSP, São Paulo, Brazil

School of Arts, Sciences and Humanities, University of São Paulo, São Paulo, Brazil

Matheus de Melo Rodrigues Computational Intelligence in Psychometrics and Epidemiology Research – Mackenzie Presbyterian University, São Paulo, Brazil

Marione Azevedo de Oliveira Druid Creative Gaming, São Paulo, Brazil

Roseli Lage de Oliveira Mental Health Department, Santa Casa de São Paulo School of Medical Sciences, São Paulo, Brazil

Felipe de Oliveira Aguiar Mental Health Department, Santa Casa de São Paulo School of Medical Sciences, São Paulo, Brazil

Artificial Intelligence Research Division, Infinity Doctor's Inc., Dover, DE, USA

Alexandre Luiz de Oliveira Serpa Computational Intelligence in Psychometrics and Epidemiology Research – Mackenzie Presbyterian University, São Paulo, Brazil

Giovanna de Paula Vidigal Faculty of Medicine, University of São Paulo – FMUSP, São Paulo, Brazil

Ana Paula Almeida de Pereira Federal University of Paraná, Curitiba, Brazil

Paul H. Delfabbro School of Psychology, The University of Adelaide, Adelaide, SA, Australia

Eduardo Dati Dias School of Arts, Sciences and Humanities, University of São Paulo, São Paulo, Brazil

Yuan Dong Duke University, Durham, NC, USA

Dayane Antunes dos Santos Federal University of Paraná, Curitiba, Brazil

Carolina Ribeiro Ferraz Mental Health Department, Santa Casa de São Paulo School of Medical Sciences, São Paulo, Brazil

Matheus Figueiredo Mental Health Department, Santa Casa de São Paulo School of Medical Sciences, São Paulo, Brazil

Jesse Fox Ohio State University, Columbus, OH, USA

Sara Barbosa Franco Mental Health Department, Santa Casa de São Paulo School of Medical Sciences, São Paulo, SP, Brazil

Luiz Antônio Vesco Gaiotto Mental Health Department, Santa Casa de São Paulo School of Medical Sciences, São Paulo, Brazil

Melina Houlis Mental Health Department, Santa Casa de São Paulo School of Medical Sciences, São Paulo, Brazil

Gabriel Kadri Mental Health Department, Santa Casa de São Paulo School of Medical Sciences, São Paulo, Brazil

Brendan Kelleher GEMINI Research Center, Duke University School of Medicine, Durham, NC, USA

Massachusetts General Hospital, Boston, MA, USA

Vanderbilt University School of Medicine, Nashville, TN, USA

Joshua Kim Purdue University, West Lafayette, IN, USA

Daniel L. King College of Education, Psychology, & Social Work, Flinders University, Adelaide, SA, Australia

Danilo Camargo Lapiccirella Electric Monkeys Games, São Paulo, Brazil

Katerina Lukasova Universidade Federal do ABC, Santo André, Brazil

Lucas Murrins Marques Mental Health Department, Santa Casa de São Paulo School of Medical Sciences, São Paulo, Brazil

Artificial Intelligence Research Division, Infinity Doctor's Inc., Dover, DE, USA

Jessica Mayumi Maruyama Computational Intelligence in Psychometrics and Epidemiology Research – Mackenzie Presbyterian University, São Paulo, Brazil

Gian Luigi Martins Menegazzo Mental Health Department, Santa Casa de São Paulo School of Medical Sciences, São Paulo, Brazil

Thomás Garboggini Moretti Mental Health Department, Santa Casa de São Paulo School of Medical Sciences, São Paulo, Brazil

Victor Namur Mental Health Department, Santa Casa de São Paulo School of Medical Sciences, São Paulo, Brazil

Amanda Yumi Ambriola Oku Universidade Federal do ABC, Santo André, Brazil

Camilo Ernesto Subenko Olalla Universidade Federal do ABC, Santo André, Brazil

Victor Henrique Oyamada Otani Mental Health Department, Santa Casa de São Paulo School of Medical Sciences, São Paulo, Brazil

Artificial Intelligence Research Division, Infinity Doctor's Inc., Dover, DE, USA

Gabriel Branquinho Pedone Mental Health Department, Santa Casa de São Paulo School of Medical Sciences, São Paulo, Brazil

Felipe Almeida Picon Universidade Federal Do Rio Grande Do Sul (UFRGS), Porto Alegre, Brazil

ADHD Outpatient Program, Clinical Research Center, Department of Psychiatry, Hospital de Clínicas de Porto Alegre, Universidade Federal do Rio Grande do Sul, Porto Alegre, Brazil

Eneyse Dayane Pinheiro Universidade Federal do ABC, Santo André, Brazil

Lara Poor Pontifícia Universidade Católica de Campinas, Campinas, São Paulo, Brazil

Gabriel Rêgo Human Developmental Sciences Graduate Program, Mackenzie Presbyterian University, São Paulo, Brazil

National Institute of Science and Technology in Social and Affective Neuroscience, INCT-SANI, São Paulo, Brazil

Nathasha Ribeiro Public Health, University of Guelph, Guelph, Canada

Natália Rocha Mental Health Department, Santa Casa de São Paulo School of Medical Sciences, São Paulo, Brazil

Raphael Santos Mental Health Department, Santa Casa de São Paulo School of Medical Sciences, São Paulo, Brazil

João Ricardo Sato Universidade Federal do ABC, Santo André, Brazil

Brett Sherrick Purdue University, West Lafayette, IN, USA

Matthew W. R. Stevens School of Biomedicine, The University of Adelaide, Adelaide, SA, Australia

College of Education, Psychology, & Social Work, Flinders University, Adelaide, SA, Australia

Christianne Fernandes Valente Takeda Harvard University, Boston, MA, USA

Pedro Makoto Uchida Pontifícia Universidade Católica de Campinas, Campinas, São Paulo, Brazil

Ricardo Riyoiti Uchida Mental Health Department, Santa Casa de São Paulo School of Medical Sciences, São Paulo, Brazil

Artificial Intelligence Research Division, Infinity Doctor's Inc., Dover, DE, USA

Ana Luiza Urbini Mental Health Department, Santa Casa de São Paulo School of Medical Sciences, São Paulo, Brazil

Jeel Vasa Association for Socially Applicable Research (ASAR), Pune, Maharashtra, India

Department of Psychiatry, All India Institute of Medical Sciences, Nagpur, Maharashtra, India

Jordan Vieira Mental Health Department, Santa Casa de São Paulo School of Medical Sciences, São Paulo, Brazil

Juliana Perez Weingartner School of Arts, Sciences and Humanities, University of São Paulo, São Paulo, Brazil

Sam Woodland Purdue University, West Lafayette, IN, USA

Siddhesh Zadey Association for Socially Applicable Research (ASAR), Pune, Maharashtra, India

Department of Epidemiology, Columbia University Mailman School of Public Health, New York, NY, USA

GEMINI Research Center, Duke University School of Medicine, Durham, NC, USA

Hao Zhang The University of Hong Kong, Pokfulam, Hong Kong

List of Abbreviations

ACC	Anterior Cingulate Cortex
ADHD	Attention Deficit Hyperactivity Disorder
AG	Angular Gyrus
AI	Anterior Insula
AI	Artificial Intelligence
APA	American Psychiatric Association
ASD	Autism Spectrum Disorder
AVG	Action Video Games
BCI	Brain-Computer Interface
BG	Biofeedback Games
BGames	Brain Games
BIS-11	Barratt Impulsivity Scale-11
BTG	Brain Training Games
CBT	Cognitive-Behavioral Therapy
CBT-IA	Cognitive-Behavioral Therapy for Internet Addiction
CC	Cingulate Cortex
CCT	Computerized Cognitive Training
DIA	Diagnostic Interview for Internet Addiction
dlPFC	Dorsolateral Prefrontal Cortex
DMN	Default Mode Network
DNA	Digital Netizen Alliance
DS	Dorsal Striatum
DSM-5	Diagnostic and Statistical Manual of Mental Disorders-5th Edition
DTI	Diffusion Tensor Imaging
ECN	Executive Control Network
EEG	Electroencephalography
EF	Executive Functions
EI	Emotional Intelligence
ESP	Embodied Social Presence
EXG	Exergames
FA	Fractional Anisotropy

FC	Functional Connectivity
fMRI	Functional Magnetic Resonance Imaging
fNIRS	Functional Near-Infrared Spectroscopy
GADIT	Gaming Disorder Identification Test
GAS-7	Game Addiction Scale-7 Items
GBG	Good Behavior Game
GD	Gaming Disorder
GMI	Gaming Motivation Inventory
GNGT	Go/No-Go Task
GSR	Galvanic Skin Response
HC	Hippocampus
HRV	Heart Rate Variability
ICD-11	International Classification of Diseases-11th Edition
ID	Intellectual Disability
IFG	Inferior Frontal Gyrus
IGA	Internet Gaming Addiction
IGD	Internet Gaming Disorder
IGDS9-SF	Internet Gaming Disorder Scale-Short Form
IGDT-10	Ten-Item Internet Gaming Disorder Test
I-PACE	Interaction of Person-Affect-Cognition-Execution
IPL	Inferior Parietal Lobule
LGBTQIA+	Lesbian, Gay, Bisexual, Transgender, Queer or Questioning, Intersex, Asexual and others
MCI	Mild Cognitive Impairment
MFG	Medial Frontal Gyrus
MMO	Massively Multiplayer Online
MMORPG	Massively Multiplayer Online Role-Playing Game
MMSE	Mini-Mental State Examination
MOBA	Multiplayer Online Battle Arena
MOG	Multiplayer Online Game
mPFC	Medial Prefrontal Cortex
MRI	Magnetic Resonance Imaging
MTG	Medial Temporal Gyrus
NAA	N-Acetyl-Aspartate
NAc	Nucleus Accumbens
NPC	Non-Playable Character
NVGP	Non-Video Game Players
OFC	Orbitofrontal Cortex
PASAT	Paced Auditory Serial Addition Test
PCC	Posterior Cingulate Cortex
PECS	Picture Exchange Communication System
PEGI	Pan European Game Information
PET	Positron Emission Tomography
PFC	Prefrontal Cortex
PTSD	Post-Traumatic Stress Disorder

PUS	Perception & Understanding of Self
PVT	Psychomotor Vigilance Task
RCT	Randomized Controlled Trial
RI	Response Inhibition
RPG	Role-Playing Game
SAM	Self-Assessment Manikin
SC	Social Cognition
SCI-IGD	Structured Clinical Interview for Internet Gaming Disorder
SDMT	Symbol Digit Modalities Test
SDT	Self-Determination Theory
SEL	Social and Emotional Learning
SG	Serious Game
SMA	Supplementary Motor Area
SPARX	Smart, Positive, Active, Realistic, X-factor thoughts
SR	Self-Regulation
SST	Stop Signal Task
STG	Superior Temporal Gyrus
STM	Short-Term Memory
SUD	Substance Use Disorder
SVG	Strategy Video Game
TAP	Test of Attentional Performance
TPJ	Temporoparietal Junction
TUD	Tobacco Use Disorder
TVA	Theory of Visual Attention
UFOV	Useful Field of View
VBM	Voxel-Based Morphometry
VGP	Video Game Players
VR	Virtual Reality
VS	Ventral Striatum
VSTM	Visual Short-Term Memory
WHO	World Health Organization
WIT	Wise IT-Use
WM	Working Memory

List of Figure

List of Tables

Part I
Neuroscience Fundamentals of Gaming

Chapter 1
Social and Affective Neuroscience of Gaming in Children's Education

Eneyse Dayane Pinheiro, Camilo Ernesto Subenko Olalla, Raimundo da Silva Soares Jr., Amanda Yumi Ambriola Oku, Katerina Lukasova, and João Ricardo Sato

Abstract The world of games has grown considerably in recent decades, consolidating itself as a powerful tool beyond entertainment. However, the potential for integrating them in healthy and effective ways in the context of early childhood, especially in education, is still little explored. With this in mind, this chapter proposes an interdisciplinary approach that connects games, neuroscience, and education, highlighting how the mechanisms of engagement and motivation present in games can transform learning. We analyze how well-designed cognitive games can strengthen learning, support emotional regulation, and promote essential social skills, and bring examples such as games based on brain-computer interfaces (BCI), which illustrate how emerging technologies can personalize learning in real-time and how design guided by cognitive load helps to balance pedagogical challenges, optimizing student engagement. On the other hand, we also discuss the associated risks, such as cyberbullying, exposure to inappropriate content, and digital addiction, highlighting the need for a careful and balanced application of game technologies. Additionally, we provide practical guidance for educators and researchers on aligning game mechanics with learning objectives, integrating assessments into gamified experiences, and creating a safe and inclusive environment. We also suggest strategies for balancing challenges and skills, fostering collaboration among students, and monitoring individual progress. These tips are intended to optimize the implementation of evidence-based gamification practices.

Keywords Games · Neuroscience · Education · Social Interactions · Children

E. D. Pinheiro · C. E. S. Olalla · A. Y. A. Oku · K. Lukasova · J. R. Sato (✉)
Universidade Federal do ABC, Santo André, Brazil
e-mail: joao.sato@ufabc.edu.br

R. da Silva Soares Jr.
Instituto D'Or de Pesquisa e Ensino, Rio de Janeiro, Brazil

L. Murrins Marques et al. (eds.), *Social and Affective Neuroscience of Gaming*,
https://doi.org/10.1007/978-3-032-00325-6_1

Introduction to the Use of Games in Education

Games can be conceptualized as artificial environments that prioritize player enjoyment over the representation of reality. While engaging in atypical or potentially harmful behaviors in the real world can result in enduring negative consequences, games create a secure space for experimentation and exploration of such behaviors (Ganter-Argast et al., 2024). Over the years, the value of commercial video games in society has increased, paralleling the development of games designed for education (Papastergiou, 2009; Noroozi et al., 2020). Video games used for educational purposes have gradually gained attention from educators and scholars. Research indicates a growing recognition of the positive impact of video games on learning (Abbott et al., 2022; Wouters et al., 2013; Ganter-Argast et al., 2024).

The integration of video games into education gained traction in the early 2000s (Manzano-León et al., 2021). Since that time, researchers have acknowledged the potential of games as valuable and versatile tools for teaching, owing to their unique ability to merge educational content with entertainment (Kalogiannakis et al., 2021). A fundamental concept in the design of educational activities that utilize gamification is the notion of transfer. Theories of transfer can be categorized into two main types: specific transfer and general transfer (Singley & Anderson, 1989). Specific transfer theory suggests that learning one skill (Skill A) will only assist in the development of another skill (Skill B) if both skills share common elements (Flores-Gallegos & Mayer, 2022). Consequently, playing a particular game is expected to enhance performance in that game but not necessarily in other areas. For instance, a study found that a group playing a game designed to improve memory outperformed a control group in a memory task but did not perform better on a visuospatial task (Wells et al., 2021). In contrast, general transfer theory proposes that learning Skill A can improve overall cognitive abilities, enhancing performance across various tasks (Singley & Anderson, 1989). For example, another study demonstrated that individuals trained in action video games displayed faster learning in two cognitive domains—perception and working memory—than those trained in non-action games (Zhang et al., 2021). From both perspectives, games can be a valuable opportunity to promote learning and interventions.

Research on game-based interventions for educational purposes has shown promising cognitive benefits. A meta-analysis found that playing action video games enhances performance on tasks that require perceptual processing, particularly in spatial cognition (Bediou et al., 2018). Another meta-analysis highlighted that game-based learning strategies have a more significant impact on individuals, especially children, compared to traditional teaching methods, particularly in terms of learning effectiveness and motivation (Wouters et al., 2013). Additionally, another study explored a year-long role-playing game therapy approach (Abbott et al., 2022). This study evaluated participant feedback, the key concepts of the model, and insights for therapists interested in this approach. Participants reported increased confidence in social situations, particularly in setting boundaries and managing potential errors, indicating that the skills acquired through the game were applicable in real-life scenarios.

Games can also help pedagogy to offer open learning for various learners who could step beyond the boundaries of conventional learning (Rahimi & Shute, 2021; Ganter-Argast et al., 2024). Gamification enhances student motivation and engagement, fostering the development of cognitive skills such as critical thinking and problem-solving, which are essential in today's world (Prensky, 2009). Adding game elements to other tasks and situations highlights the wide range of options and scenarios that games can offer, showing us how much potential there is to explore.

Neuroscience research has shown promising results regarding the use of video games. For instance, a neuroimaging study examined adults who played Super Mario 64 for 30 min daily over a 2-month period (Kühn & Gallinat, 2014). The findings indicated that the intervention led to structural changes in brain areas. Specifically, the right hippocampus, associated with memory, and the dorsolateral prefrontal cortex are involved in decision-making, problem-solving, and regulating attention and behavior. Participants demonstrated increased volume in both the right hippocampus and the dorsolateral prefrontal cortex, suggesting that neuroplasticity occurred due to the game intervention (Kühn & Gallinat, 2014). Thus, video games uniquely integrate various forms of stimuli into a single interactive experience. This combination, along with the active engagement required by players, makes video games an excellent tool for fostering learning.

In contrast, a systematic review highlighted that not all video games are capable of enhancing cognition (Rahimi & Shute, 2021): in fact, certain game genres are more effective than others. Specifically, video games that foster a state of flow, encourage player co-creation, and enhance intrinsic motivation are particularly effective at promoting creativity (Ganter-Argast et al., 2024). Given that co-creation and motivation are important aspects of learning, games can provide a valuable opportunity to capture students' attention during educational activities designed to develop specific skills. For instance, a geometry class that students typically find "boring" or "difficult" could transform into an engaging challenge through an educational game designed to enhance visuospatial skills. Additionally, students can interact and assist each other in completing the challenges, which fosters both spatial cognition and social skills. This emphasizes the importance of assessing the characteristics of games to determine which elements are beneficial or irrelevant for intentional interventions or educational purposes.

While games can offer various benefits, including enhanced cognitive control, improved memory, and stronger social connections (Anguera et al., 2013; Abbott et al., 2022; Kleschnitzki et al., 2022), there are also valid concerns regarding their negative aspects. These concerns include the potential for addiction, the promotion of violence, increased cognitive load, and a lack of effective transfer of intended learning outcomes from games (Garneli et al., 2017; Hess & Gunter, 2013; Weng et al., 2013). These factors have been the subject of scientific research and should be considered in discussions about the most effective ways to utilize games for teaching and developing important cognitive skills, such as memory and attention, among others.

This chapter aims to provide a thorough overview of relevant research in cognitive and social Neuroscience as it relates to games. Our goal is to offer insights into the underlying principles and methodologies that guide research in this area, preparing readers for an in-depth exploration of how games affect the brain and behavior. We will introduce the complex interactions among games, social dynamics, neuroscientific concepts, and the challenges associated with game applications. Additionally, we will share insights that could lead to best practices and future directions for using games to promote cognitive and social development.

Neuroscientific Principles Supporting Games

One of the key neuroscientific aspects relevant to gamification is brain plasticity. The human brain has a remarkable ability to adapt and modify its structure and function in response to experiences and challenges. This plasticity allows the brain to create new synaptic connections and reorganize its neural networks, facilitating the acquisition of knowledge and skills (Kolb et al., 2017). The gaming environment provides complex scenes, an engaging experience, and a variety of goals, creating ideal stimuli for what Bavelier et al. (2012) refer to as the "learning to learn" framework. This concept suggests that video games enhance learning skills, such as improved probabilistic inference, pattern recognition, and resource allocation, which further support the adaptation to new and diverse tasks.

To understand the relationship between the use of video games and the functional and structural brain changes they may produce, Palaus et al. (2017) conducted a systematic review on this topic. They were able to establish some consistent links between gaming and neural function in the following circuits. Notable changes were reported in the parietal cortex, which supports visual-spatial processing and sensory integration—skills sharpened by action and strategy games requiring spatial awareness and quick responses. Similarly, fronto-parietal activity was linked to executive functions like decision-making and planning, often involving the hippocampus for memory-related tasks. The cerebellum and motor cortices were also activated in games that demand precise coordination, short reaction times, and motor skills. Furthermore, the occipital and temporal regions, particularly the visual cortex, showed enhanced visual processing (West et al., 2017; Kühn & Gallinat, 2014) and attention-related activity, especially in fast-paced games (Palaus et al., 2017). The regions mentioned here are depicted below (Fig. 1.1):

One practical application of these findings is the intentional design of games that align with specific educational objectives. For example, games focusing on visual-spatial skills and sensory integration could be tailored to subjects like mathematics or geometry. Age-appropriate games, such as *Minecraft* and *NanoRoboMatch*, have been used to enhance numerical and geometric skills. The efficacy of these games has been explored in a review by Hui and Mahmud (2023). Similarly, games emphasizing executive functions such as decision-making and planning, linked to fronto-parietal activity and hippocampal engagement, could enhance critical thinking and

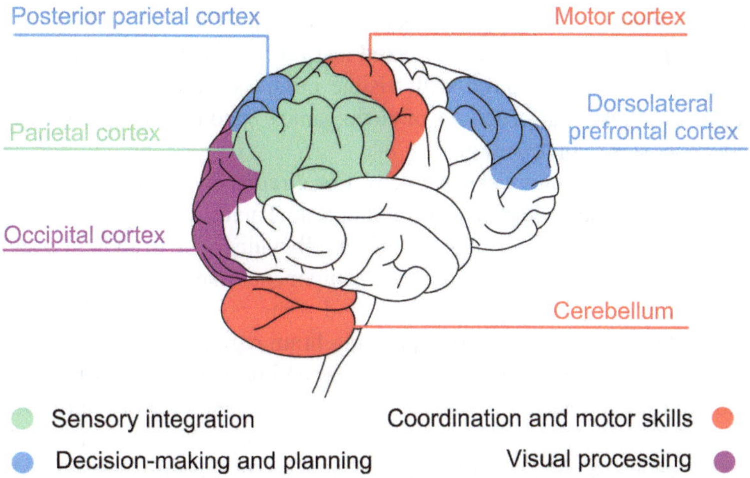

Fig. 1.1 Regions highlighted throughout the chapter and related functions. (Illustration by Falconi-Souto)

strategic reasoning in disciplines like science or history. For this purpose, a game such as *Nano Legends* (Kaipal & Figg, 2009) may be indicated.

On the neurofunctional level, digital games have long been associated with the activation of the brain's reward systems, releasing neurotransmitters like dopamine, which are associated with motivation, pleasure, and learning (Koepp et al., 1998). The impact of video game training on the brain's reward system was evaluated in a longitudinal study with 50 healthy participants, randomly assigned to either a video game training group (TG) or a control group (CG). The study aimed to investigate both the functional predictors of video gaming experience and the functional changes in the brain resulting from video game training. To achieve this, the researchers conducted functional magnetic resonance imaging (fMRI) before and after 2 months of training with the Mario Bros video game platform, utilizing a reward task unrelated to gaming (Lorenz et al., 2015). The findings suggest that video game training may lead to significant changes in the brain's reward system, with the strongest activation occurring in the ventral striatum (VS) during reward anticipation in the TG group. The authors proposed that this striatal reward signal could reflect sustained engagement in the TG group during the post-test. The responsiveness in reward processing and motivation to complete the task at a similar level of engagement at the second time point might indicate a positive effect on the reward system. These results showcase other studies that found video games' impact on brain function (Brilliant et al., 2019) and structure (Kühn et al., 2011), through effective reward schedules, strong engagement, and satisfaction stemming from the sense of competence and autonomy (Junttila et al., 2022).

The practical application of these findings suggests that understanding reward systems can help educators optimize teaching strategies. For example, allowing students to make choices within tasks by selecting objectives, customizing avatars, or choosing strategies. It may enhance their sense of autonomy, a key driver of intrinsic motivation. Furthermore, goal-oriented learning can be promoted through clear and attainable objectives paired with visible progress indicators. These elements replicate the motivational structures of video games, encouraging sustained engagement and persistence in learning.

However, excessive play risks overactivating circuits involved in reward and control, emphasizing the importance of moderation (Palaus et al., 2017). The neural reward system of some gamers may be profoundly vulnerable to addiction and show structural and functional changes in key components like the orbitofrontal cortex (OFC), anterior cingulate cortex (ACC), and some regions of the limbic system (Haber, 2011). As well as various supporting brain regions, such as the amygdala, hippocampus (HC), posterior cingulate cortex, and prefrontal cortex mediate complex addiction-linked behaviors such as craving, emotional memory retrieval, visual attention to gaming clues, and reward-seeking motivation (Bäcklund et al., 2022).

Another aspect of games' influence on humans is related to our social brain, since online gaming communities have also been assessed for monitoring changes in an individual's ability to influence communication behaviors and interpersonal bonds formed through gaming. The role of community gaming was shown to positively impact the interpersonal bonds through communication and sense of belonging (Ni et al., 2023). Engaging in cooperative gameplay can lead to increased self-control and significantly influence emotional and behavioral outcomes in aging subjects. Successful aging is considered to involve maintaining physical health, mental health, and social engagement. The study by Allaire et al. (2013) investigated how gaming can contribute to these dimensions in a sample of 140 independently living older adults that were divided into three groups of regular, occasional gamers, and non-gamers. The results showed that regular and occasional gamers perform better for well-being, social functioning, and negative affect in general. The research indicates that gaming can provide opportunities for positive social experiences and thus enhance self-perceived quality of life, playing a beneficial role in promoting successful aging.

The impact of gaming on social intuition and decision-making has been examined extensively, particularly within games that require players to engage in interaction and make strategic decisions. In such environments, players must navigate complex social cues, adjust to opponents' behavior, and balance cooperation with competition, which can influence their performance and outcomes. Yoshida et al. (2010) explored how humans infer the strategies and beliefs of others in cooperative game settings, particularly focusing on recursive belief inference. The authors used a "stag hunt" game in which participants interacted with a computerized agent that varied its strategy complexity (ranging from simple to more sophisticated recursive strategies). The study aimed to investigate the neural basis of belief inference, especially how individuals predict others' actions based on understanding their mental states. The results suggested that different areas of the prefrontal cortex are involved in belief inference. The rostral medial prefrontal cortex, associated with mentalizing, encodes the uncertainty about the opponent's strategy, while the dorsolateral prefrontal cortex encodes the complexity of the beliefs involved in cooperation (how deeply one considers the other's beliefs about their own beliefs, and so on). This suggests a computational framework within the brain for managing cooperation that relies on managing complex beliefs about us and others (Yoshida et al., 2010).

Furthermore, the nuances of broader dynamics of competition and social interaction are often shaped by specific game mechanics and contextual features. In group decision-making settings, the interplay between collaboration and competition influences players' choices, balancing short-term individual incentives with long-term group benefits. This trade-off is regulated by complex neural processes involving the ventral prefrontal cortex, which plays a critical role in social and moral decision-making. This region helps mediate the conflict between personal rewards and collective welfare, enabling players to make choices that align with both personal goals and the needs of the group (Hampton et al., 2006).

The sophisticated neural underpinnings of social behaviors fostered in gaming suggest potential applications in enhancing real-world social skills and decision-making abilities. However, there is a need for refining the descriptions of the effect of games on cognition and neural plasticity, considering the interaction between the characteristics of the games themselves and the individual differences in players (Bavelier & Green, 2019; Dale et al., 2020). By understanding the neuroscientific principles of brain plasticity and the reward and motivation systems, educators can design gamified learning experiences that enhance engagement, the acquisition of knowledge, and skills among students. This neuroscience-based approach can contribute to the development of more effective educational strategies tailored to the needs of the current and future technological society.

Skills Developed Through Games

The development of critical thinking and problem-solving skills is a fundamental aspect of contemporary education, and the use of games as a pedagogical tool has proven effective in this context (Snyder & Snyder, 2008). Exposing students to challenging situations and the need to make strategic decisions, present in many games, significantly contributes to the enhancement of these competencies. According to Gee (2003), digital games provide a rich environment for practicing analytical skills, where players are encouraged to explore different solutions to complex problems.

Games that involve collaboration among participants have been highlighted in promoting skills such as teamwork, communication, and negotiation. The social interaction required in these games allows students to practice active listening, empathy, and respect for others' opinions. Research by Deterding et al. (2011) suggests that these collaborative experiences enhance academic performance and help prepare students for professional environments where teamwork is often essential. These social skills develop through both the gaming environment and the time spent interacting with others during gameplay. Having an environment in which you can simulate attempts, successes, and failures several times provides players with knowledge acquired through multiple experiences.

In addition to improving social interactions, the use of digital games can foster students' creativity and innovation. Online interactions enable players to gain knowledge and explore physically distant environments. By engaging with settings from various countries and regions, players can experience diverse ethnic, religious, and cultural contexts. This exposure fosters greater respect, empathy, and connection by encouraging an understanding of others' challenges, peculiarities, and environments. Additionally, it promotes awareness of historical, political, and social factors, such as economic disparities, conflicts, and climate-related issues affecting different regions. Games that encourage exploration and experimentation allow students to develop out-of-the-box thinking by testing new ideas and solutions. This creative freedom is essential in a constantly changing world, where the ability to innovate is increasingly valued. Research by Koivisto and Hamari (2019) shows that interaction with games offering a safe space for experimentation can enhance students' confidence in their creative abilities. In a scenario of creative potential, games have an advantage when compared to conventional methods: the online environment. Having friends playing simultaneously from different countries, cultures, and contexts allows the player to develop strategies, ways of relating, and self-control that could take years if he needed to visit each country to learn something new.

Integrating games into the educational environment enriches learning in terms of critical thinking and problem-solving while also promoting social and creative skills in the children's repertoire. Research on the impact of these games suggests that, when used strategically, they can be a valuable resource for developing students' social, motor, and cognitive skills. Socially, such games foster teamwork, communication, and empathy as students collaborate and engage with peers. These elements are evident in various examples, such as MMORPGs (Massively Multiplayer Online Role-Playing Games), where players must cooperate to complete quests and overcome challenges. Additionally, some games require teamwork in specific contexts. For instance, *Moving Out* and *Overcooked* demand coordination and communication to transport furniture between homes or prepare meals in a restaurant, respectively. Emotional engagement is also a key aspect in games like *It Takes Two*, where two players take on the roles of a divorcing couple striving to understand each other better while considering what is best for their daughter. Motor skills are enhanced through activities that require coordination, precise movements, and active participation. These elements are present in various motion-based games, such as *Just Dance*, as well as Virtual Reality platform games and consoles like the Wii or Kinect (Xbox). Cognitively, games challenge students to think critically, solve problems, and make strategic decisions. These abilities can be developed through various approaches but are particularly common in puzzle games like *Portal*, *Human: Fall Flat*, *Braid*, *The Witness*, and *Humanity*. For example, in *Portal*, players use a gun that shoots two portals—an entry and an exit—to solve challenges based on physics concepts such as gravity and inertia. Similarly, in *Human: Fall Flat*, players rely solely on their malleable bodies to overcome obstacles, using only the movement and gripping capabilities of their upper limbs to interact with objects. By integrating these multidimensional experiences, games contribute to academic knowledge and to a well-rounded development that supports various aspects of student growth.

Challenges and Considerations

Digital games have significant potential to transform the learning experience, making education more engaging and effective. However, their implementation faces notable challenges, such as the lack of adequate infrastructure in schools (Prioste & Raiça, 2017). These challenges include the availability of technological devices, a stable internet connection, appropriate physical spaces, technical support, teacher training, and funding. In schools with lower socioeconomic resources, these issues are even more pronounced, exacerbating educational inequalities and limiting many students' access to the benefits of digital games (Van Eck, 2006).

These characteristics are further magnified when comparing educational contexts across different countries, particularly between developing and developed nations. In developing countries, schools often struggle with fundamental issues such as poor connectivity and inadequate access to devices, limiting the potential for game-based learning. In contrast, while infrastructure may be more readily available in developed countries, the integration of digital games can still face resistance due to traditional educational mindsets or insufficient teacher training (Lavy, 2010).

Teacher training is an essential aspect of effectively integrating digital games into education (da Cunha & Bizelli, 2016). Beyond merely learning to use the tools, educators must develop competencies to align games with pedagogical objectives, ensuring they are applied in a meaningful and relevant manner (Sefton & Galini, 2022; Araújo, 2020). Without proper training, teachers may lack confidence and utilize these technologies. Research shows that continuous training programs, workshops, and adequate technical support are vital to ensuring that educators feel prepared to use games effectively in the classroom (Koehler & Mishra, 2005). Examples include integrating multiplayer games to develop teamwork and communication skills or utilizing simulations to deepen understanding in science topics (Arango et al., 2008). For instance, Stone et al. (2019) demonstrated that online multiplayer games provided platforms for students with autism to use speech to engage in reciprocal conversations, share information, make requests, give commands, and direct others. This highlights the importance of equipping teachers with strategies to incorporate such games in ways that foster meaningful participation and assess learning for all students.

Building on the necessity for well-structured teacher training and inclusive game integration strategies, another critical challenge is assessing the learning outcomes achieved through game-based education. It is essential to balance the playful nature of games with the measurement of educational objectives. Traditional assessment tools often fail to capture the development of skills such as creativity, collaboration, and problem-solving—key aspects frequently nurtured by digital games. To address this, complementary methods like formative assessments, which provide continuous feedback to both students and teachers, are vital (Shute, 2008). Moreover, successful case studies demonstrate that implementing real-time feedback mechanisms in digital games can effectively engage students and promote active learning (Erhel & Jamet, 2013). For instance, schools that have integrated adaptive gaming platforms report improved student motivation and enhanced skill acquisition, showcasing the potential for overcoming these assessment barriers.

Parental involvement is equally critical in this process. By recognizing the positive role of digital games in education, parents can encourage the constructive use of these tools, helping to select age-appropriate games aligned with learning objectives (Ramos & Pimentel, 2021). This collaboration between schools and families strengthens the incorporation of games into children's daily educational activities that can transform their learning experience, making it more engaging and effective. By combining principles from neuroscience and social development, both educators and parents can foster children's cognitive skills and address their emotional and social growth (von Gillern & Nash, 2024). However, for this integration among games, family, and school education to be effective, it is necessary to adopt practices, such as providing proper feedback, using adaptive games, and measuring the emotional and social impact of the game's children play.

To engage parents more effectively, schools can implement a series of strategies. First, they can host workshops and informational sessions that explain the benefits of digital games and how they align with school curricula. These sessions could include practical demonstrations of educational games. Additionally, schools can create online discussion groups where parents can share experiences and recommendations about games, fostering a supportive community. Another effective approach is to send regular newsletters with game suggestions, usage tips for the home, and success stories from other families. By providing these engagement opportunities, schools can build a strong partnership with parents, enriching children's educational experiences.

Beyond the use of digital games, the creation of educational games has also gained prominence, particularly in the development of games designed to teach while entertaining (Löytömäki et al., 2024). These games require a collaborative approach between developers and educators, who must create engaging experiences aligned with pedagogical needs. Research shows that serious games can promote skills such as critical thinking, problem-solving, and cognitive engagement while reinforcing curricular content (Annetta, 2010). This intersection between technology and education underscores the need for an integrated approach, where all stakeholders—teachers, parents, students, and developers—work together to build an inclusive and innovative educational ecosystem.

Trends and Perspectives

With the advancement of neuroscience techniques in educational research, it has become feasible to conduct experiments that empirically and naturalistically assess the effectiveness of different teaching methodologies and approaches. This progress allows researchers to move beyond traditional observational methods, providing objective insights into how various pedagogical strategies impact cognitive and neural processes involved in learning (Brockington et al., 2018). Recent research by Oku et al. (2023) compares the traditional classes' approach with interactive, game-based learning environments. The study utilized functional near-infrared spectroscopy (fNIRS), a non-invasive technique that measures brain activity indirectly by detecting changes in blood oxygenation levels. Additionally, the hyperscanning modality enabled simultaneous data acquisition from both teacher and student, allowing the researchers to capture real-time interactions between dyads. Using graph theory for data representation, results indicated that game-based interactive activities fostered significantly more connections between participants compared to theoretical lessons. These findings underscore the potential of incorporating games into educational practices to enhance engagement and strengthen teacher-student relationships at a neurobiological level. The utilization of hyperscanning methods to investigate the neural correlates of social interactions and gaming is a promising but still scarce field of research.

Proper feedback is a fundamental element for effective learning. The literature suggests that this approach can significantly increase a student's motivation, particularly when feedback is given immediately after the completion of a task and with high frequency (Barkela & Leuchter, 2024). In this context, neurofeedback emerges as a powerful tool that can provide students with real-time data about their brain activity, thus increasing their ability to self-regulate learning processes. According to Hammond (2011), neurofeedback can facilitate immediate adjustments in cognitive and emotional responses, ultimately leading to improved performance and motivation. Thus, by allowing students to visualize their brain activity and receive feedback on their mental states, it is possible to increase the effectiveness of traditional feedback methods, further promoting a growth mindset and fostering deeper learning experiences (Thibault et al., 2018). Thus, integrating neurofeedback into educational environments can significantly amplify the benefits of immediate feedback, making learning more effective and engaging.

In addition, the adaptability of games to the gamer—that is, their ability to adjust to each student's skill level—is crucial to maintaining interest and promoting continuous learning. Adaptive games modify their difficulty based on the player's performance, offering personalized challenges that meet each child's cognitive capacity (Gee, 2003). This personalization of learning is crucial to avoid frustration caused by difficult tasks and to prevent demotivation resulting from challenges that are too simple. This approach creates a zone of proximal development, a fundamental concept for effective learning, as proposed by Vygotsky (1978), where the game would play the role of the teacher during this interaction. That said, the neuroscience of adaptability offers valuable insights into how the brain responds to different levels of challenge and how this affects motivation and learning. Studies show that the brain's reward circuits are closely linked to frustration and motivation. The neuroscience of adaptability provides valuable insights into how the brain responds to different levels of challenge and how this impacts motivation and learning. Research suggests that the brain's reward circuits are closely linked to frustration and motivation. When students face challenges far beyond their capabilities, frustration can arise, leading to demotivation. On the other hand, appropriately matched challenges can activate these reward circuits, fostering engagement and motivation (Schunk & Zimmerman, 2012). However, the brain continuously seeks a state of balance—known as homeostasis. Challenges that are either too difficult or too simple can disrupt this balance, resulting in frustration or disengagement. Therefore, offering balanced challenges that are stimulating yet achievable is crucial for the development of both cognitive and emotional skills. This approach enables students to manage frustration constructively and build their ability to adapt to new situations. Moreover, frequent and timely feedback is an effective strategy to support this process. By integrating the principles of the neuroscience of adaptability into education, we can promote more effective and meaningful learning while fostering a positive and engaging learning environment.

Furthermore, online games offer clear benefits, such as improved collaboration and enhanced problem-solving skills. However, these environments also present risks, such as cyberbullying and toxic social interactions. While multiplayer online games provide opportunities for social engagement, they can expose children to negative behaviors. Therefore, educators and parents must be aware of the social risks associated with online gaming, promoting respectful interactions and providing guidance on how to handle harmful behaviors (Kowert & Quandt, 2015). Additionally, monitoring social dynamics in online games can provide insights into social skills such as empathy, cooperation, and conflict resolution. However, without proper guidance, indiscriminate use of these platforms can reinforce negative behaviors. The debate about the relationship between violent games and aggressive behavior has persisted for decades. Yet, a growing body of research suggests that there is no direct causal relationship between violent games and violent behaviors, especially when children play age-appropriate games. A comprehensive study conducted by Zhang et al. (2020), involving thousands of Chinese children, found no significant link between violent games and aggressive behavior. The study highlighted that context is an important factor: children who played games inappropriate for their age or who lacked parental supervision were more likely to exhibit problematic behaviors. This underscores the importance of ensuring that children play content suitable for their age and that there is adult guidance during the process (Zhang et al., 2020).

Therefore, measuring the emotional and social impact of games is essential to ensure that they promote not only cognitive development but also emotional regulation and positive social interactions (von Gillern & Nash, 2024). As demonstrated by the works cited, neuroscience can play a crucial role in this endeavor, providing tools to assess emotional and social impacts from both cognitive and neurofunctional perspectives.

Conclusion

In short, integrating games into education presents a powerful opportunity to engage students across cognitive, emotional, and social dimensions. When applied strategically, educational games can strengthen neural pathways for learning, support emotional regulation, and foster the development of essential social skills. However, it is equally important for educators and caregivers to remain mindful of the challenges involved, such as the risks of cyberbullying, overexposure to violent games, and potential over-reliance on digital tools.

Emerging innovations, such as brain-computer interface (BCI) games, represent a promising frontier in educational technology. These tools leverage real-time neural feedback to personalize learning experiences, adapting to each student's cognitive state and optimizing their engagement. Similarly, cognitive load-based games are designed to balance the mental demands placed on students, ensuring that challenges remain stimulating but manageable. Such approaches align with evidence-based practices, enabling students to achieve deeper learning while minimizing frustration or cognitive overload. As these technologies evolve, it becomes essential to maintain a balance between digital immersion and real-world social interactions. While virtual environments and gamified platforms can enhance learning experiences, opportunities for face-to-face collaboration and meaningful peer connections should not be overlooked. This holistic approach ensures that students not only acquire knowledge and skills but also develop the interpersonal competencies needed to navigate an increasingly interconnected world.

When thoughtfully designed and implemented, games can serve as a transformative tool in modern education, promoting academic success alongside emotional and social growth. The adoption of gamified practices—augmented by emerging technologies like BCI and cognitive load-based designs—marks a critical step in preparing students for the complexities of the twenty-first century. By leveraging the full potential of these innovations, educators and parents can equip students with the adaptability, critical thinking, and resilience needed to thrive in a rapidly changing world.

Acknowledgements EDP is supported by a scholarship from CEBRASPE/CARREFOUR. AYAO is supported by São Paulo Research Foundation (FAPESP, #24/20442-2). KL is support by São Paulo Research Foundation (FAPESP, #22/03206-8). JRS is supported by São Paulo Research Foundation (FAPESP, # 23/18337-3; 24/09675-5, 21/05332-8 and 23/16997-6) and Allan Faconi-Souto for the illustration in Figure 1.

References

Abbott, M. S., Stauss, K. A., & Burnett, A. F. (2022). Table-top role-playing games as a therapeutic intervention with adults to increase social connectedness. *Social Work With Groups, 45*(1), 16–31. https://doi.org/10.1080/01609513.2021.1932014

Allaire, J. C., McLaughlin, A. C., Trujillo, A., Whitlock, L. A., LaPorte, L., & Gandy, M. (2013). Successful aging through digital games: Socioemotional differences between older adult gamers and non-gamers. *Computers in Human Behavior, 29*(4), 1302–1306.
Anguera, J. A., Boccanfuso, J., Rintoul, J. L., Al-Hashimi, O., Faraji, F., Janowich, J., Kong, E., Larraburo, Y., Rolle, C., Johnston, E., & Gazzaley, A. (2013). Video game training enhances cognitive control in older adults. *Nature, 501*(7465), 97–101. https://doi.org/10.1038/nature12486
Annetta, L. A. (2010). The "i's" have it: A framework for serious educational game design. *Review of General Psychology, 14*(2), 105–112. https://doi.org/10.1037/a0018985
Arango, F., Aziz, E. S., Esche, S. K., & Chassapis, C. (2008). A review of applications of computer games in education and training. In *2008 38th annual frontiers in education conference* (pp. T4A–1). IEEE.
Araújo, L. C. (2020). *Jogos educacionais digitais como ferramentas promissoras no processo de aprendizagem dos campos de experiências da educação infantil* [Master's thesis]. Universidade Federal do Rio Grande do Norte.
Bäcklund, C., Elbe, P., Gavelin, H. M., Sörman, D. E., & Ljungberg, J. K. (2022). Gaming motivations and gaming disorder symptoms: A systematic review and meta-analysis. *Journal of Behavioral Addictions, 11*(3), 667–688. https://doi.org/10.1556/2006.2022.00053. PMID: 36094861; PMCID: PMC9872536.
Barkela, V., & Leuchter, M. (2024). Effectiveness of automated formative feedback in an online tutorial for promoting summarizing. *Journal of Educational Technology Development and Exchange (JETDE), 17*(1), 67–95.
Bavelier, D., & Green, C. S. (2019). Enhancing attentional control: Lessons from action video games. *Neuron, 104*(1), 147–163.
Bavelier, D., Green, C. S., Pouget, A., & Schrater, P. (2012). Brain plasticity through the life span: Learning to learn and action video games. *Annual Review of Neuroscience, 35*(1), 391–416.
Bediou, B., Adams, D. M., Mayer, R. E., Tipton, E., Green, C. S., & Bavelier, D. (2018). Meta-analysis of action video game impact on perceptual, attentional, and cognitive skills. *Psychological Bulletin, 144*(1), 77–110. https://doi.org/10.1037/bul0000130
Brilliant, T. D., Nouchi, R., & Kawashima, R. (2019). Does video gaming have impacts on the brain: Evidence from a systematic review. *Brain Sciences, 9*(10), 251. https://doi.org/10.3390/brainsci9100251
Brockington, G., Balardin, J. B., Zimeo Morais, G. A., Malheiros, A., Lent, R., Moura, L. M., & Sato, J. R. (2018). From the laboratory to the classroom: The potential of functional near-infrared spectroscopy in educational neuroscience. *Frontiers in Psychology, 9*, 1840.
da Cunha, M. D., & Bizelli, J. L. (2016). Caminhos para TIC em sala de aula sob a perspectiva dos professores. *Revista on line de Política e Gestão Educacional, 20*, 282–300.
Dale, G., Joessel, A., Bavelier, D., & Green, C. S. (2020). A new look at the cognitive neuroscience of video game play. *Annals of the New York Academy of Sciences, 1464*(1), 192–203.
Deterding, S., Sicart, M., Nacke, L., O'Hara, K., & Dixon, D. (2011). Gamification. Using game-design elements in non-gaming contexts. In *CHI'11 extended abstracts on human factors in computing systems* (pp. 2425–2428). Association for Computing Machinery.
Erhel, S., & Jamet, E. (2013). Digital game-based learning: Impact of instructions and feedback on motivation and learning effectiveness. *Computers & Education, 67*, 156–167.
Flores-Gallegos, R., & Mayer, R. E. (2022). Learning cognitive skills by playing video games at home: Testing the specific transfer of general skills theory. *Journal of Cognitive Enhancement, 6*(4), 485–495. https://doi.org/10.1007/s41465-022-00253-8
Ganter-Argast, C., Schipper, M., Shamsrizi, M., Stein, C., & Khalil, R. (2024). The light side of gaming: Creativity and brain plasticity. *Frontiers in Human Neuroscience, 17*. https://doi.org/10.3389/fnhum.2023.1280989
Garneli, V., Giannakos, M., & Chorianopoulos, K. (2017). Serious games as a malleable learning medium: The effects of narrative, gameplay, and making on students' performance and

attitudes. *British Journal of Educational Technology, 48*(3), 842–859. https://doi.org/10.1111/bjet.12455

Gee, J. P. (2003). *What video games have to teach us about learning and literacy*. Palgrave Macmillan.

Haber, S. N. (2011). Neuroanatomy of reward: a view from the ventral striatum. *Neurobiology of sensation and reward*

Hammond, D. C. (2011). What is neurofeedback: An update. *Journal of Neurotherapy, 15*(4), 305–336.

Hampton, A. N., Bossaerts, P., & O'Doherty, J. P. (2006). The role of the ventromedial prefrontal cortex in abstract state-based inference during decision making in humans. *Journal of Neuroscience, 26*(32), 8360–8367.

Hess, T., & Gunter, G. (2013). Serious game-based and nongame-based online courses: Learning experiences and outcomes. *British Journal of Educational Technology, 44*(3), 372–385. https://doi.org/10.1111/bjet.12024

Hui, H. B., & Mahmud, M. S. (2023). Influence of game-based learning in mathematics education on the students' cognitive and affective domain: A systematic review. *Frontiers in Psychology, 14*, 1105806. https://doi.org/10.3389/fpsyg.2023.1105806

Junttila, K., Smolander, A. R., Karhila, R., Giannakopoulou, A., Uther, M., Kurimo, M., & Ylinen, S. (2022). Gaming enhances learning-induced plastic changes in the brain. *Brain and Language, 230*, 105124.

Kaipal, K., & Figg, C. (2009). Using video games in science instruction: Pedagogical, social, and concept-related aspects. *Canadian Journal of Science, Mathematics, and Technology Education, 9*, 117–134. https://doi.org/10.1080/14926150903047780

Kalogiannakis, M., Papadakis, S., & Zourmpakis, A.-I. (2021). Gamification in science education. A systematic review of the literature. *Education Sciences, 11*(1), 22.

Kleschnitzki, J. M., Beyer, L., Beyer, R., & Großmann, I. (2022). The effectiveness of a serious game (MemoreBox) for cognitive functioning among seniors in care facilities: Field study. *JMIR Serious Games, 10*(2), e33169. https://doi.org/10.2196/33169

Koehler, M. J., & Mishra, P. (2005). What happens when teachers design educational technology? The development of technological pedagogical content knowledge. *Journal of Educational Computing Research, 32*(2), 131–152.

Koepp, M. J., Gunn, R. N., Lawrence, A. D., Cunningham, V. J., Dagher, A., Jones, T., et al. (1998). Evidence for striatal dopamine release during a video game. *Nature, 393*, 266–268. https://doi.org/10.1038/30498

Koivisto, J., & Hamari, J. (2019). The rise of motivational information systems: A review of gamification research. *International Journal of Information Management, 45*, 191–210.

Kolb, B., Harker, A., & Gibb, R. (2017). Principles of plasticity in the developing brain. *Developmental Medicine & Child Neurology, 59*(12), 1218–1223.

Kowert, R., & Quandt, T. (2015). *The video game debate: Unravelling the physical, social, and psychological effects of video games*. Routledge.

Kühn, S., & Gallinat, J. (2014). Amount of lifetime video gaming is positively associated with entorhinal, hippocampal and occipital volume. *Molecular Psychiatry, 19*(7), 842–847.

Kühn, S., Romanowski, A., Schilling, C., Lorenz, R., Mörsen, C., Seiferth, N., et al. (2011). The neural basis of video gaming. *Translational Psychiatry, 1*(11), e53.

Lavy, V. (2010). *Do differences in school's instruction time explain international achievement gaps in math, science, and reading?: Evidence from developed and developing countries*. National Bureau of Economic Research.

Lorenz, R. C., Gleich, T., Gallinat, J., & Kühn, S. (2015). Video game training and the reward system. *Frontiers in Human Neuroscience, 9*, 76760.

Löytömäki, J., Ohtonen, P., & Huttunen, K. (2024). Serious game the Emotion Detectives helps to improve social–emotional skills of children with neurodevelopmental disorders. *British Journal of Educational Technology, 55*(3), 1126–1144.

Manzano-León, A., Camacho-Lazarraga, P., Guerrero, M. A., Guerrero-Puerta, L., Aguilar-Parra, J. M., Trigueros, R., & Alias, A. (2021). Between level up and game over: A systematic literature review of gamification in education. *Sustainability, 13*(4), 2247.

Ni, S., Dong, R., & Ueichi, H. (2023). The influence of online multiplayer games on social capital and interdependent well-being in Japan. *Entertainment Computing, 47*, 100587.

Noroozi, O., Dehghanzadeh, H., & Talaee, E. (2020). A systematic review on the impacts of game-based learning on argumentation skills. *Entertainment Computing, 35*, 100369.

Oku, A. Y. A., Pinheiro, E. D., da Silva Soares, R., Jr., & Sato, J. R. (2023). Analyzing teacher-student interactions through graph theory applied to hyperscanning fNIRS data. *Progress in Brain Research, 282*, 123–143.

Palaus, M., Marron, E. M., Viejo-Sobera, R., & Redolar-Ripoll, D. (2017). Neural basis of video gaming: A systematic review. *Frontiers in Human Neuroscience, 11*, 231323.

Papastergiou, M. (2009). Digital game-based learning in high school computer science education: Impact on educational effectiveness and student motivation. *Computers & Education, 52*(1), 1–12.

Prensky, M. (2009). H. sapiens digital: From digital immigrants and digital natives to digital wisdom. *Innovate: Journal of Online Education, 5*(3), 1–6

Prioste, C., & Raiça, D. (2017). Inclusão digital e os principais desafios educacionais brasileiros. *Revista on line de Política e Gestão Educacional, 21*, 860–880.

Rahimi, S., & Shute, V. J. (2021). The effects of video games on creativity: A systematic review. In S. W. Russ, J. D. Hoffmann, & J. C. Kaufman (Eds). *Handbook of lifespan development of creativity (Vol. 37)*. Cambridge University Press

Ramos, D. K., & Pimentel, F. S. C. (2021). *Cognição, aprendizagem e jogos digitais* (p. 13). BG Business Graphics Editora.

Schunk, D. H., & Zimmerman, B. J. (2012). Self-regulation and learning. In *Handbook of psychology* (2nd ed., p. 7). Wiley.

Sefton, A. P., & Galini, M. E. (2022). *Metodologias ativas: desenvolvendo aulas ativas para uma aprendizagem significativa*. Freitas Bastos.

Shute, V. J. (2008). Focus on formative feedback. *Review of Educational Research, 78*(1), 153–189. https://doi.org/10.3102/0034654307313795

Singley, M. K., & Anderson, J. R. (1989). *The transfer of cognitive skill*. Harvard University Press.

Snyder, L. G., & Snyder, M. J. (2008). Teaching critical thinking and problem solving skills. *The Journal of Research in Business Education, 50*(2), 90.

Söderström, S., & Palm, T. (2024). Feedback in mathematics education research: A systematic literature review. *Research in Mathematics Education*, 1–22. https://doi.org/10.1080/14794802.2024.2401488

Stone, B. G., Mills, K. A., & Saggers, B. (2019). Online multiplayer games for the social interactions of children with autism spectrum disorder: A resource for inclusive education. *International Journal of Inclusive Education, 23*(2), 209–228.

Thibault, R. T., MacPherson, A., Lifshitz, M., Roth, R. R., & Raz, A. (2018). Neurofeedback com fMRI: Uma revisão sistemática crítica. *NeuroImage, 172*, 786–807.

Van Eck, R. (2006). Digital game-based learning: It's not just the digital natives who are restless. *Educause Review, 41*(2), 16–30.

von Gillern, S., & Nash, B. (2024). Video games and literacy learning: Exploring the research on serious and entertainment games. *Technology, Knowledge and Learning*. https://doi.org/10.1007/s10758-024-09787-6

Vygotsky, L. S. (1978). *Mind in society: The development of higher psychological processes*. Harvard University Press.

Wells, A., Mayer, R. E., Plass, J. L., & Homer, B. D. (2021). Playing a video game and learning to think: What's the connection? *Journal of Cognitive Enhancement, 5*, 459–467. https://doi.org/10.1007/s41465-021-00214-7

Weng, C.-B., Qian, R.-B., Fu, X.-M., Lin, B., Han, X.-P., Niu, C.-S., & Wang, Y.-H. (2013). Gray matter and white matter abnormalities in online game addiction. *European Journal of Radiology, 82*(8), 1308–1312. https://doi.org/10.1016/j.ejrad.2013.01.031

West, G. L., Zendel, B. R., Konishi, K., Benady-Chorney, J., Bohbot, V. D., Peretz, I., & Belleville, S. (2017). Playing Super Mario 64 increases hippocampal grey matter in older adults. *PLoS One, 12*(12), e0187779.

Wouters, P., Van Nimwegen, C., Van Oostendorp, H., & Van Der Spek, E. D. (2013). A meta-analysis of the cognitive and motivational effects of serious games. *Journal of Educational Psychology, 105*(2), 249.

Yoshida, W., Seymour, B., Friston, K. J., & Dolan, R. J. (2010). Neural mechanisms of belief inference during cooperative games. *Journal of Neuroscience, 30*(32), 10744–10751.

Zhang, Q., Tian, J., & Chen, L. (2020). Violent video game effects on aggressive behavior among children: The role of aggressive motivation and trait-aggressiveness in China. *Journal of Aggression, Maltreatment & Trauma, 30*(2), 175–192. https://doi.org/10.1080/10926771.2020.1866135

Zhang, R. Y., Chopin, A., Shibata, K., Lu, Z. L., Jaeggi, S. M., Buschkuehl, M., et al. (2021). Action video game play facilitates "learning to learn". *Communications Biology, 4*(1), 1154.

Chapter 2
Neurobiological Basis of Gaming Behavior

Felipe de Oliveira Aguiar, Diogo de Lacerda Veiga, Luiz Antônio Vesco Gaiotto, Thales Marcon Almeida, Roseli Lage de Oliveira, and Ricardo Riyoiti Uchida

Abstract This chapter presents an integrative overview of the neurobiological foundations underlying gaming behavior. It explores how decision-making, reward processing, cognitive control, impulsivity, compulsivity, emotional regulation, and neuroplasticity are involved in engagement and immersion in gaming. Structural and functional neuroimaging findings are reviewed, highlighting alterations in brain anatomy and connectivity associated with both recreational and pathological gaming behaviors, particularly Internet gaming disorder (IGD). The chapter also addresses the neurobiological intersections between IGD and other psychiatric conditions, such as substance use disorders (SUD) and attention-deficit/hyperactivity disorder (ADHD). Special attention is given to the role of craving, impulsiveness, compulsivity, and emotional dysregulation as core mechanisms that sustain problematic gaming behavior. These insights provide a deeper understanding of the neural processes involved in gaming and their implications for mental health, prevention, and intervention strategies.

Keywords Gaming behavior · Neurobiology · Internet Gaming Disorder · Reward system · Impulsivity

F. de Oliveira Aguiar · R. R. Uchida (✉)
Mental Health Department, Santa Casa de São Paulo School of Medical Sciences, São Paulo, SP, Brazil

Artificial Intelligence Research Division, Infinity Doctor's Inc., Dover, DE, USA
e-mail: ricardouchida@scioed.com

D. de Lacerda Veiga · L. A. V. Gaiotto · T. M. Almeida · R. L. de Oliveira
Mental Health Department, Santa Casa de São Paulo School of Medical Sciences, São Paulo, SP, Brazil

L. Murrins Marques et al. (eds.), *Social and Affective Neuroscience of Gaming*,
https://doi.org/10.1007/978-3-032-00325-6_2

Introduction

This chapter explores the fundamental neurobiological concepts underlying gaming behavior, addressing the neural mechanisms involved in engagement and immersion. It also examines individual differences in performance and the development of pathological patterns, such as Internet gaming disorder (IGD). Furthermore, the chapter analyzes the neurobiological similarities between IGD and other psychiatric disorders, such as substance use disorders (SUD) and attention-deficit/hyperactivity disorder (ADHD). Finally, we discuss the implications of these findings for understanding the impact of gaming on cognition, mental health, and human behavior.

Electronic games have become a widely disseminated form of entertainment and social interaction, affecting millions of individuals worldwide. Advances in neuroscience have enabled the investigation of the cognitive and behavioral effects of gaming, encompassing both the benefits of recreational engagement and the risks associated with excessive use. The neurobiology of gaming behavior examines the neural substrates involved in decision-making, reward processing, cognitive control, impulsivity, motivation, and other fundamental processes that regulate the player's interaction with the virtual environment.

Neurobiology is the area of science that investigates the behavior of the central nervous system, including the brain's responses to environmental stimuli, how it represents the environment, the influence of external interactions, and the role of genetics in behavior. It also studies how we learn, forget, the evolution of the brain and nervous system, and behaviors that may indicate disorders, among other relevant aspects.

The neuroscience behind gaming behavior plays a fundamental role in understanding the structural and functional changes of the brain in response to stimuli received. Visual and auditory stimuli, psychomotor activities and social interactions during gaming can have positive impacts on the lives of people of different ages, helping them to cope with stressors (Jones et al., 2014), treat depression and anxiety (Kowal et al., 2021) and improve cognitive function in people with dementia (Saragih et al., 2022). However, gaming can also have negative effects, such as loneliness, depression, and a loss of clear identity, especially when used as a form of escapism (Marques et al., 2023).

In humans, neuroimaging represents the primary set of techniques used in neurobiological investigation and can be classified into structural and functional modalities, each with distinct characteristics and purposes.

Structural neuroimaging is used to visualize and assess morphological aspects of the brain, allowing, for example, the evaluation of the volume of specific brain structures or regions, typically through techniques such as voxel-based morphometry applied to magnetic resonance imaging (MRI) data. In MRI studies, diffusion tensor imaging (DTI) can be employed to assess the directional organization of cerebral white matter through tractography, using indices such as fractional anisotropy (FA), which estimates molecular directionality within white matter.

Functional neuroimaging, on the other hand, is used to assess brain activity. Functional magnetic resonance imaging (fMRI) and positron emission tomography (PET) are examples of functional techniques that infer neural activity by evaluating changes in cerebral blood flow or estimated brain metabolism. The conceptual differences between these methods influence how the data are interpreted. In PET, virtually any radioactive ligand can be produced—from glucose to neurotransmitters and pharmacological agents. However, it involves the use of radioactive substances, and the production of short half-life isotopes requires an expensive cyclotron located near the imaging facility. These techniques have made it possible to identify patterns of brain activation during the gaming experience.

The magnetic resonance imaging (MRI), in contrast, does not involve ionizing radiation, offers higher spatial resolution, and allows for multiple types of acquisitions using the same scanner. For example, by assessing brain activity over a given period while the individual is at rest (resting-state), a low-frequency rhythmic pattern of activation can be observed in specific regions of interest. Such studies allow for the estimation of functional connectivity (FC). Additionally, and simultaneously, structural connectivity can be evaluated through DTI, thereby providing bimodal data on white matter connectivity.

In summary, while structural neuroimaging provides a static view (trait) of brain anatomy, functional neuroimaging offers a dynamic perspective (state) of brain activity.

Understanding the neurochemistry and neurobiology involved in gaming behavior allows for a deeper comprehension of the complexity surrounding this behavior, as well as the underlying mechanisms involved, such as the reward system, impulsivity to compulsivity spectrum, salience processing, cognitive control, decision-making, and emotional regulation.

Neurochemistry and Neurobiology of Gaming Behavior

The neurochemistry involved in gaming behavior is complex. Norepinephrine is responsible for arousal and hyperactivation; dopamine is linked to reward and reinforcement systems; opioids are associated with pleasure and impulse control; serotonin plays a role in the initiation or cessation of gaming behavior—impulse control (Potenza, 2008; Potenza, 2013), decision-making, behavioral control, and inhibition (Yau & Potenza, 2015); cortisol is related to stress responsivity; and glutamate is involved in cognitive functioning, including cognitive flexibility (Potenza, 2013). In individuals with problematic gaming behavior or gaming disorder, significant alterations in these aspects have been observed.

The reward system is closely tied to decision-making and motivation. Structures such as the nucleus accumbens and the orbitofrontal cortex play a central role in reward processing, reinforcing behaviors that generate pleasure and satisfaction. This system is involved in both everyday activities and in risk behaviors and disorders such as addiction and pathological gaming.

When evaluating sensitivity to reward (SR) and sensitivity to punishment (SP) between men and women through a monetary incentive delay task assessed via fMRI, the study concluded that women exhibited higher SP, with lower neuronal responses to both high and low magnitude wins. Men, on the other hand, exhibited higher SR, regardless of the incentive magnitude, with greater activation in the bilateral orbitofrontal cortex and visual areas, while monetary losses activated the bilateral orbitofrontal cortex (Dhingra et al., 2021). This study demonstrated that men and women have different sensitivities to reward and punishment.

A review investigating reward mechanisms in individuals with Gambling Disorder (GD) through fMRI revealed increased ventral striatal activity during reward anticipation in pathological gamblers, concluding that there is an anticipatory vigilance sustained by waiting for rewards, accompanied by distorted reward expectation evaluations and increased dopamine regulation, though without structural brain alterations (Clark et al., 2019).

Impulsivity is characterized by a tendency to act quickly without considering consequences. It is modulated by a balance between the reward system and inhibitory control by the prefrontal cortex. High levels of impulsivity are associated with difficulties in self-control and greater vulnerability to risk behaviors, such as substance abuse and pathological gambling.

Impulsivity can be observed across different subcomponents, such as: motor impulsivity—the tendency to act without forethought, associated with deficits in inhibitory control; cognitive impulsivity—the difficulty in weighing risks and rewards carefully, which can lead to poor decision-making; and choice impulsivity—the preference for immediate gratification over delayed rewards (Punia & Balodis, 2019).

In a study with university students, Yan et al. (2016) assessed whether participants exhibited problematic gambling behavior and its relationship with impulsive behavior facets through the Impulsive Behaviors Scale (BIS-11), such as sensation seeking, lack of perseverance, lack of planning, negative urgency, and positive urgency (Smith et al., 2007), as well as impulsivity subtraits via the Barratt Impulsiveness Scale, including motor, non-planning, and attentional impulsivity (Stanford et al., 2009).

Their results identified that men were more likely to be problematic gamblers than women. Furthermore, when compared to recreational gamblers, there was a significant increase in both positive and negative urgency among pathological gamblers, whereas at-risk gamblers showed a significant increase only in positive urgency, with no significant difference in negative urgency relative to recreational gamblers (Yan et al., 2016).

Negative urgency refers to acting impulsively under distress and is similar to the impulsivity facet of neuroticism (Smith et al., 2007). Positive urgency is the tendency to act impulsively during exceptionally positive mood states (Tan & Tam, 2024). Thus, both positive and negative urgency are important personality facets related to impulsive behavior.

One of the main contributions of the study was that negative urgency predicted both pathological and at-risk gamblers compared to recreational gamblers, suggesting that negative urgency may represent a specific trait predisposing individuals to the development of psychopathology, thereby serving as a potential vulnerability marker for developing Gambling Disorder. Negative urgency is associated with a strong impulse tendency in contexts of negative affect, reflecting deficits in inhibitory control processes. The authors concluded that impulsivity is a central feature in Gambling Disorder (Yan et al., 2016).

In a study with recreational gamblers, the association between impulsivity, gambling-related cognitive distortions, cognitive reappraisal, and problematic gambling was investigated. The results demonstrated that elevated levels of impulsivity traits, gambling-related cognitive distortions, and cognitive reappraisal were significant moderating variables that strengthened the association between cognitive reappraisal and problematic gambling. Thus, individuals with high levels of impulsivity showed a positive correlation between cognitive reappraisal and problematic gambling, perpetuating their gambling behavior, whereas those with low impulsivity exhibited a near-zero correlation (Tan & Tam, 2024).

Another important finding identified by the authors was that among individuals with low levels of gambling-related cognitive distortions, cognitive reappraisal became an adaptive coping and emotional regulation strategy. Instead of engaging in maladaptive gambling behaviors to escape negative emotions, these individuals were able to restructure their cognitions to better cope with emotional experiences. In this way, cognitive reappraisal may serve as a protective factor against problematic gambling (Tan & Tam, 2024).

While impulsivity involves immediate, unplanned responses without a real evaluation of risks or negative consequences, compulsivity, on the other hand, is characterized by persistent and rigid behaviors often driven by negative reinforcement, such as avoiding negative emotions or discomfort states. Unlike impulsivity, which involves quick, reactive responses, compulsivity is marked by resistance to change even when the behavior is no longer advantageous (Bottesi et al., 2015; Lee et al., 2019).

Salience refers to the degree of importance or relevance attributed to specific stimuli, directly influencing attention and behavior. In the context of the reward system, stimuli that intensely activate the dopaminergic circuit become highly salient, potentially leading to a compulsive craving for certain activities, such as gaming or substance use. While this mechanism is essential for motivation, when dysregulated, it can contribute to compulsive behaviors and addiction disorders.

Cognitive control refers to the ability to regulate thoughts and actions in order to achieve goals. It is related to executive function and depends on the activity of the dorsolateral prefrontal cortex. Cognitive control enables the inhibition of impulsive responses, planning, and cognitive flexibility, which are essential for adapting to different contexts.

Decision-making is a neurocognitive process that involves evaluating options, risks, and rewards before making a choice. This process depends on the interaction between the prefrontal cortex, which regulates logical reasoning and planning, and the striatum, which contributes to the evaluation of rewards and punishments. In individuals with gambling disorder (GD), decision-making tasks have shown reduced activation in prefrontal regions, such as the orbitofrontal cortex, which is associated with risk evaluation (Clark et al., 2019; Miedl et al., 2010).

Another important aspect concerns inhibitory control. Pathological gamblers show reduced activity in the dorsolateral prefrontal cortex when performing inhibitory control tasks, making them more vulnerable to intense urges to persist in gambling and impulsive behavior (Cosenza et al., 2024). Chasing behavior may be driven by dopamine release in the ventral striatum, sustaining the cycle of disadvantageous decision-making and, consequently, compulsive gambling (Spagnolo et al., 2019).

Craving encompasses both the anticipation of pleasure (e.g., winning bets)—positive reinforcement—and the expectation of relief from negative affect (e.g., continued gambling)—negative reinforcement—through the alleviation of stress, anxiety, and depressive symptoms. Individuals who exhibit a more dependent decision-making style tend to have greater difficulty making decisions and persisting in gambling, which may serve as a protective factor against the continuation of chasing behavior (Cosenza et al., 2024).

Craving can be induced either by cues or by stress. Cue-induced craving occurs when players are exposed to images, sounds, or situations associated with gambling, leading to activation of brain areas linked to the urge to gamble. fMRI studies have shown greater activation in the orbitofrontal cortex and ventral striatum, even in individuals who had abstained from gambling for extended periods (Spagnolo et al., 2019). Stress-induced craving, frequently observed among pathological gamblers, involves the use of gambling as a means to alleviate negative emotions such as anxiety and frustration. Pathological gamblers exposed to stressful situations exhibit greater activation in the amygdala and anterior cingulate cortex, brain regions involved in emotional processing and decision-making (Cosenza et al., 2024). Under stress, craving intensity likely increases through the action of the noradrenaline and cortisol systems, which amplify feelings of urgency and loss of control over impulsivity (Spagnolo et al., 2019).

Emotions and emotion regulation play a critical role in gambling behavior, and two main aspects can be identified. On one hand, the emotional function of gambling is observed, where players use gambling to manage specific emotional states such as anxiety, anger, and frustration (Marchica et al., 2020). On the other hand, studies have discussed deficits in emotion regulation that may predispose individuals to maladaptive gambling behavior and the development of gambling disorder (Rogier & Velotti, 2018). Emotions exert an ambivalent influence on gambling behavior. Recreational gamblers tend to report pleasure, excitement, and euphoria during gambling sessions, which may reinforce gambling behavior and contribute to its social acceptability (Rogier & Velotti, 2018). However, among pathological gamblers, gambling may serve as an emotional regulation strategy to avoid negative emotional states (Marchica et al., 2020).

A meta-analysis conducted with 38 studies sought to examine the association between specific emotion regulation difficulties and emotion regulation strategies among gamblers. The study found that pathological gamblers had significantly higher levels of emotion regulation difficulties compared to individuals without the disorder. Among specific emotion regulation difficulties, a positive relationship was observed with a small effect size for the dimensions of mindfulness, clarity, goals, and impulse, and a medium effect size for non-acceptance and accessibility/flexibility in pathological gamblers, highlighting the role of emotion dysregulation in the onset and maintenance of the disorder (Velotti et al., 2021).

Pathological gamblers may exhibit impairments in various components of emotion regulation, such as low ability to identify and understand their own emotions, difficulties in tolerating unpleasant emotions, and a tendency to use gambling as a strategy for emotional relief (Marchica et al., 2020). The excessive use of maladaptive strategies, such as emotional suppression or experiential avoidance, has also been discussed in other studies (Maniaci et al., 2017; Marchica et al., 2020; Rogier & Velotti, 2018).

Performance Studies with Gamers

When comparing high-performing gamers to those with average or low performance (Wang et al., 2018), the former showed greater activation in the declive, a cerebellar region involved in motor control and motor skill learning, as well as increased activation in the postcentral gyrus and the striatum. Regarding brain connectivity, high-performing gamers demonstrated greater functional connectivity between the declive, the precentral gyrus, and the thalamus, whereas low-performing gamers exhibited reduced connectivity in the middle frontal gyrus. Furthermore, early skill acquisition and performance improvement were positively associated with dorsal striatal volume, but not with hippocampal volume (Erickson et al., 2010).

These findings suggest that better performance in online gaming is associated with coordination within the brain's sensorimotor pathways, including the precentral gyrus (motor responses), the occipital gyrus (visual processing), and the thalamus, which functions as an information-processing hub and is connected to the prefrontal cortex (PFC) (Georgopoulos, 2000; de Haas et al., 2021; Cassel et al., 2013). The relationship between striatum and performance may reflect the relevance of the limbic network in learning, reinforcement, reward, motivation, and cognitive processes related to strategy and planning in gameplay (Cox & Witten, 2019; Shohamy, 2011). Additionally, the striatum is strongly connected to areas involved in executive control and decision-making, such as the middle and inferior frontal gyri (Brown, 2011).

A study examining professional online gamers found a positive correlation between cortical thickness in the right superior frontal gyrus, right superior parietal gyrus, and right precentral gyrus and career duration. Additionally, cortical thickness in the prefrontal cortex was associated with win rates (Hyun et al., 2013). Compared to amateur players, expert gamers showed greater gray matter volume in the right posterior parietal cortex (Tanaka et al., 2013), which correlated with superior performance in visual working memory tasks. They also exhibited increased gray matter volume in the insula and enhanced functional connectivity within attentional and sensorimotor networks, including the left insula (Gong et al., 2015). Taken together, these findings indicate that superior gaming performance is associated with enhanced visuospatial, attentional, and motor connectivity (Anguera et al., 2013; Green & Bavelier, 2006).

Moreover, professional gamers exhibit distinct patterns of brain morphology compared to individuals with online gaming addiction. While professional gamers show increased gray matter volume in the left cingulate gyrus, individuals with gaming addiction display reduced gray matter volume in the thalamus, a region associated with dopaminergic reinforcement and reward processes (Han et al., 2012; Sánchez-González et al., 2005). Increased gray matter in the left cingulate is positively correlated with lower impulsivity and fewer perseverative errors. Additionally, this structure is associated with better executive function and attentional control (MacDonald 3rd et al., 2000), potentially serving as a marker of successful performance in online gaming.

Internet Gaming Disorder

Internet gaming disorder (IGD) is a condition recognized by the World Health Organization (WHO) under code 6C51.0 in the ICD-11, and it is included in the "Conditions for Further Study" section of the Diagnostic and Statistical Manual of Mental Disorders, Fifth Edition (DSM-5), by the American Psychiatric Association (APA). In both classifications, it is described as a disorder that can lead to significant impairments in daily life due to excessive online gaming.

In the DSM-5 Revised (APA, 2022), the following diagnostic criterion is proposed:

Persistent and recurrent use of the Internet to engage in games, often with other players, leading to clinically significant impairment or distress as indicated by five (or more) of the following 12-month period:

1. Preoccupation with Internet games. (The individual thinks about previous gaming activity or anticipates playing the next game; Internet gaming becomes the dominant activity in daily life.)

Note: This disorder is distinct from Internet gambling, which is included under gambling disorder.

2. Withdrawal symptoms when Internet gaming is taken away. (These symptoms are typically described as irritability, anxiety, or sadness, but there are no physical signs of pharmacological withdrawal.)
3. Tolerance—the need to spend increasing amounts of time engaged in Internet games.
4. Unsuccessful attempts to control participation in Internet games.
5. Loss of interest in previous hobbies and entertainment as a result of, and with the exception of, Internet games.

6. Continued excessive use of Internet games despite knowledge of psychosocial problems.
7. Has deceived family members, therapists, or others regarding the amount of Internet gaming.
8. Use of Internet games to escape or relieve a negative mood (e.g., feelings of helplessness, guilt, anxiety).
9. Has jeopardized or lost a significant relationship, job, or educational or career opportunity because of participation in Internet games.

Note: Only nongambling Internet games are included in this disorder. Use of the Internet for required activities in a business or profession is not included; nor is the disorder intended to include other recreational or social Internet use. Similarly, sexual Internet sites are excluded.

Specify current severity: Internet gaming disorder can be mild, moderate, or severe depending on the degree of disruption of normal activities. Individuals with less severe Internet gaming disorder may exhibit fewer symptoms and less disruption of their lives. Those with severe Internet gaming disorders will have more hours spent on the computer and more severe loss of relationships or career or school opportunities.

The etiology of Internet gaming disorder (IGD) is complex, involving internal factors such as poor mood regulation and dysregulation of the reward system, as well as external factors including family environment and social skills. Longitudinal studies indicate that IGD is associated with reduced psychosocial well-being, negatively impacting self-esteem, social support, and life satisfaction. Prevalence rates of IGD vary, but studies suggest that it may affect between 1% and 9% of the population, depending on factors such as age and geographic location (Teng et al., 2020; Gentile et al., 2017).

Studies have demonstrated gray matter reduction in several regions: bilaterally in the anterior cingulate cortex (ACC), supplementary motor area (SMA), superior parietal cortex, left dorsolateral prefrontal cortex (DLPFC), left insula, and bilaterally in the cerebellum. The occurrence of incongruent responses in the Stroop task was negatively correlated with gray matter volume in the ACC (Wang et al., 2015).

Individuals with Internet gaming disorder (IGD) show increased volume in both the ventral striatum (nucleus accumbens) and dorsal striatum (caudate), a pattern similar to that observed in studies on psychoactive substance addiction. This increase in right caudate volume, along with resting-state functional connectivity alterations within the fronto-striatal pathway (DLPFC–caudate), is also correlated with greater Stroop task incongruence, reflecting impaired cognitive control (Yuan et al., 2017).

Gray matter reduction in prefrontal cortex regions and bilateral amygdala has been correlated with increased impulsivity among individuals with IGD, as assessed by the Barratt Impulsiveness Scale (Lin et al., 2015). Findings of gray matter reduction in the ACC, SMA, insula, and inferior temporal gyrus have been replicated across multiple studies and summarized in meta-analyses (Weinstein et al., 2017).

Interestingly, individuals with IGD exhibit increased fractional anisotropy in several brain regions, which correlates with the severity of their gaming addiction. These changes suggest enhanced white matter integrity, potentially due to increased myelination or other structural changes in response to prolonged gaming activities. These regions include the bilateral anterior thalamic radiation, anterior limb of the internal capsule, corticospinal tract, inferior fronto-occipital fasciculus, corpus callosum, and inferior longitudinal fasciculus. These changes may be linked to the reward circuitry and sensory and motor control systems, which are often engaged during gaming (Dong et al., 2018; Jeong et al., 2016; Dong et al., 2012).

Resting-state fMRI studies reveal significant alterations in brain activity and connectivity in individuals with IGD. These findings underscore the importance of understanding the neural mechanisms to develop targeted interventions and improve treatment outcomes.

Individuals with persistent IGD (PER-IGD) show decreased activity in brain regions associated with reward processing and inhibitory control, such as the orbitofrontal cortex (OFC), precuneus, and dorsolateral prefrontal cortex (DLPFC), compared to those who have recovered (RE-IGD) (Liu et al., 2023). These differences may influence natural recovery from IGD.

Adolescents with internet gaming addiction (IGA) exhibit altered functional connectivity in the default-mode network (DMN), with increased connectivity in the bilateral cerebellum posterior lobe and middle temporal gyrus, and decreased connectivity in the bilateral inferior parietal lobule and right inferior temporal gyrus (Ding et al., 2013). These patterns are partially consistent with those observed in substance addiction, suggesting shared neurobiological abnormalities.

By examining white matter tracts, it is possible to analyze the efficiency of brain connectivity at different levels: global efficiency, which assesses the number of parallel connections available along a pathway; local efficiency, which evaluates how communication occurs between nearby regions of interest following the removal of a hub; and regional efficiency, which reflects the average shortest path length from a given hub to all other nodes. Individuals with IGD show reduced global and local efficiency, along with an increase in the shortest communication path length. The severity of IGD is negatively correlated with global efficiency. Impairments in regional efficiency have been observed in the frontal cortex, anterior cingulate cortex, and the globus pallidus (Zhai et al., 2017).

In IGD, there is impaired functional connectivity within executive control networks, as assessed through the activity of predefined regions such as the ventromedial prefrontal cortex, dorsolateral prefrontal cortex, and parietal cortex. This reduction in functional connectivity (FC) shows a negative correlation with performance accuracy on the Stroop task (Dong et al., 2015).

The assessment of insular FC is of particular interest in IGD, as this region is associated with the salience network, which may render individuals with IGD more susceptible to assigning increased value to gaming-related cues and maintaining engagement in gaming activities. Individuals with IGD exhibit increased FC between the anterior insula and the ACC—a key component of the salience network—compared to healthy controls. These individuals also show increased connectivity between the insula and the angular gyrus, precuneus, and inferior parietal lobe, which are components of the Default Mode Network (DMN). The insula appears to function as a modulator between the DMN and the Executive Control Network (ECN). While the DMN is involved in sustaining habitual and self-referential behaviors, the ECN is responsible for generating goal-directed behavior and enhancing cognitive control. Functional connectivity alterations between the insula and ACC have been correlated with the duration of IGD. These findings suggest not only a possible predisposition in individuals with such neural alterations to develop IGD, but also that the disorder itself may negatively affect the functioning of these neural pathways (Zhang et al., 2016).

The assessment of insular functional connectivity (FC) has proven particularly relevant in the context of IGD, as the insula is associated with the salience network. This network may increase susceptibility in individuals with IGD to assign greater value to gaming-related cues and to sustain engagement with such stimuli. Studies have shown increased FC between the anterior insula and the ACC—core component of the salience network—in individuals with IGD compared to healthy controls. Additionally, these individuals also exhibit enhanced connectivity between the insula and the angular gyrus, precuneus, and inferior parietal lobe, which are regions associated with the default mode network (DMN). The insula appears to function as a modulator between the DMN and the executive control network (ECN). While the DMN is involved in maintaining habitual and self-referential behaviors, the ECN is responsible for generating goal-directed behaviors and modulating cognitive control. Alterations in functional connectivity between the insula and ACC have been correlated with the duration of IGD. These findings suggest not only a potential predisposition in individuals with such neural alterations to develop IGD, but also that the disorder itself may negatively affect the functioning of these neural pathways (Zhang et al., 2016).

This set of findings points to alterations in the balance of activation between the insula and the ECN. In a study conducted by Han et al. (2018a), the dynamic functional connectivity (FC) of the DLPFC—a cortical structure critically involved in executive control—was investigated. Dynamic FC is an approach that examines interactions between regions of interest on a shorter temporal scale than that typically used in studies based on blood-oxygen-level dependent (BOLD) signals. This technique allows for increased sensitivity to temporal fluctuations in neural activation. The study revealed a reduction in dynamic FC between the right DLPFC and the left insula, a finding that showed a negative correlation with the severity of IGD. Additional findings included reduced dynamic FC between the right DLPFC and the right putamen, as well as with the left precentral gyrus, along with increased connectivity with the precuneus.

Impulse control, assessed using the Go/No-Go task, does not show significant performance differences between the IGD group and the control group. However, individuals with IGD performed more poorly on inhibitory responses (No-Go trials) compared to execution responses (Go trials), a pattern not observed in the control group. fMRI assessments during No-Go tasks showed that the IGD group, compared to controls, exhibited hyperactivation in the left superior medial frontal gyrus, right ACC, right superior and medial frontal gyri, precentral gyrus, as well as the left precuneus and cuneus. In contrast, hypoactivation was observed in the bilateral middle temporal gyrus, bilateral inferior temporal gyrus, and right superior parietal lobule. Activation in the left medial frontal gyrus was positively correlated with the severity of internet addiction, as measured by the CIAS, and with impulsivity scores assessed using the BIS-11 scale. This pattern of hyperactivation in frontoparietal regions and areas involved in inhibitory control suggests that, even in the absence of evident behavioral performance differences, individuals with IGD exhibit reduced cognitive efficiency, requiring greater neural recruitment to perform the same task (Ding et al., 2014; Rypma et al., 2006).

When individuals with addiction to Massive Multiplayer Online Role-Playing Games (MMORPGs) were exposed to images of themselves, strangers, and their avatars, and asked to evaluate these images using the body image questionnaire (FKB-20) and a visual analog scale assessing attractiveness, likability, and gender identification, they exhibited hypoactivation in the bilateral angular gyrus and the middle occipital gyrus while evaluating their own image. In contrast, when evaluating their avatar's, increased activation was observed in the left angular gyrus (Leménager et al., 2014). In a subsequent study, it was demonstrated that activation of the left angular gyrus during reflection on one's avatar was positively correlated with the severity of IGD (Leménager et al., 2016).

Taken together, neurobiological findings in IGD reveal impairments in cognitive control, reward processing, and self-referential networks, indicating both predisposing vulnerabilities and the impact of excessive gaming. These alterations set the stage for understanding how IGD may intersect with other conditions characterized by dysregulation in similar systems, such as substance use disorder (SUD) and attention-deficit/hyperactivity disorder (ADHD).

Substance Use Disorder and Internet Gaming Disorder

Weinstein and Weizman (2012) summarize the possible psychological pathways for the development of IGD in three hypotheses: (1) games are used as a way to cope with frustration, stress, and fears; (2) the skill-based and reinforcement mechanisms present in games influence certain affective dimensions, particularly excitement, physiological arousal, and frustration; (3) the stimulatory effects on reward and sensitization systems may sustain excessive gaming, in a mechanism similar to that observed in substance use disorders. Understanding the shared characteristics between IGD and substance use disorder (SUD) may enhance insight into the underlying mechanisms of addictive behaviors and problematic online gaming.

The prefrontal regions are central to understanding the similarities between IGD and SUD. Both disorders show alterations in the medial frontal gyrus, which extend to the ACC and the inferior frontal gyrus, areas critical for executive and inhibitory control (Klugah-Brown et al., 2021). Han et al. (2007) observed that individuals with IGD exhibit higher reward dependence scores compared to healthy individuals—a pattern also observed in cases of alcohol, nicotine, and opioid abuse.

In SUDs, the processes of "wanting" and "liking" can become dissociated. "Wanting" often becomes more dominant in the progression of the disorder, possibly due to an incentive-sensitization mechanism. In this context, individuals with IGD and tobacco use disorder (TUD) showed similar functional connectivity within the "wanting" circuitry—including the substantia nigra, amygdala, caudate nucleus, putamen, and DLPFC—when exposed to substance- or gaming-related images. This suggests a neural mechanism of sensitization to reward stimuli, contributing to compulsive behaviors in both IGD and TUD (Ma et al., 2024).

Both disorders also share neural connectivity patterns between subcortical and motor networks, including abnormal resting-state functional connectivity between the right thalamus and the right postcentral gyrus, indicating common disruptions in brain circuits across these disorders (Chen et al., 2023).

Some common mechanisms between SUD and IGD have been discussed in this text. However, there is a fundamental difference between them: the presence of a psychoactive substance. Even so, in both IGD and Tobacco Use Disorder (TUD), regions within the fronto-limbic network are activated when individuals are exposed to game- or tobacco-related cues (Ko et al., 2013).

Although typical substance-related effects, such as the reduction in cerebellar and medial frontal cortex volume observed in alcohol use disorder (AUD), are absent in IGD, excessive gaming leads to other alterations. These include increased volume of the hippocampus/amygdala and right precuneus, regions involved in memory and attention, as well as stronger connectivity between the hippocampus/amygdala and the left ventromedial prefrontal cortex, when compared to individuals with AUD. Therefore, even in the absence of neurotoxic effects from substances, IGD-related alterations may be associated with abnormally heightened memory processes and exaggerated emotional responses to gaming-related cues (Yoon et al., 2017).

Internet Gaming Disorder and Attention Deficit and Hyperactivity Disorder

One of the interesting contradictions about IGD is that playing games demands high levels of concentration and hours of continuous engagement, so, how is possible that individuals with attention deficit and hyperactivity disorder (ADHD), who exhibit hyperactivity, short attention span, motor restlessness, and impulsivity, can be so prone to developing IGD?

A possible answer is that online games are highly captivating and stimulating, offering quick rewards and a sense of progress when reaching the "next level." The rapidly changing visual stimuli require minimal attention or working memory, areas in which individuals with ADHD typically struggle. ADHD is associated with increased reward-seeking behavior, and games activate the brain's reward system through dopamine release in the striatum.

ADHD and IGD may have a bidirectional relationship: ADHD symptoms make gaming particularly appealing, while gaming, in turn, exacerbates core ADHD traits such as impulsivity, the need for immediate gratification, and inattention (Weiss et al., 2011).

Despite this counterintuitive association, an emerging epidemiological relationship has been identified. In a five-year cohort study involving patients seeking treatment for IGD, 32.7% of the 755 individuals evaluated were diagnosed with ADHD. Furthermore, those who scored higher on the Korean ADHD Rating Scale were less likely to achieve recovery from IGD (Han et al., 2018b).

In another longitudinal study conducted in 2018, it was found that having ADHD at age 20 increased the risk of developing IGD at age 25, and likewise, having IGD at age 20 increased the risk of ADHD at age 25. However, only inattention symptoms showed a bidirectional relationship between ADHD and IGD; hyperactivity symptoms did not reach statistical significance for this bidirectional association (Marmet et al., 2018). In a separate longitudinal study, it was also observed that gaming behavior was influenced by ADHD-related inattention (Schoenmacker et al., 2020).

A more recent meta-analysis identified a moderate association between IGD and ADHD across both symptom domains: inattention and hyperactivity/impulsivity. Regarding symptom severity, individuals with more severe IGD were more likely to simultaneously exhibit symptoms of both inattention and hyperactivity related to ADHD. Furthermore, the presence of ADHD was positively associated with greater IGD severity (Koncz et al., 2023).

Individuals with comorbid IGD and ADHD were less likely to achieve recovery than those with IGD alone. In a three-year follow-up study involving 128 individuals, it was found that 93% of the IGD-only group recovered, whereas only 60% of the ADHD+IGD group achieved recovery, indicating a poorer outcome in the management of IGD when ADHD is present as a comorbidity (Lee et al., 2021).

A cross-sectional study showed that children with both ADHD and IGD tend to have significantly poorer interpersonal relationships compared to children with ADHD but without IGD (Tzang et al., 2022).

Supporting these epidemiological findings, IGD and ADHD share neurobiological similarities. It is well established that online gaming activates the brain's dopamine-related reward system. This system plays a central role in both ADHD and IGD. The increase in dopamine during online gaming may act as a compensatory mechanism for the reduced dopamine levels typically observed in individuals with untreated ADHD (Weinstein & Weizman, 2012).

The neural structures and pathways shared by both disorders generally involve deficits in reward and executive function circuits. Both Internet gaming disorder (IGD) and attention-deficit/hyperactivity disorder (ADHD) are associated with alterations in the prefrontal cortex (PFC) and the striatum. The primary cortical regions within the PFC linked to reward and executive functions include the dorsolateral cortex, the prefrontal cortex, the anterior cingulate cortex, and the orbitofrontal cortex.

The IGD+ADHD group showed evidence of frontal neurochemical abnormalities, characterized by reduced levels of N-acetyl-aspartate (NAA), indicating neuronal dysfunction and impairment in regions associated with impulse control and attention. Additionally, this group exhibited lower levels of glutamate+glutamine (Glu+Gln) compared to the ADHD-only group, suggesting further neurochemical alterations linked to problematic online gaming. This reduction may reflect an adaptive mechanism to dopaminergic hyperactivity induced by constant gaming stimulation. While isolated ADHD is associated with glutamatergic hyperactivity, the IGD+ADHD combination appears to neutralize this characteristic due to the chronic impact of gaming activity. These alterations differ from the pattern observed in individuals with ADHD without IGD, highlighting the influence of problematic gaming on the neurobiology of ADHD and its clinical manifestations, such as attentional deficits and increased symptom severity (Bae et al., 2016).

When considering shared alterations, the structures within the prefrontal-striatal circuit play a crucial role. This network includes the anterior cingulate cortex, the orbitofrontal prefrontal cortex, the ventral striatum, the ventral pallidum, the dorsal prefrontal cortex, the amygdala, and the hippocampus.

In a study comparing males with IGD to healthy controls, atrophy was observed in brain regions related to decision-making, behavioral inhibition, and emotional regulation, such as the inferior frontal gyrus, cingulate gyrus, insula, and precuneus. Reduced gray matter volume in the cingulate gyrus, a region suggested to be part of the executive control system, makes it more difficult to regulate compulsive behavior or inhibit impulsive urges to play, even in the face of negative consequences (Lin et al., 2015).

A recent meta-analysis also demonstrated that patients with both IGD and ADHD frequently exhibited reduced gray matter volume in the ACC and the insula, with hyperactivation of the former and hypoactivation of the latter. The main shared alterations between the two conditions were observed in the striatum: both showed reduced gray matter volume in this region. However, only IGD was associated with decreased functional connectivity, while no significant changes were found in fMRI assessments of patients with ADHD (Gao et al., 2021).

Both disorders are associated with impaired inhibitory control. In a case-control study, participants with Internet gaming disorder (IGD) showed reduced gray matter volume in regions involved in executive control, such as the anterior cingulate cortex and the supplementary motor area. These lower volumes were associated with higher scores on self-report measures of impulsivity, such as the Barratt Impulsiveness Scale (Lee et al., 2018).

Another important aspect observed is that both disorders exhibit characteristic cortical asymmetries, although each presents distinct features. Compared to recreational gamers, patients with IGD showed rightward asymmetry in gray matter volume in the orbitofrontal cortex (OFC) and middle temporal gyrus, and leftward asymmetry in the medial prefrontal cortex (mPFC), precuneus, superior parietal lobule (SPL), and the anterior and posterior inferior temporal gyrus (ITG) (Chen et al., 2024).

The analysis of cortical development in right-handed neurotypical individuals reveals an increase in the relative thickness of the right orbitofrontal cortex and right inferior frontal cortex, along with a relative increase in cortical thickness in the left occipital regions (Shaw et al., 2009). Similarly, in right-handed individuals with ADHD, an increase in cortical thickness is also observed in the left occipital regions; however, there is no corresponding trend of increased relative thickness in the right frontal cortex. Instead, there was a relative increase in cortical thickness in the right anterior superior temporal gyrus (Shaw et al., 2009).

A resting-state fMRI study indicated that individuals with both IGD and ADHD showed a similar reduction in functional connectivity between cortical and subcortical regions. The IGD+ADHD group exhibited decreased functional connectivity between cortical areas, specifically the middle frontal cortex and the cingulate gyrus, and the subcortical caudate nucleus. Similarly, the IGD-only group showed reduced functional connectivity between the dorsolateral prefrontal cortex and subcortical areas associated with behavioral regulation (Han et al., 2021). These findings suggest that both disorders are linked to deficits in impulse and behavioral control, even when examined independently.

However, individuals with IGD who had a childhood diagnosis of ADHD showed expanded functional connectivity between the posterior cingulate cortex, a central region of the default mode network, and the cerebellum. This increased connectivity was positively associated with impulsivity, as measured by self-report scales. This alteration was not observed in IGD individuals without a history of ADHD, suggesting that inefficient executive control in children with ADHD may serve as a predisposing factor for the development of IGD (Lee et al., 2017).

Partially answering the question of why individuals with ADHD play videogames, a quantitative electroencephalogram (QEEG) study showed lower relative power in the delta band and higher relative power in the beta band in the temporal regions of individuals with comorbid IGD and ADHD, compared to those with ADHD only. Additionally, the severity of IGD symptoms was positively correlated with intra-hemispheric coherence in the delta, theta, alpha, and beta bands in the parieto-occipital regions of the IGD+ADHD group. The increased beta activity in the temporal regions may reflect repeated stimulation of the dopaminergic system during gaming, similar to the effect of methylphenidate in ADHD. Thus, internet video game use may represent a form of self-medication in IGD patients who experience attentional difficulties.

Treating ADHD may help reduce internet addiction and IGD. In a clinical trial, children treated with extended-release methylphenidate for 8 weeks showed a reduction in problematic internet use, provided that ADHD symptoms were also effectively managed (Han et al., 2009). Another study found that treatment with extended-release methylphenidate or atomoxetine for 3 months reduced internet addiction scores in individuals with both ADHD and IGD (Park et al., 2016). However, ADHD medication does not appear to reduce the risk of developing IGD in these individuals (Isorna Folgar et al., 2024).

Conclusion

The evidence gathered throughout this work points to a robust set of neurobiological alterations involved in IGD, encompassing structural and functional changes in cortical and subcortical regions, as well as neurochemical imbalances in reward and executive control systems. Hyperactivation of dopamine-related circuits, such as the striatum and PFC, combined with dysfunction in regions like the ACC and the insula, suggests a neural pattern that favors compulsive behavior, immediate reward-seeking, and impaired inhibitory control. These alterations manifest in reduced cognitive and functional efficiency, even in the absence of overt behavioral deficits, as seen in tasks involving inhibitory control, decision-making, and emotional regulation.

The integrated analysis of IGD, ADHD, and SUD revealed shared neurobiological mechanisms, including dysfunction of the fronto-limbic network, heightened sensitivity to reward-related stimuli, and involvement of cortico-striatal pathways. The overlap between these disorders, particularly between IGD and ADHD, was demonstrated by structural, functional, and electrophysiological findings, as well as epidemiological and comorbidity data. The presence of asymmetric cortical development patterns altered functional connectivity in executive and default mode networks, and the potential influence of gaming as a form of self-medication reinforces the need for a more integrated and transdiagnostic diagnostic and therapeutic approach. These findings not only enhance our understanding of the neurobiological underpinnings of gaming behavior but also pave the way for more targeted prevention, intervention, and treatment strategies.

References

American Psychiatric Association – APA. (2022). *Diagnostic and statistical manual of mental disorders – DSM-5-TR* (5th ed., text revision). APA.

Anguera, J. A., Boccanfuso, J., Rintoul, J. L., Al-Hashimi, O., Faraji, F., Janowich, J., Kong, E., Larraburo, Y., Rolle, C., Johnston, E., & Gazzaley, A. (2013). Video game training enhances cognitive control in older adults. *Nature, 501*(7465), 97–101.

Bae, S., Han, D. H., Kim, S. M., Shi, X., & Renshaw, P. F. (2016). Neurochemical correlates of internet game play in adolescents with attention deficit hyperactivity disorder: A proton magnetic resonance spectroscopy (MRS) study. *Psychiatry Research – Neuroimaging, 254*, 10–17.

Bottesi, G., Ghisi, M., Ouimet, A. J., Tira, M. D., & Sanavio, E. (2015). Compulsivity and impulsivity in pathological gambling: Does a dimensional–transdiagnostic approach add clinical utility to DSM-5 classification? *Journal of Gambling Studies, 31*, 825–847.

Brown, J. W. (2011). Medial prefrontal cortex activity correlates with time-on-task: What does this tell us about theories of cognitive control? *NeuroImage, 57*, 314–315.

Cassel, J. C., Pereira de Vasconcelos, A., Loureiro, M., Cholvin, T., Dalrymple-Alford, J. C., & Vertes, R. P. (2013). The reuniens and rhomboid nuclei: Neuroanatomy, electrophysiological characteristics and behavioral implications. *Progress in Neurobiology, 111*, 34–52.

Chen, H., Zha, R., Lai, X., Liu, Y., Wei, Z., Wang, M., Zuo, H., Hong, W., Fan, C., Jin, C., Cui, G., Tao, R., Liang, P., & Zhang, X. (2023). Internet gaming disorder and tobacco use disorder share neural connectivity patterns between the subcortical and the motor network. *Human Brain Mapping, 44*(6), 2607–2619.

Chen, S., Yan, J., Lock, M., Wang, T., Wang, M., Wang, L., Yuan, L., Zhuang, Q., & Dong, G.-H. (2024). Alterations of gray matter asymmetry in internet gaming disorder. *Scientific Reports, 14*(1), 28282.

Clark, L., Boileau, I., & Zack, M. (2019). Neuroimaging of reward mechanisms in Gambling disorder: An integrative review. *Molecular Psychiatry, 24*(5), 674–693.

Cosenza, M., Sacco, M., Ciccarelli, M., Pizzini, B., Jiménez-Murcia, S., Fernández-Aranda, F., & Nigro, G. (2024). Getting even: Chasing behavior, decision-making, and craving in habitual gamblers. *BMC Psychology, 12*(1), 445.

Cox, J., & Witten, I. B. (2019). Striatal circuits for reward learning and decision-making. *Nature Reviews. Neuroscience, 20*(8), 482–494.

de Haas, B., Sereno, M. I., & Schwarzkopf, D. S. (2021). Inferior occipital gyrus is organized along common gradients of spatial and face-part selectivity. *The Journal of Neuroscience, 41*(25), 5511–5521.

Dhingra, I., Zhang, S., Zhornitsky, S., Wang, W., Le, T. M., & Li, C. S. R. (2021). Sex differences in neural responses to reward and the influences of individual reward and punishment sensitivity. *BMC Neuroscience, 22*, 1–14.

Ding, W., Sun, J., Sun, Y., Zhou, Y., Li, L., Xu, J., & Du, Y. (2013). Altered default network resting-state functional connectivity in adolescents with internet gaming addiction. *PLoS One, 8*, e59902.

Ding, W. N., Sun, J. H., Sun, Y. W., et al. (2014). Trait impulsivity and impaired prefrontal impulse inhibition function in adolescents with internet gaming addiction revealed by a Go/No-Go fMRI study. *Behavioral and Brain Functions, 10*, 20.

Dong, G., DeVito, E., Huang, J., & Du, X. (2012). Diffusion tensor imaging reveals thalamus and posterior cingulate cortex abnormalities in internet gaming addicts. *Journal of Psychiatric Research, 46*(9), 1212–1216.

Dong, G., Lin, X., & Potenza, M. N. (2015). Decreased functional connectivity in an executive control network is related to impaired executive function in Internet gaming disorder. *Progress in Neuro-Psychopharmacology and Biological Psychiatry, 3*(57), 76–85.

Dong, G., Wu, L., Wang, Z., Wang, Y., Du, X., & Potenza, M. (2018). Diffusion-weighted MRI measures suggest increased white-matter integrity in Internet gaming disorder: Evidence from the comparison with recreational Internet game users. *Addictive Behaviors, 81*, 32–38.

Erickson, K. I., Boot, W. R., Basak, C., Neider, M. B., Prakash, R. S., Voss, M. W., Graybiel, A. M., Simons, D. J., Fabiani, M., Gratton, G., & Kramer, A. F. (2010). Striatal volume predicts level of video game skill acquisition. *Cerebral Cortex, 20*(11), 2522–2530.

Gao, X., Zhang, M., Yang, Z., Wen, M., Huang, H., Zheng, R., Wang, W., Wei, Y., Cheng, J., Han, S., & Zhang, Y. (2021). Structural and functional brain abnormalities in internet gaming disorder and attention-deficit/hyperactivity disorder: A comparative meta-analysis. *Frontiers in Psychiatry, 12*, 679437.

Gentile, D. A., Bailey, K., Bavelier, D., et al. (2017). Internet gaming disorder in children and adolescents. *Pediatrics, 140*(Suppl 2), S81–S85.

Georgopoulos, A. P. (2000). Neural aspects of cognitive motor control. *Current Opinion in Neurobiology, 10*(2), 238–241.

Gong, D., He, H., Liu, D., Ma, W., Dong, L., Luo, C., & Yao, D. (2015). Enhanced functional connectivity and increased gray matter volume of insula related to action video game playing. *Scientific Reports, 5*, 9763.

Green, C. S., & Bavelier, D. (2006). Effect of action video games on the spatial distribution of visuospatial attention. *Journal of Experimental Psychology. Human Perception and Performance, 32*(6), 1465–1478.

Han, D. H., Lee, Y. S., Yang, K. C., Kim, E. Y., Lyoo, I. K., & Renshaw, P. F. (2007). Dopamine genes and reward dependence in adolescents with excessive internet video game play. *Journal of Addiction Medicine, 1*(3), 133–138.

Han, D. H., Lee, Y. S., Na, C., Ahn, J. Y., Chung, U. S., Daniels, M. A., Haws, C. A., & Renshaw, P. F. (2009). The effect of methylphenidate on Internet video game play in children with attention-deficit/hyperactivity disorder. *Comprehensive Psychiatry, 50*(3), 251–256.

Han, D. H., Lyoo, I. K., & Renshaw, P. F. (2012). Differential regional gray matter volumes in patients with on-line game addiction and professional gamers. *Journal of Psychiatric Research, 46*(4), 507–515.

Han, X., Wu, X., Wang, Y., et al. (2018a). Alterations of resting-state static and dynamic functional connectivity of the dorsolateral prefrontal cortex in subjects with internet gaming disorder. *Frontiers in Human Neuroscience, 12*, 41.

Han, D. H., Yoo, M., Renshaw, P. F., & Petry, N. M. (2018b). A cohort study of patients seeking Internet gaming disorder treatment. *Journal of Behavioral Addictions, 7*(4), 930–938.

Han, D. H., Bae, S., Hong, J., Kim, S. M., Son, Y. D., & Renshaw, P. (2021). Resting-state fMRI study of ADHD and internet gaming disorder. *Journal of Attention Disorders, 25*(8), 1080–1095.

Hyun, G. J., Shin, Y. W., Kim, B. N., Cheong, J. H., Jin, S. N., & Han, D. H. (2013). Increased cortical thickness in professional on-line gamers. *Psychiatry Investigation, 10*(4), 388–392.

Isorna Folgar, M., Faílde Garrido, J. M., Dapía Conde, M. D., & Braña Rey, F. (2024). Evaluation of problematic video game use in adolescents with ADHD and without ADHD: New evidence and recommendations. *Behavioral Sciences, 14*(7), 524.

Jeong, B., Han, D., Kim, S., Lee, S., & Renshaw, P. (2016). White matter connectivity and Internet gaming disorder. *Addiction Biology, 21*, 732–742.

Jones, C., Scholes, L., Johnson, D., Katsikitis, M., & Carras, M. C. (2014). Gaming well: Links between video games and flourishing mental health. *Frontiers in Psychology, 5*, 260.

Klugah-Brown, B., Zhou, X., Pradhan, B. K., Zweerings, J., Mathiak, K., Biswal, B., & Becker, B. (2021). Common neurofunctional dysregulations characterize obsessive–compulsive, substance use, and gaming disorders—An activation likelihood meta-analysis of functional imaging studies. *Addiction Biology, 26*(4), e12997.

Ko, C. H., Liu, G. C., Yen, J. Y., Yen, C. F., Chen, C. S., & Lin, W. C. (2013). The brain activations for both cue-induced gaming urge and smoking craving among subjects comorbid with Internet gaming addiction and nicotine dependence. *Journal of Psychiatric Research, 47*(4), 486–493.

Koncz, P., Demetrovics, Z., Takacs, Z. K., Griffiths, M. D., Nagy, T., & Király, O. (2023). The emerging evidence on the association between symptoms of ADHD and gaming disorder: A systematic review and meta-analysis. *Clinical Psychology Review, 106*, 102343. Elsevier Inc.

Kowal, M., Conroy, E., Ramsbottom, N., Smithies, T., Toth, A., & Campbell, M. (2021). Gaming your mental health: A narrative review on mitigating symptoms of depression and anxiety using commercial video games. *JMIR Serious Games, 9*(2), e26575.

Lee, D., Lee, J., Lee, J. E., & Jung, Y.-C. (2017). Altered functional connectivity in default mode network in Internet gaming disorder: Influence of childhood ADHD. *Progress in Neuro-Psychopharmacology and Biological Psychiatry, 75*, 135–141.

Lee, D., Namkoong, K., Lee, J., & Jung, Y. C. (2018). Abnormal gray matter volume and impulsivity in young adults with Internet gaming disorder. *Addiction Biology, 23*(5), 1160–1167.

Lee, R. S., Hoppenbrouwers, S., & Franken, I. (2019). A systematic meta-review of impulsivity and compulsivity in addictive behaviors. *Neuropsychology Review, 29*, 14–26.

Lee, J., Bae, S., Kim, B. N., & Han, D. H. (2021). Impact of attention-deficit/hyperactivity disorder comorbidity on longitudinal course in Internet gaming disorder: A 3-year clinical cohort study. *Journal of Child Psychology and Psychiatry, 62*(9), 1110–1119.

Leménager, T., Dieter, J., & Hill, H. (2014). Neurobiological correlates of physical self-concept and self-identification with avatars in addicted players of Massively Multiplayer Online Role-Playing Games (MMORPGs). *Addictive Behaviors, 39*(12), 1789–1797.

Leménager, T., Dieter, J., Hill, H., et al. (2016). Exploring the neural basis of avatar identification in pathological Internet gamers and of self-reflection in pathological social network use. *Journal of Behavioral Addictions, 5*(3), 485–499.

Lin, X., Dong, G., Wang, Q., et al. (2015). Abnormal gray matter and white matter volume in Internet gaming addicts. *Addictive Behaviors, 40*, 137–143.

Liu, X., Zheng, Y., Niculescu, M., Liang, Q., Yang, A., Dong, G., Gao, Z., Lin, P., Liu, Y., Chen, L., & Xu, D. (2023). The involvement of spontaneous brain activity in natural recovery from internet gaming disorder: A resting-state fMRI study. *Frontiers in Psychiatry, 14*, 1093784.

Ma, X., Wang, M., Zhou, W., Zhang, Z., Ni, H., Jiang, A., Zheng, Y., Du, X., Potenza, M. N., & Dong, G. H. (2024). Wanting-liking dissociation and altered dopaminergic functioning: Similarities between internet gaming disorder and tobacco use disorder. *Journal of Behavioral Addictions, 13*(2), 596–609.

MacDonald, A. W., 3rd, Cohen, J. D., Stenger, V. A., & Carter, C. S. (2000). Dissociating the role of the dorsolateral prefrontal and anterior cingulate cortex in cognitive control. *Science, 288*(5472), 1835–1838. https://doi.org/10.1126/science.288.5472.1835. PMID: 10846167.

Maniaci, G., Picone, F., van Holst, R. J., Bolloni, C., Scardina, S., & Cannizzaro, C. (2017). Alterations in the emotional regulation process in gambling addiction: The role of anger and alexithymia. *Journal of Gambling Studies, 33*, 633–647.

Marchica, L. A., Keough, M. T., Montreuil, T. C., & Derevensky, J. L. (2020). Emotion regulation interacts with gambling motives to predict problem gambling among emerging adults. *Addictive Behaviors, 106*, 106378.

Marmet, S., Studer, J., Grazioli, V. S., & Gmel, G. (2018). Bidirectional associations between self-reported gaming disorder and adult attention deficit hyperactivity disorder: Evidence from a sample of young swiss men. *Frontiers in Psychiatry, 9*, 649.

Marques, L. M., et al. (2023). Escaping through virtual gaming—What is the association with emotional, social, and mental health? A systematic review. *Frontiers in Psychiatry, 14*, 1257685.

Miedl, S. F., Fehr, T., Meyer, G., & Herrmann, M. (2010). Neurobiological correlates of problem gambling in a quasi-realistic blackjack scenario as revealed by fMRI. *Psychiatry Research: Neuroimaging, 181*, 165–173.

Park, J. H., Lee, Y. S., Sohn, J. H., & Han, D. H. (2016). Effectiveness of atomoxetine and methylphenidate for problematic online gaming in adolescents with attention deficit hyperactivity disorder. *Human Psychopharmacology: Clinical and Experimental, 31*(6), 427–432.

Potenza, M. N. (2008). The neurobiology of pathological gambling and drug addiction: An overview and new findings. *Philosophical Transactions of the Royal Society B: Biological Sciences, 363*(1507), 3181–3189.

Potenza, M. N. (2013). Neurobiology of gambling behaviors. *Current Opinion in Neurobiology, 23*(4), 660–667.

Punia, K., & Balodis, I. M. (2019). An overview of the neurobiology of impulsivity in gambling and gaming disorder. *Current Behavioral Neuroscience Reports, 6*, 209–218.

Rogier, G., & Velotti, P. (2018). Conceptualizing gambling disorder with the process model of emotion regulation. *Journal of Behavioral Addictions, 7*(2), 239–251.

Rypma, B., et al. (2006). Neural correlates of cognitive efficiency. *NeuroImage, 33*(3), 969–979.

Sánchez-González, M. A., García-Cabezas, M. A., Rico, B., & Cavada, C. (2005). The primate thalamus is a key target for brain dopamine. *The Journal of Neuroscience, 25*(26), 6076–6083.

Saragih, I. D., Everard, G., & Lee, B. O. (2022). A systematic review and meta-analysis of randomized controlled trials on the effect of serious games on people with dementia. *Ageing Research Reviews, 82*, 101740.

Schoenmacker, G. H., Groenman, A. P., Sokolova, E., Oosterlaan, J., Rommelse, N., Roeyers, H., Oades, R. D., Faraone, S. V., Franke, B., Heskes, T., Arias Vasquez, A., Claassen, T., & Buitelaar, J. K. (2020). Role of conduct problems in the relation between Attention-Deficit Hyperactivity disorder, substance use, and gaming. *European Neuropsychopharmacology, 30*, 102–113.

Shaw, P., Lalonde, F., Lepage, C., Rabin, C., Eckstrand, K., Sharp, W., Greenstein, D., Evans, A., Giedd, J. N., & Rapoport, J. (2009). Development of cortical asymmetry in typically developing children and its disruption in attention-deficit/hyperactivity disorder. *Em Archives of General Psychiatry, 66*(8), 888–896.

Shohamy, D. (2011). Learning and motivation in the human striatum. *Current Opinion in Neurobiology, 21*(3), 408–414.
Smith, G. T., Fischer, S., Cyders, M. A., Annus, A. M., & Spillane, N. S. (2007). On the validity and utility of discriminating among impulsivity-like traits. *Assessment, 14*, 155–170.
Spagnolo, P. A., Gómez Pérez, L. J., Terraneo, A., Gallimberti, L., & Bonci, A. (2019). Neural correlates of cue-and stress-induced craving in gambling disorders: Implications for transcranial magnetic stimulation interventions. *European Journal of Neuroscience, 50*(3), 2370–2383.
Stanford, M. S., Mathias, C. W., Dougherty, D. M., Lake, S. L., Anderson, N. E., & Patton, J. H. (2009). Fifty years of the Barratt Impulsiveness Scale: An update and review. *Personality and Individual Differences, 47*(5), 385–395.
Tan, G. S. L., & Tam, C. L. (2024). Impulsivity, gambling-related cognitions, cognitive reappraisal and gambling behaviour in a Malaysian sample. *Journal of Gambling Studies, 40*(2), 475–492.
Tanaka, S., Ikeda, H., Kasahara, K., Kato, R., Tsubomi, H., Sugawara, S. K., Mori, M., Hanakawa, T., Sadato, N., Honda, M., & Watanabe, K. (2013). Larger right posterior parietal volume in action video game experts: A behavioral and voxel-based morphometry (VBM) study. *PLoS One, 8*(6), e66998.
Teng, Z., Pontes, H. M., Nie, Q., et al. (2020). Internet gaming disorder and psychosocial well-being: A longitudinal study of older-aged adolescents and emerging adults. *Addictive Behaviors, 110*, 106530.
Tzang, R.-F., Chang, C.-H., & Chang, Y.-C. (2022). Structural equation modeling (SEM): Gaming disorder leading untreated attention-deficit/hyperactivity disorder to disruptive mood dysregulation. *International Journal of Environmental Research and Public Health, 19*(11), 6648.
Velotti, P., Rogier, G., Zobel, S. B., & Billieux, J. (2021). Association between gambling disorder and emotion (dys) regulation: A systematic review and meta-analysis. *Clinical Psychology Review, 87*, 102037.
Wang, H., Jin, C., Yuan, K., Shakir, T. M., Mao, C., Niu, X., Niu, C., Guo, L., & Zhang, M. (2015). The alteration of gray matter volume and cognitive control in adolescents with internet gaming disorder. *Frontiers in Behavioral Neuroscience, 9*, 64.
Wang, M., Dong, G., Wang, L., Zheng, H., & Potenza, M. N. (2018). Brain responses during strategic online gaming of varying proficiencies: Implications for better gaming. *Brain and Behavior: A Cognitive Neuroscience Perspective, 8*(8), e01076.
Weinstein, A., & Weizman, A. (2012). Emerging association between addictive gaming and attention-deficit/hyperactivity disorder. *Em Current Psychiatry Reports, 14*(5), 590–597.
Weinstein, A., Livny, A., & Weizman, A. (2017). New developments in brain research of internet and gaming disorder. *Neuroscience & Biobehavioral Reviews, 75*, 314–330.
Weiss, M. D., Baer, S., Allan, B. A., Saran, K., & Schibuk, H. (2011). The screens culture: Impact on ADHD. *ADHD Attention Deficit and Hyperactivity Disorders, 3*(4), 327–334.
Yan, W. S., Zhang, R. R., Lan, Y., Li, Y. H., & Sui, N. (2016). Comparison of impulsivity in non-problem, at-risk and problem gamblers. *Scientific Reports, 6*(1), 39233.
Yau, Y. H., & Potenza, M. N. (2015). Gambling disorder and other behavioral addictions: Recognition and treatment. *Harvard Review of Psychiatry, 23*(2), 134–146.
Yoon, E. J., Choi, J. S., Kim, H., Sohn, B. K., Jung, H. Y., Lee, J. Y., Kim, D. J., Park, S. W., & Kim, Y. K. (2017). Altered hippocampal volume and functional connectivity in males with Internet gaming disorder comparing to those with alcohol use disorder. *Scientific Reports, 7*(1), 5744.
Yuan, K., Yu, D., Cai, C., Feng, D., Li, Y., Bi, Y., Liu, J., Zhang, Y., Jin, C., Li, L., Qin, W., & Tian, J. (2017). Frontostriatal circuits, resting state functional connectivity and cognitive control in internet gaming disorder. *Addiction Biology, 22*(3), 813–822.
Zhai, J., Luo, L., Qiu, L., et al. (2017). The topological organization of white matter network in internet gaming disorder individuals. *Brain Imaging and Behavior, 11*(6), 1769–1778.
Zhang, J. T., Yao, Y. W., Li, C. S., et al. (2016). Altered resting-state functional connectivity of the insula in young adults with Internet gaming disorder. *Addiction Biology, 21*(3), 743–751.

Chapter 3
Cognitive Processes in Gaming

Amanda Douat Cardoso, Matheus de Melo Rodrigues, Alexandre Luiz de Oliveira Serpa, and Jessica Mayumi Maruyama

Abstract Interactive entertainment, particularly video and board games, has become an integral aspect of contemporary recreation and cognitive engagement. With nearly 40% of the global population engaging in gaming activities, research has increasingly focused on the cognitive processes involved in gameplay. This chapter explores the complex interplay between gaming and cognitive functions, including attention, memory, problem-solving, executive functions, and decision-making. Empirical studies suggest that gaming enhances selective attention, visuospatial processing, working memory, and cognitive flexibility. Moreover, video games have been investigated for their potential as cognitive training tools, with promising applications in educational and clinical settings. However, the relationship between gaming and cognition is bidirectional, with individual cognitive abilities influencing game engagement and vice versa. While well-structured gaming experiences may foster cognitive development across different life stages, excessive or unregulated gameplay can pose risks, such as diminished academic performance and cognitive overload. This chapter also examines methodological challenges in assessing gaming's cognitive benefits, the impact of different game genres, and the role of serious games in cognitive interventions. Future research should refine experimental designs and explore the long-term cognitive implications of interactive entertainment in diverse populations.

Keywords Cognitive Functions · Video Games · Executive Functions · Cognitive Training · Neuropsychology of Gaming

---

A. D. Cardoso · M. de Melo Rodrigues · A. L. de Oliveira Serpa (✉) · J. M. Maruyama
Computational Intelligence in Psychometrics and Epidemiology Research, Mackenzie Presbyterian University, São Paulo, Brazil

L. Murrins Marques et al. (eds.), *Social and Affective Neuroscience of Gaming*,
https://doi.org/10.1007/978-3-032-00325-6_3

Introduction

Interactive entertainment, including video and board games, has evolved into a prevalent medium for both recreation and interpersonal connections in contemporary life. Literature indicates that almost 40% of people worldwide engage with games in their leisure time, and it is predicted that this industry will reach a market value exceeding U$521 billion in 2027 (Goh et al., 2023). This demonstrates how gaming activities occupy a substantial portion of recreational time and mental engagement for numerous individuals. Researchers exploring the implications of this widespread gaming phenomenon have particularly focused on understanding how gameplay influences and potentially strengthens various mental capabilities or cognitive functions.

These cognitive demands of gaming exemplify the complex interplay of mental processes that constitute our broader cognitive functions. Cognitive functions are higher-order mental processes that promote the acquisition, manipulation, and usage of information from an individual's environment. These functions comprise an array of neuropsychological operations, including but not limited to perception, attention, memory, language, problem-solving, decision-making, executive functions, processing speed, and perception.

Cognitive Functions and Games

Gaming, whether electronic or non-electronic, requires players to engage multiple cognitive systems simultaneously. Players must pay attention to dynamic visual and auditory information, hold rules and strategies in working memory, make rapid decisions, and adapt their approach based on changing game states. This multifaceted cognitive engagement has led researchers to investigate how different types of games may impact various aspects of cognition, including attention, memory, problem-solving, and executive functions (Bediou et al., 2018). To understand the potential effects of gaming on these cognitive processes, it is essential to briefly define each function.

Attention represents an ability that enables people to consciously direct their awareness toward particular elements of their surroundings while disregarding other stimuli. It includes selective attention, sustained attention, and divided attention (Petersen & Posner, 2012). In the context of gaming, attention is crucial for monitoring dynamic visual and auditory stimuli, tracking multiple objects, and responding to relevant game events (Cutting & Cairns, 2022). Among the cognitive areas extensively researched in gaming studies, attention mechanisms stand out prominently. Action-oriented digital games particularly challenge players' visual processing systems. Such games typically present fast-paced visual scenarios requiring players to swiftly identify and act upon important cues while disregarding irrelevant information (Green & Bavelier, 2012). Research has demonstrated that individuals who regularly play action games exhibit superior selective visual attention and enhanced spatial awareness compared to those who don't play game (Bediou et al., 2018). Additionally, findings indicate that these players excel at tracking multiple moving objects and demonstrate superior visual search capabilities (Green & Bavelier, 2003). A systematic review further found that video games positively impact the development of focused, sustained, and divided attention, with evidence suggesting that these benefits transfer to other tasks, such as improved pseudoword reading tasks performance and shorter reaction times (Sampalo et al., 2023).

Memory refers to processes of encoding, storing, and retrieving information from the brain, allowing organisms to acquire, retain, and later recall experiences, knowledge, and learned behaviors. Working memory involves the temporary maintenance and manipulation of information necessary for complex cognitive tasks (Baddeley, 2012). Long-term memory, in contrast, involves the storage and retrieval of information over extended periods of time. Both working memory and long-term memory play critical roles in gaming, from remembering game rules and strategies to recalling past game events and experiences (Martinez et al., 2022). Memory systems are also heavily engaged during gameplay. Working memory is constantly taxed as players must maintain and update information about game states, rules, and strategies. Strategy games and role-playing games often require players to hold large amounts of information in memory while planning future moves. Studies have found that gaming experience is associated with improvements in visual and spatial working memory and short-term memory recall (Blacker et al., 2014; Clemenson et al., 2020). Additionally, the immersive narratives in many games may engage episodic memory systems as players remember story elements and past game events.

Problem-solving is a complex cognitive function that entails recognizing issues, formulating possible solutions, and executing and assessing those solutions. This process demands the coordination of various cognitive abilities, such as reasoning, planning, and making decisions, into complex scenarios (Knauff & Wolf, 2010). Within the realm of gaming, problem-solving plays a pivotal role in tackling obstacles, crafting strategies, and adjusting to new scenarios presented by the game. Problem-solving and decision-making processes are central to most gaming experiences. Players must analyze complex situations, generate potential solutions, and make rapid choices that impact game outcomes. Puzzle games directly challenge problem-solving abilities, while strategy games require long-term planning and decision-making under uncertainty. Bavelier et al. (2012) have suggested that gaming may enhance cognitive flexibility and improve problem-solving skills that transfer to non-gaming contexts. Jordan and Dhamala (2022) evaluated gamers and non-gamers using a sensorimotor decision-making task and investigated brain and behavior responses within an fMRI framework. The results showed an overall increase in decision accuracy and a decrease in decision reaction time favoring gamers, which aligned with the neural activation and connection patterns observed between the groups.

Executive functions are a set of abilities that regulate, control, and manage behavior or thought. Components include inhibition, shifting, and updating (Miyake et al., 2000). Executive functions are critical for goal-directed behavior, planning, and adapting to changing task demands. In gaming, executive functions are engaged when players must manage multiple tasks, switch between different game objectives, and inhibit automatic responses in favor of more strategic actions. Executive functions, including cognitive control, task switching, and inhibition, are heavily recruited during gameplay. Most games require players to manage multiple goals, shift between different task sets, and inhibit prepotent responses. For instance, in a first-person shooter game, players must frequently alternate between attacking and defending while suppressing the impulse to shoot at irrelevant objects. Research indicates that individuals who play action video games exhibit superior task-switching skills and greater cognitive control compared to those who do not play such games, although the results vary with regard to the specific game investigated (Colzato et al., 2010; Dale et al., 2020; Dobrowolski et al., 2021).

These cognitive functions do not operate in isolation but interact in complex ways to support effective gameplay. For instance, effective problem-solving in a game frequently demands prolonged focus, optimal utilization of working memory, and the integration of various executive functions. Understanding how these cognitive processes are engaged and potentially enhanced through gaming is a key focus of current research in cognitive psychology and neuroscience (Bediou et al., 2018; Chen et al., 2024). While much of the research has focused on action video games, other genres and board games also engage in important cognitive processes. Board games that emphasize strategy, such as chess, have been linked to enhancements in executive functions and problem-solving skills (Aciego et al., 2012). Similarly, cooperative board games may engage in social cognition and theory of mind as players must reason about others' knowledge and intentions.

The connection between gaming and cognitive function is intricate and likely operates in both directions. On one hand, gaming might contribute to cognitive improvements; on the other hand, individuals with higher cognitive skills may naturally gravitate toward games that require greater mental engagement. Moreover, excessive gaming can have adverse effects on cognition, especially when it disrupts essential activities such as sleep, physical exercise, or other daily responsibilities (Palaus et al., 2017). The cognitive demands of gaming have led researchers to explore potential applications in educational and clinical contexts. Game-based learning approaches leverage the engaging nature of games to improve student motivation and cognitive skills (Bediou et al., 2018). In clinical settings, specially designed "serious games" are being used to assess and potentially improve cognitive functions in various populations, from toddlers to elderly people (Neugnot, 2015), and several mental health conditions, including individuals with attention deficit disorders (Anguera & Gazzaley, 2015), intellectual disability (Shapoval et al., 2022), autism spectrum disorder (Jiménez-Muñoz et al., 2022), and others. As gaming technology advances, incorporating innovations like virtual and augmented reality, the cognitive demands of gameplay are expected to grow increasingly complex and diverse. Future studies will need to adapt to these changes to explore how these emerging gaming experiences influence cognition and brain activity.

Cognitive Functions and Video Games Interventions

Electronic and non-electronic games have shown promise as tools for cognitive intervention, particularly in clinical settings. Shapoval et al. (2022) reviewed 573 studies published since 2015, identifying 10 that focused specifically on executive function training in adults with intellectual disabilities. The targeted executive function domains included planning, decision-making, working memory, and behavioral regulation. Although the improvements observed were modest, these studies provided valuable insights into the technical and procedural aspects of game design for cognitive interventions. Video games have also been explored as interventions for mitigating age-related cognitive decline, particularly in memory. Clemenson et al. (2020) demonstrated that 4 weeks of immersive 3D video gaming significantly enhanced hippocampal-based memory in older adults, replicating findings previously observed in younger populations. Notably, these improvements persisted for up to 4 weeks post-intervention, suggesting that video games may act as a form of environmental enrichment for the aging brain.

Action video games (AVG) have received particular attention for their cognitive benefits. Podlogar and Podlesek (2022) reported that AVG players outperformed non-gamers in tasks assessing mental rotation, attentional capacity, and task-switching speed, though no differences were observed in cognitive flexibility. Similarly, Bediou et al. (2023) conducted meta-analyses revealing significant cognitive benefits associated with AVG play, with intervention studies demonstrating causal improvements in visuospatial skills, attention, and processing speed. These findings highlight the potential of AVG to enhance specific cognitive domains, particularly those related to perception and attention.

Dobrowolski et al. (2021) demonstrated that training in real-time strategy games like StarCraft II led to enhancements in perceptual and attentional abilities. Moreover, in-game behavioral changes were directly correlated with cognitive performance gains, underscoring the importance of detailed behavioral analyses in understanding gaming's cognitive effects. However, factors such as competitiveness and aggression may modulate these outcomes. For instance, Pichon et al. (2019) found that competitive personality traits were strongly linked to aggressive behaviors among AVG players, suggesting that personality traits may confound the observed cognitive benefits.

Methodological considerations play a critical role in shaping the observed outcomes of these studies. Waris et al. (2019) explored the relationship between gaming and working memory (WM) by employing both extreme groups, comparing gamers and non-gamers. While gamers outperformed non-gamers in verbal WM, visuospatial WM, and n-back tasks in the extreme groups analysis, the effects were significantly weaker when the full participant spectrum was included. This finding highlights the importance of balanced study designs for generating accurate and generalizable insights into the cognitive effects of gaming.

Another key challenge in the use of video games for cognitive training is the limited transfer of skills to untrained tasks. (Flores-Gallegos and Mayer 2022) found that while gaming could lead to task-specific improvements, broad cognitive benefits were not observed. This aligns with broader critiques of cognitive training paradigms, which often struggle to demonstrate far transfer effects. Nonetheless, the observed domain-specific gains indicate that video games may still serve as effective tools for targeted cognitive training, with potential for refinement in design and application.

A growing body of research underscores the multifaceted relationship between video games and cognition. Evidence suggests that gaming can enhance specific cognitive domains, including memory, attention, and visuospatial skills, though these effects are often constrained by methodological limitations and the specificity of the tasks studied. However, when designed without careful consideration of the context in which they will be implemented, their intended objectives, and the overall design, these games risk becoming ineffective (Clark et al., 2016; Howard-Jones et al., 2015; Palaus et al., 2017). As emphasized by Plass et al. (2015), it is crucial to incorporate key elements such as mechanics, aesthetics, narratives, rewards, and feedback in a way that aligns with the game's specific learning objectives or addresses the specific needs and unique characteristics of the target population, including traits associated with disorders.

Games and Cognitive Development Across the Lifespan

As previously discussed, existing evidence suggests that gaming can enhance specific cognitive skills development, offering distinct benefits at different stages of life. From childhood to old age, games engage abilities such as attention, memory, and problem-solving, adapting to the cognitive and social demands of each phase of the lifespan (Alotaibi, 2024; Dell'Osso et al., 2024). The impact of video games varies across age groups, offering potential cognitive benefits while also posing unique risks, and scientific evidence on these age-specific effects is still evolving.

In childhood and adolescence, playing video or board games is expected to have an influential role in shaping cognitive and social abilities, leveraging the brain's high plasticity during these stages (Alotaibi, 2024). Compared to non-video gamers, children engaged in games, particularly those focused on problem-solving and memory, have been shown to enhance executive functions such as working memory, attention, response inhibition, and cognitive flexibility (Chaarani et al., 2022; Lamb et al., 2018). Multiplayer and collaborative games further support the development of social skills, encouraging teamwork, communication, metacognition, and conflict resolution (Alotaibi, 2024; Ricker & Richert, 2021; Xu et al., 2023). However, the benefits depend heavily on the context and moderation of gameplay; while structured use can foster learning and interaction, excessive or unregulated gaming may lead to issues such as reduced academic performance and social withdrawal (Alotaibi, 2024; Chaarani et al., 2022). During the COVID-19 pandemic, for example, increased reliance on video games provided virtual socialization opportunities but also raised concerns about prolonged screen time and its impact on mental health (Chaarani et al., 2022). These findings highlight the dual nature of gaming in childhood and adolescence, where its effects are shaped by how and why the games are played.

In adulthood, games served as useful tools for maintaining and enhancing cognitive abilities during a period of relative cognitive stability. Different genres of games target specific cognitive domains: action and strategy games can improve attention, decision-making, and processing speed, while puzzle and role-playing games (RPGs) are linked to better problem-solving skills and memory retention (Zioga et al., 2024). Beyond entertainment, games are increasingly integrated into professional and educational contexts, offering platforms for continuous learning and skill development. For instance, serious games designed for workplace training or simulations provide immersive environments to build practical skills while reinforcing cognitive functions (Dahalan et al., 2023; Ferreira et al., 2017). Moreover, gaming can also offer stress relief and mental stimulation, contributing to overall well-being (Wang et al., 2017). In addition, adults who were introduced to video games later in life may experience cognitive benefits differently compared to digital natives who have grown up immersed in a technology-rich environment (Marston & del Carmen Miranda Duro, 2020). Future research should explore how generational differences in early exposure to video games influence cognitive development and adaptability across the lifespan.

In older adults, serious games have shown significant promise as non-pharmacological interventions to mitigate cognitive decline and improve quality of life (Dell'Osso et al., 2024; Sokolov et al., 2020; Vázquez et al., 2018). Research highlights that video games, particularly those designed for cognitive training, can enhance memory, attention, and executive functions, addressing challenges associated with aging such as mild cognitive impairment (Sokolov et al., 2020; Vázquez et al., 2018). Puzzle and strategy games have been linked to improvements in working memory and mental flexibility, while action games can boost processing speed and visuospatial skills (Dell'Osso et al., 2024). Moreover, gaming, particularly through board games and in-person activities, can actively promote social interaction and strengthen community bonds, effectively addressing the pervasive issue of isolation often faced by older populations (Allaire et al., 2013; Bavelier et al., 2012; Pallavicini et al., 2018). Exergames, which combine physical activity with interactive gameplay, offer a dual benefit by promoting both cognitive and physical health (Vázquez et al., 2018). By creating stimulating and engaging environments, games provide older adults with tools to maintain independence, support daily functioning, and enhance overall well-being (Allaire et al., 2013; Bavelier et al., 2012; Pallavicini et al., 2018).

Overall, gaming plays a significant role in cognitive development across all ages, delivering unique advantages while also carrying potential challenges based on usage and circumstances. From supporting executive function development in childhood to maintaining cognitive health in old age, their potential is multifaceted and still under evaluation in scientific literature (Alotaibi, 2024; Dell'Osso et al., 2024). The impact of gaming is highly dependent on the type of game, the context in which it is played, and the balance between engagement and moderation (Lamb et al., 2018). Future research should explore further into the long-term effects of different game genres and investigate how targeted interventions can maximize cognitive and social benefits while minimizing potential risks (Alotaibi, 2024; Dell'Osso et al., 2024). Furthermore, exploring culturally and demographically diverse populations can provide a broader understanding of gaming's role in development (Alotaibi, 2024; Dell'Osso et al., 2024). In the end, adopting a thoughtful and deliberate approach to gaming can maximize its potential as a powerful tool for improving cognitive health and enriching quality of life throughout every stage of human development.

As a mentally stimulating activity, gaming engages various cognitive processes, such as problem-solving and executive functions. While further research is essential to fully understand the long-term impacts of gaming on cognitive development and performance—including potential risks—current findings indicate that moderate gaming can foster measurable improvements in certain abilities. Given its widespread presence in modern life, investigating the cognitive effects of gaming is a vital and dynamic area of study with promising implications for education, cognitive enhancement, and public health. By bridging these insights with innovative applications, future research can unlock the full potential of gaming as a tool for personal and societal growth.

References

Aciego, R., García, L., & Betancort, M. (2012). The benefits of chess for the intellectual and social-emotional enrichment in schoolchildren. *The Spanish Journal of Psychology, 15*(2), 551–559. https://doi.org/10.5209/rev_sjop.2012.v15.n2.38866

Allaire, J. C., McLaughlin, A. C., Trujillo, A., Whitlock, L. A., LaPorte, L., & Gandy, M. (2013). Successful aging through digital games: Socioemotional differences between older adult gamers and Non-gamers. *Computers in Human Behavior, 29*(4), 1302–1306. https://doi.org/10.1016/j.chb.2013.01.014

Alotaibi, M. S. (2024). Game-based learning in early childhood education: A systematic review and meta-analysis. *Frontiers in Psychology, 15*, 1307881. https://doi.org/10.3389/fpsyg.2024.1307881

Anguera, J. A., & Gazzaley, A. (2015). Video games, cognitive exercises, and the enhancement of cognitive abilities. *Current Opinion in Behavioral Sciences, 4*, 160–165. https://doi.org/10.1016/j.cobeha.2015.06.002

Baddeley, A. (2012). Working memory: theories, models, and controversies. *Annual Review of Psychology, 63*(1), 1–29. https://doi.org/10.1146/annurev-psych-120710-100422

Bavelier, D., Green, C. S., Pouget, A., & Schrater, P. (2012). Brain plasticity through the life span: Learning to learn and action video games. *Annual Review of Neuroscience, 35*(1), 391–416. https://doi.org/10.1146/annurev-neuro-060909-152832

Bediou, B., Adams, D. M., Mayer, R. E., Tipton, E., Green, C. S., & Bavelier, D. (2018). Meta-analysis of action video game impact on perceptual, attentional, and cognitive skills. *Psychological Bulletin, 144*(1), 77–110.

Bediou, B., Rodgers, M. A., Tipton, E., Mayer, R. E., Green, C. S., & Bavelier, D. (2023). Effects of action video game play on cognitive skills: A meta-analysis. *Technology, Mind, and Behavior, 4*(1). https://doi.org/10.1037/tmb0000102

Blacker, K. J., Curby, K. M., Klobusicky, E., & Chein, J. M. (2014). Effects of action video game training on visual working memory. *Journal of Experimental Psychology. Human Perception and Performance, 40*(5), 1992–2004. https://doi.org/10.1037/a0037556

Chaarani, B., Ortigara, J., Yuan, D., Loso, H., Potter, A., & Garavan, H. P. (2022). Association of video gaming with cognitive performance among children. *JAMA Network Open, 5*(10), e2235721. https://doi.org/10.1001/jamanetworkopen.2022.35721

Chen, T., Ou, J., Li, G., & Luo, H. (2024). Promoting mental health in children and adolescents through digital technology: A systematic review and meta-analysis. *Frontiers in Psychology, 15*, 1356554. https://doi.org/10.3389/fpsyg.2024.1356554

Clark, D. B., Tanner-Smith, E. E., & Killingsworth, S. S. (2016). Digital games, design, and learning: A systematic review and meta-analysis. *Review of Educational Research, 86*(1), 79–122. https://doi.org/10.3102/0034654315582065

Clemenson, G. D., Stark, S. M., Rutledge, S. M., & Stark, C. E. L. (2020). Enriching hippocampal memory function in older adults through video games. *Behavioural Brain Research*, 112667. https://doi.org/10.1016/j.bbr.2020.112667

Colzato, L. S., van Leeuwen, P. J. A., van den Wildenberg, W. P. M., & Hommel, B. (2010). DOOM'd to switch: Superior cognitive flexibility in players of First Person Shooter games. *Frontiers in Psychology, 1*, 8. https://doi.org/10.3389/fpsyg.2010.00008

Cutting, J., & Cairns, P. (2022). Investigating game attention using the Distraction Recognition Paradigm. *Behaviour & Information Technology, 41*(5), 981–1001. https://doi.org/10.1080/0144929x.2020.1849402

Dahalan, F., Alias, N., & Shaharom, M. S. N. (2023). Gamification and game based learning for vocational education and training: A Systematic Literature Review. *Education and Information Technologies, 29*(2), 1–39. https://doi.org/10.1007/s10639-022-11548-w

Dale, G., Joessel, A., Bavelier, D., & Green, C. S. (2020). A new look at the cognitive neuroscience of video game play. *Annals New York Academy of Sciences, 1464*, 192–203. https://doi.org/10.1111/nyas.14295

Dell'Osso, L., Nardi, B., Massoni, L., Battaglini, S., De Felice, C., Bonelli, C., Pini, S., Cremone, I. M., & Carpita, B. (2024). Video gaming in older people: What are the implications for cognitive functions? *Brain Sciences, 14*(7), 731. https://doi.org/10.3390/brainsci14070731

Dobrowolski, P., Skorko, M., Myśliwiec, M., Kowalczyk-Grębska, N., Michalak, J., & Brzezicka, A. (2021). Perceptual, attentional, and executive functioning after real-time strategy video game training: Efficacy and relation to in-game behavior. *Journal of Cognitive Enhancement, 5*(4), 397–410. https://doi.org/10.1007/s41465-021-00211-w

Ferreira, A. T., Araújo, A. M., Fernandes, S., & Miguel, I. C. (2017). Gamification in the workplace: A systematic literature review. In *Advances in intelligent systems and computing* (pp. 283–292). Springer International Publishing. https://doi.org/10.1007/978-3-319-56541-5_29

Flores-Gallegos, R., & Mayer, R. E. (2022). Learning cognitive skills by playing video games at home: Testing the specific transfer of general skills theory. *Journal of Cognitive Enhancement*.https://doi.org/10.1007/s41465-022-00253-8

Goh, E., Al-Tabbaa, O., & Khan, Z. (2023). Unravelling the complexity of the video game industry: An integrative framework and future research directions. *Telematics and Informatics Reports, 100100*, 100100. https://doi.org/10.1016/j.teler.2023.100100

Green, C. S., & Bavelier, D. (2003). Action video game modifies visual selective attention. *Nature, 423*(6939), 534–537.https://doi.org/10.1038/nature01647

Green, C. S., & Bavelier, D. (2012). Learning, attentional control, and action video games. *Current Biology: CB, 22*(6), R197–R206. https://doi.org/10.1016/j.cub.2012.02.012

Howard-Jones, P. A., Jay, T., Mason, A., & Jones, H. (2015). Gamification of learning deactivates the default mode network. *Frontiers in Psychology, 6*, 1891. https://doi.org/10.3389/fpsyg.2015.01891

Jiménez-Muñoz, L., Peñuelas-Calvo, I., Calvo-Rivera, P., Díaz-Oliván, I., Moreno, M., Baca-García, E., & Porras-Segovia, A.(2022). Video games for the treatment of autism spectrum disorder: A systematic review. *Journal of Autism and Developmental Disorders, 52*, 169–188. https://doi.org/10.1007/s10803-021-04934-9

Jordan, T., & Dhamala, M. (2022). Video game players have improved decision-making abilities and enhanced brain activities. *Neuroimage. Reports, 2*(3), 100112. https://doi.org/10.1016/j.ynirp.2022.100112

Knauff, M., & Wolf, A. G. (2010). Complex cognition: The science of human reasoning, problem-solving, and decision-making. *Cognitive Processing, 11*(2), 99–102. https://doi.org/10.1007/s10339-010-0362-z

Lamb, R. L., Annetta, L., Firestone, J., & Etopio, E. (2018). A meta-analysis with examination of moderators of student cognition, affect, and learning outcomes while using serious educational games, serious games, and simulations. *Computers in Human Behavior, 80*, 158–167. https://doi.org/10.1016/j.chb.2017.10.040

Marston, H. R., & del Carmen Miranda Duro, M. (2020). Revisiting the twentieth century through the lens of Generation X and digital games: A scoping review. *The Computer Games Journal, 9*(2), 127–161. https://doi.org/10.1007/s40869-020-00099-0

Martinez, L., Gimenes, M., & Lambert, E. (2022). Entertainment video games for academic learning: A systematic review. *Journal of Educational Computing Research, 60*(5), 1083–1109. https://doi.org/10.1177/07356331211053848

Neugnot, M. (2015). The use of games in paediatric cognitive intervention: A systematic review. *International Journal of Physical Medicine & Rehabilitation, 03*(04). https://doi.org/10.4172/2329-9096.1000286

Palaus, M., Marron, E. M., Viejo-Sobera, R., & Redolar-Ripoll, D. (2017). Neural basis of video gaming: A systematic review. *Frontiers in Human Neuroscience, 11*, 248. https://doi.org/10.3389/fnhum.2017.00248

Pallavicini, F., Ferrari, A., & Mantovani, F. (2018). Video games for well-being: A systematic review on the application of Computer Games for cognitive and emotional training in the adult population. *Frontiers in Psychology, 9*, 2127. https://doi.org/10.3389/fpsyg.2018.02127

Petersen, S. E., & Posner, M. I. (2012). The attention system of the human brain: 20 years after. *Annual Review of Neuroscience, 35*(1), 73–89. https://doi.org/10.1146/annurev-neuro-062111-150525

Pichon, S., Antico, L., Chanal, J., Singer, T., & Bavelier, D. (2019). The link between competitive personality, aggressive and altruistic behaviors in action video game players. In *PsyArXiv*. https://doi.org/10.31234/osf.io/te83n

Plass, J. L., Homer, B. D., & Kinzer, C. K. (2015). Foundations of game-based learning. *Educational Psychologist, 50*(4), 258–283. https://doi.org/10.1080/00461520.2015.1122533

Podlogar, N., & Podlesek, A. (2022). Comparison of mental rotation ability, attentional capacity and cognitive flexibility in action video gamers and non-gamers. *Cyberpsychology, 16*(2). https://doi.org/10.5817/cp2022-2-8

Ricker, A. A., & Richert, R. A. (2021). Digital gaming and metacognition in middle childhood. *Computers in Human Behavior, 115*(106593), 106593. https://doi.org/10.1016/j.chb.2020.106593

Sampalo, M., Lázaro, E., & Luna, P.-M. (2023). Action video gaming and attention in young adults: A systematic review. *Journal of Attention Disorders, 27*(5), 530–538. https://doi.org/10.1177/10870547231153878

Shapoval, S., Gimeno-Santos, M., Mendez Zorrilla, A., Garcia-Zapirain, B., Guerra-Balic, M., Signo-Miguel, S., & Bruna-Rabassa, O. (2022). Serious games for executive functions training for adults with intellectual disability: Overview. *International Journal of Environmental Research and Public Health, 19*(18), 11369. https://doi.org/10.3390/ijerph191811369

Sokolov, A. A., Collignon, A., & Bieler-Aeschlimann, M. (2020). Serious video games and virtual reality for prevention and neurorehabilitation of cognitive decline because of aging and neurodegeneration. *Current Opinion in Neurology, 33*(2), 239–248. https://doi.org/10.1097/WCO.0000000000000791

Vázquez, F. L., Otero, P., García-Casal, J. A., Blanco, V., Torres, Á. J., & Arrojo, M. (2018). Efficacy of video game-based interventions for active aging. A systematic literature review and meta-analysis. *PLoS One, 13*(12), e0208192. https://doi.org/10.1371/journal.pone.0208192

Wang, P., Zhu, X.-T., Liu, H.-H., Zhang, Y.-W., Hu, Y., Li, H.-J., & Zuo, X.-N. (2017). Age-related cognitive effects of videogame playing across the adult life span. *Games for Health, 6*(4), 237–248. https://doi.org/10.1089/g4h.2017.0005

Waris, O., Jaeggi, S. M., Seitz, A. R., Lehtonen, M., Soveri, A., Lukasik, K. M., Söderström, U., Hoffing, R. C., & Laine, M. (2019). Video gaming and working memory: A large-scale cross-sectional correlative study. *Computers in Human Behavior, 97*(March), 94–103. https://doi.org/10.1016/j.chb.2019.03.005

Xu, K., Geng, S., Dou, D., & Liu, X. (2023). Relations between video game engagement and social development in children: The mediating role of executive function and age-related moderation. *Behavioral Sciences, 13*(10), 833. https://doi.org/10.3390/bs13100833

Zioga, T., Ferentinos, A., Konsolaki, E., Nega, C., & Kourtesis, P. (2024). Video game skills across diverse genres and cognitive functioning in early adulthood: Verbal and visuospatial short-term and working memory, hand-eye coordination, and empathy. *Behavioral Sciences, 14*(10). https://doi.org/10.3390/bs14100874

Part II
Developmental Aspects of Gaming

Chapter 4
The Role of Gaming in Cognitive Development and Learning

Victor Namur, Danilo Camargo Lapiccirella, Marione Azevedo de Oliveira, and Gian Luigi Martins Menegazzo

Abstract Digital gaming has emerged as a powerful context for cognitive development and learning, engaging diverse mental processes through interactive and immersive environments. This chapter explores how different video game genres, particularly action and strategy games, contribute to enhancing cognitive skills such as visuospatial abilities, problem-solving, cognitive flexibility, and executive functions. It also addresses the impact of player profiles and motivation on cognitive outcomes and highlights the educational potential of active learning embedded within games. Additionally, the chapter discusses the cognitive dimensions of E-sports and emphasizes the importance of balancing gaming with offline activities for holistic development. Finally, it outlines strategies for players, developers, and researchers to optimize the cognitive benefits of gaming while critically examining methodological challenges and future research directions.

Keywords Cognitive Development · Active Learning · Action Video Games · Strategy Games · Educational Gaming

V. Namur (✉) · G. L. M. Menegazzo
Mental Health Department, Santa Casa de São Paulo School of Medical Sciences, São Paulo, SP, Brazil

D. C. Lapiccirella
Electric Monkeys Games, São Paulo, SP, Brazil

M. A. de Oliveira
Druid Creative Gaming, São Paulo, SP, Brazil

L. Murrins Marques et al. (eds.), *Social and Affective Neuroscience of Gaming*,
https://doi.org/10.1007/978-3-032-00325-6_4

Introduction

Digital games have become an integral part of daily life for billions of individuals around the globe. According to the Global Games Market Report, the number of players worldwide is projected to reach 3.42 billion, with approximately 2.8 billion being mobile gamers, marking a 4.5% increase from the previous year (Newzoo, 2024). This significant growth is reflected in the uprising attention video games receive from researchers, educators, and policymakers. The enthusiasm for exploring the multifaceted impacts of gaming on human cognition and behavior is reflected in the increasing body of literature dedicated to this subject.

Despite their widespread presence, video games are often simplistically treated as a single entity in popular media and even in some scientific discussions. However, this perspective overlooks the complexity and diversity inherent in video game genres. Just as different genres of music can evoke distinct emotional responses and cultural significances—consider the profound differences between a Beethoven symphony and an energetic samba track—different types of video games can activate a different range of cognitive skills and learning processes. This diversity claims a nuanced approach to understanding the impacts of gaming on cognitive development and brain function.

Research indicates that the cognitive skills engaged during video game play are as varied as the games themselves (Sala et al., 2018). For instance, action video games may enhance visuospatial abilities and reflexive responses, while strategy games could foster problem-solving skills and strategic thinking. The effects of gaming are thus not only dependent on the individual games but also on the particular cognitive domains they engage. This heterogeneity challenges the view that video games can be assessed uniformly regarding their cognitive impacts.

In this chapter, while we do not aim to exhaustively classify or define every type of video game available today, we acknowledge the significance of distinguishing between game categories and their respective cognitive influences. By examining the most commonly studied genres, we aim to provide insights into how these digital platforms may affect cognitive development and learning. Through a critical evaluation of current research, we hope to illuminate the potential educational and cognitive benefits of video games, while also considering the broader implications of their pervasive presence in modern society.

Cognitive Skills and Gaming

Cognitive Skills

Cognitive skills encompass a wide range of mental processes that are crucial for acquiring knowledge, understanding, and interacting with the world. Unlike static knowledge that is stored in the mind, cognitive skills are dynamic performances that are elicited in specific contexts (Vallotton & Fischer, 2008). These skills are task-specific, yet they possess the potential to be transferred to new situations through building connections and adapting learned strategies to novel problems. This adaptability is a hallmark of cognitive development, allowing individuals to apply learned skills flexibly across various domains.

In the realm of digital gaming, the diversity of video game genres plays a significant role in eliciting different sets of cognitive skills. Video games offer rich, interactive environments that engage a multitude of cognitive processes, including visuospatial navigation, strategic planning, and even social cognition (Bediou et al., 2018). The specific skills developed can vary widely depending on the game category, with action games often enhancing reflexes and spatial reasoning, while puzzle and strategy games may foster problem-solving and critical thinking abilities (Pallavicini et al., 2018). By examining the impact of gaming on cognitive skills, researchers are uncovering how these digital experiences can contribute to cognitive growth and adaptation, demonstrating the potential of games as valuable tools for cognitive skill development.

Video Game Classifications

Video games can be categorized in a multitude of ways depending on adopted reference points, such as quantity of engaging players, general theme, or structure, but it is important to look for a base standard that is contemplated by scientific literature. In a systematic review from 2018, Pallavicini et al. establish the distinction between serious games (SGs) and commercial games. The first category comprises games which are designed from the ground up as cognitive improvement or rehabilitation tools specifically, while the latter pertains to video games designed with the main goal of entertaining its player base. Commercial games can be subdivided into further categories, such as action video games (AVG) and strategy video games (SVG).

While SGs, due to their design philosophy, are expected to improve cognitive abilities and social skills of their participants, the potential of commercial games continues to be a less explored territory, as they are mostly scrutinized by medicine practitioners for their addictive potential and negative impact on quality of life. It is important to note that, once again according to Pallavicini et al., there's a large gap in literature regarding these classifications, as AVG is much more discussed than other types of commercial games, to the point that, due to the lack of standardized classification of gamers, players who prefer other genres of games than AVG are often listed in studies as non-VG players.

Action Video Games (AVG)

Action video games (AVGs), predominantly defined as first and third-person games (with the first-person mostly consisting of shooting games), have been one of the main focuses of research investigating the cognitive effects of gaming (REF). These games present distinctive elements, such as vast visual fields where enemies can appear from any direction, visually cluttered scenes, and numerous rapidly and independently moving targets, which inherently demand extensive engagement of perceptual, attentional, and cognitive skills (Bediou et al., 2023).

Wang et al. (2016) conducted a comprehensive meta-analysis to assess the impact of AVG training on cognitive abilities. The findings revealed that healthy adults experienced moderate improvements in overall cognitive ability and specific cognitive domains, such as visuospatial ability, processing speed/attention, and memory function. Notably, young adults reaped more significant benefits from AVG training than older adults, particularly in processing speed. Despite these positive outcomes, the authors caution that the underlying mechanisms driving these cognitive enhancements remain insufficiently understood, and thus, the results warrant careful interpretation.

Adding to this body of work, Bediou et al. (2023) offered a nuanced perspective through a recent meta-analysis that separately evaluated cross-sectional and intervention studies. Their research focused exclusively on young, healthy adults with no prior gaming experience. In intervention studies, participants trained on AVGs typically showed more extensive cognitive improvements than those engaged with control video games. However, the magnitude of these effects is generally smaller compared to cross-sectional studies.

In intervention studies, the results demonstrate that individuals trained in action video games tend to exhibit broader cognitive improvements compared to their peers trained in control video games. However, the effects observed in these studies are usually smaller than those found in cross-sectional studies and thus should be interpreted with caution.

The limited number of intervention studies and their methodological variability pose challenges to forming a cohesive understanding of AVG's cognitive benefits and the sustainability of these effects. This variability highlights the need for further research to clarify the effects of AVGs on cognitive skills and to establish more robust methodologies for future studies.

Furthermore, there is a smaller number of such studies available in the literature, and those that have been published display considerable variability in sample size and methodology, which complicates a more cohesive interpretation of video game training, as well as its effects and their duration (Bediou et al., 2023).

Strategy Video Games (SVG)

Strategy video games have gained significant attention for their potential to enhance cognitive skills and contribute to learning outcomes. These games require players to engage in complex decision-making, resource management, and long-term planning, which can translate into real-world cognitive benefits. The cognitive demands imposed by strategy games make them an intriguing subject for researchers exploring the malleability of cognitive abilities through digital gaming.

Glass et al. (2013) conducted a study demonstrating the impact of real-time strategy game training on cognitive flexibility. Their research found that players of these games exhibited enhanced cognitive flexibility, a critical skill that allows individuals to adapt their thinking and behavior in response to changing environments. This trait is particularly beneficial in real-life scenarios that require quick adaptation and problem-solving. The study highlights how the dynamic and often unpredictable nature of real-time strategy games fosters an environment where cognitive flexibility can be developed and refined.

Adachi and Willoughby (2013) explored the longitudinal relationships between playing strategic video games and the development of problem-solving skills and academic performance. Their findings suggest that strategy games can positively influence players' problem-solving abilities, which in turn can lead to improved academic outcomes. By engaging with complex in-game scenarios that require strategic thinking and planning, players may develop enhanced cognitive skills that support their academic pursuits. This study underscores the potential of strategy games as tools for cognitive skill development that extend beyond entertainment purposes.

Strobach and Schubert (2017) further investigated the impact of video game training on executive functions, which are crucial for managing cognitive processes such as task-switching, updating, and inhibition. Although their study encompasses various game genres, the insights gained are relevant to strategy games as well. The research indicates that strategy games can contribute to the enhancement of executive functions, providing players with improved cognitive control and better decision-making capabilities. These findings suggest that the cognitive demands of strategy games may foster the development of essential cognitive skills, making them valuable for both educational and everyday cognitive challenges.

In summary, strategy video games present a promising avenue for cognitive skill enhancement. Through engaging and challenging gameplay, players can develop vital cognitive abilities, such as cognitive flexibility, problem-solving skills, and executive functions, which have practical applications beyond the gaming environment. This growing body of research highlights the potential for strategy games to be leveraged as effective tools for cognitive development and learning.

E-sports

Highly prevalent in the current historical moment, especially in the post-COVID-19 pandemic landscape, is the spread of E-sports—competitive gaming that generates millions of dollars in market revenue, whether through major competition prize pools, athlete contracts, or spectator viewership. There is a growing interest in the mental, physical, and cognitive health of those engaged in this competitive environment, primarily, but not exclusively, the athletes. However, this field remains underexplored within science, with a primary focus on enhancing players' performance and exploring the interface between this activity and mental health. This interest is particularly high due to the rise in diagnosis of Internet Gaming Disorder in the general population and investigations into whether E-sports players are especially affected by this condition.

Studies examine the quality of life of players (Li et al., 2023), noticing negative impacts on sleep quality and reduced engagement in physical activities.

It is safe to assume that E-sports as a practice will continue to evolve and grow, which should lead to greater interest in researching its practitioners through the scope of cognitive development, since, similar to physical sports, athletes tend to be exposed to long hours of video game engagement from a very young age.

Player Profile Classifications

Players can be categorized into several profiles. Based on player's motivations, Frye developed on his doctoral thesis a theoretical framework for player profiles composed of seven player profiles (Frye, 2013). Below is a table with the profiles of players, detailing each with a brief description (Table 4.1):

Despite the lack of consensus within the scientific community regarding the delineation of rigid, specific player profiles, Frye's categorization is widely employed in the commercial gaming industry. This framework significantly contributes to designing games and identifying market demands and emerging trends. Frye's model aids developers and designers in creating more targeted and engaging gaming experiences by recognizing varying player motivations and behaviors.

Learning in Gaming

The exploration of learning within gaming environments sheds light on a dynamic intersection between traditional educational methods and interactive digital experiences. As digital games become increasingly integral to educational strategies, understanding the mechanisms of learning they facilitate is essential.

Table 4.1 Player profiles according to motivational orientations in gaming

Player profile	Description
Competitor	Motivated by showcasing skills and achieving victories, they excel in competitive settings.
Explorer	Enjoys discovering all aspects of a game, especially expansive worlds and hidden secrets.
Socializer	Sees games as a social extension and is motivated by interactions with others.
Achiever	Pursues overcoming challenges and reaching goals, finding satisfaction in game progress.
Fantasist	Engages with fantasy worlds for immersion, valuing the escape into alternate realities.
Escapist	Uses games for relaxation and mental breaks, seeking comfort from real-world pressures.
Thrill seeker	Seeks excitement and adrenaline, drawn to fast-paced and intense gaming experiences.

Research consistently demonstrates that active learning approaches, which are inherent in gaming, significantly enhance cognitive processes more effectively than passive learning methods (Michel et al. 2009).

Active learning encompasses instructional techniques that place students at the center of the learning experience, fostering engagement and accountability (Michel et al. 2009). The context of gaming naturally embodies these principles as they provide interactive, problem-solving scenarios where players must apply knowledge actively, make decisions, and experience consequences within a simulated environment.

In contrast, passive learning is characterized by activities such as reading, lecturing, and note-taking, and provides limited opportunities for engagement and reflection (Harris and Bacon et al 2019). Therefore, this learning style lacks the interactive and adaptive elements that active learning and gaming environments offer.

Games tend to provide players with immediate feedback, the ability to make choices, and the opportunity to learn from mistakes, all of which are crucial elements of active learning. Studies show that such interactive environments contribute to both lower- and higher-order cognitive skills, including critical thinking, problem-solving, and adaptability, surpassing the cognitive outcomes typically associated with passive learning (Green & Bavelier, 2019).

Modulating Factors of Learning Through Gaming

When referring to VG and learning, one of the most important factors to consider is player expertise. It has been demonstrated that different players will use different approaches and learn different aspects connected to the game played, depending on prior experience with said game (Pallavicini et al., 2018).

For example, in a multiplayer online battle arena (MOBA) game, like league of legends, experienced players tend to focus on learning new strategies and positioning, while inexperienced ones tend to focus on improving mechanical execution of many aspects of the game, such as movement and character specific skills (Gong et al., 2019). Age also plays an important factor, as younger adults potentially acquire more skills through cognitive enhancement than older individuals (Wang et al., 2016). Baseline cognitive function (e.g., reasoning skill, attentional skill) also correlates to improved learning through gaming, since it influences the choice of VG engagement and the extent of the benefits of gaming (Bediou et al., 2018).

Finally, motivation can also be a modulating factor for enhancements through VG, as research shows that highly motivated gamers present more functional changes in brain regions related to cognition and reward (Gleich et al., 2017).

Improving Cognitive Skill Development and Learning in Gaming

The intersection of gaming and cognitive skill development presents a unique opportunity to harness digital platforms for educational and cognitive enhancement. Both players and developers play crucial roles in optimizing these benefits.

Notes for Players

Selecting games that challenge and engage specific cognitive domains can be a good strategy. For instance, action games can enhance visuospatial skills and reaction times, while strategy games can improve problem-solving and planning abilities (Green & Bavelier, 2019). Players can intentionally choose games that align with cognitive skills they wish to develop and engage with these games regularly to reinforce skill acquisition.

Also, balancing gaming with other cognitive-enhancing activities can lead to more holistic cognitive development. Engaging in physical exercise, participating in social interactions, and pursuing hobbies that stimulate creativity can help complement the cognitive benefits gained from gaming (Staiano et al., 2013).

Lastly, players can reflect on their gaming experiences to maximize learning outcomes. This involves analyzing in-game decisions, outcomes, and strategies. To foster metacognitive awareness and enhance learning transfer to real-life situations, players can reflect on in-game decisions, outcomes, and strategies to maximize learning outcomes (Sitzmann, 2011). Additionally, to prevent negative outcomes associated with excessive gaming, players should benefit from mindful gaming strategies that help set limits on playtime.

Notes for Developers

Game developers have a significant role in designing games that not only entertain but also promote cognitive skill development. One fundamental strategy is incorporating educational elements subtly into gameplay. By embedding cognitive tasks within the narrative or mechanics of the game, developers can create immersive learning environments that engage players without compromising entertainment value (Wouters et al., 2013).

Creating adaptive game environments that cater to the individual needs of players by including variable difficulty levels, personalized feedback, and challenges that adjust to the player's skill level, ensuring that the game remains engaging and educational for a broad audience (Plass et al., 2015). This level of adaptability can sustain player interest and still promote continuous cognitive engagement.

Also, the incorporation of collaborative and social elements within games can enhance cognitive and social skills. Multiplayer games that require players to work together to solve complex problems can foster social interaction and cooperation skills that are transferable to real-world scenarios (Adachi & Willoughby, 2013).

The balance between challenge and skill level to achieve a state of flow, where players are fully immersed and motivated in the game, is crucial for maximizing cognitive engagement and ensuring that players experience enjoyment and educational benefits from gaming. We recommend developers prioritize finding this balance while developing games and testing user's experiences. Further research is needed for the development of tools that can help developers find reliable measures of *such* balance, as it will depend on game type and players' experience.

Notes for Researchers

One issue found during research for this chapter is the fact that, among many papers describing or proposing gamer profiles, most of these studies do not cite each other or take root in a common psychometric ground. This gives the impression that gamer profiles and their assessment tests may be arbitrarily designed and proposed as the market evolves.

The scarcity of common psychometric variables among scales and profile assessment tools shows how fragile these concepts are now in regard to being used as serious psychological research tools, and points out an interesting niche that can be focused on in the future.

Researchers should strive to build a consensus on the core psychometric constructs that underpin gamer profiles. Establishing a standardized set of variables will not only enhance the robustness of these profiles but will also facilitate comparisons across studies.

Conclusion

This chapter explored the intricate relationship between gaming and cognitive development, highlighting the potential benefits and considerations necessary for maximizing these effects. Throughout our examination, we have identified that digital gaming environments, particularly those using active learning frameworks, can significantly enhance cognitive processes, including visuospatial abilities, problem-solving skills, and strategic thinking. The diversity among game categories provides unique opportunities for players to engage in complex cognitive tasks within immersive and interactive contexts. As evidenced by research, action video games, in particular, foster improvements in specific cognitive domains, although the mechanisms behind these enhancements remain an area for further exploration.

As gaming practices increase and evolve, so too should the research and methodologies aimed at assessing their impact, ensuring that both players and developers can harness the full potential of games as tools for cognitive skill and learning development. Through these strategies and insights, the chapter aims to contribute to the growing discourse on gaming and cognitive enhancement, advocating for informed and intentional use of games in educational and cognitive contexts.

A substantial portion of research into learning and cognitive development through gaming has focused primarily on the type of game rather than the player's profile. While understanding the influence of different game genres is crucial, it may be more beneficial for scientific research to also aim at validating existing player profiles commonly used in the gaming industry. Utilizing these profiles as a variable could provide insightful perspectives on how they influence learning and development facilitated by games.

By considering player profiles, researchers could bridge the existing gap between scientific methodologies and the gaming community, ensuring that academic studies remain relevant and effectively applied within the context of game design and player engagement. Without this alignment, there is a risk of continuing to overlook the potential variations in player interactions and learning experiences that these profiles could reveal, thus maintaining a disparity between scientific findings and practical gaming industry applications.

References

Adachi, P. J. C., & Willoughby, T. (2013). More than just fun and games: The longitudinal relationships between strategic video games, self-reported problem solving skills, and academic grades. *Journal of Youth and Adolescence*, 42(7), 1041–1052.https://doi.org/10.1007/s10964-013-9913-9

Bavelier, D., & Green, C. S. (2019). Enhancing attentional control: Lessons from action video games. *Neuron, 104*, 147–163.

Bediou, B., Adams, D. M., Mayer, R. E., Tipton, E., Green, C. S., & Bavelier, D. (2018). Meta-analysis of action video game impact on perceptual, attentional, and cognitive skills. *Psychological Bulletin, 144*(1), 77–110.

Bediou, B., Rodgers, M. A., Tipton, E., Mayer, R. E., Green, C. S., & Bavelier, D. (2023). Effects of action video game play on cognitive skills: A meta-analysis. *Technology, Mind, and Behavior, 4*(1). https://doi.org/10.1037/tmb0000102

Frye, J. M. (2013). Video game player profiles: Bridging industry, game studies and social science perspectives(Doctoral dissertation). New York University, Steinhardt School of Culture, *Education, and Human Development*. ProQuest Dissertations Publishing.https://www.proquest.com/dissertations-theses/video-game-player-profilesbridging-industry-game/docview/1461167281

Glass, B. D., Maddox, W. T., & Love, B. C. (2013). Real-time strategy game training: Emergence of a cognitive flexibility trait. *PLoS ONE*, 8(8), e70350.https://doi.org/10.1371/journal.pone.0070350

Gleich, T., Lorenz, R. C., Gallinat, J., & Kühn, S. (2017). Functional changes in the reward circuit in response to gaming-related cues after training with a commercial video game. *NeuroImage, 152*, 467–475.

Gong, D., Ma, W., Liu, T., Yan, Y., & Yao, D. (2019). Electronic-sports experience related to functional enhancement in central executive and default mode areas. *Neural Plasticity, 2019*, 1940123.

Harris, N. A., & Welch Bacon, C. E. (2019). Developing cognitive skills through active learning: A systematic review of health care professions. *Athletic Training Education Journal*, 14(2), 135–148.https://doi.org/10.4085/1402135

Li, T., Xu, Q., Wang, S., Qi, K., Su, P., Silva, R. M., & Clemente, F. M. (2023). Effects of recreational small-sided games from different team sports on the improvement of aerobic fitness in youth sedentary populations: A systematic review. *Heliyon*, 9(11).https://doi.org/10.1016/j.heliyon.2023.e22041

Newzoo International B.V. (2024). The Global Games Market Report 2024 [Preview of the Games Market Reports and Forecasts subscription]. *Newzoo*.https://newzoo.com/resources/blog/globalgames-market-revenue-estimates-and-forecasts-in-2024

Michel, N. J., Cater, J., & Varela, O. E. (2009). Active versus passive teaching styles: An empirical study of student learning outcomes. *Human Resource Development Quarterly*, 20(4), 397–418.https://doi.org/10.1002/hrdq.20025

Pallavicini, F., Ferrari, A., & Mantovani, F. (2018). Video games for well-being: A systematic review on the application of computer games for cognitive and emotional training in the adult population. *Frontiers in Psychology, 9*, Article 2127.

Plass, J. L., Homer, B. D., & Kinzer, C. K. (2015). Foundations of Game-Based Learning.*Educational Psychologist*,50(4), 258–283.https://doi.org/10.1080/00461520.2015.1122533

Sala, G., Tatlidil, K. S., & Gobet, F. (2018). Video game training does not enhance cognitive ability: A comprehensive meta-analytic investigation. *Psychological Bulletin, 144*, 111–139.

Sitzmann, T. (2011). A Meta-Analytic Examination of the Instructional Effectiveness of Computer-Based Simulation Games.*Personnel Psychology* , 64, 489-528. PMID: 23592669; PMCID: PMC3473097.https://doi.org/10.1111/j.1744-6570.2011.01190.x

Staiano AE, Abraham AA, Calvert SL. (2013). Adolescent exergame play for weight loss and psychosocial improvement: a controlled physical activity intervention. *Obesity (Silver Spring)*. 21(3):598-601. PMID: 23592669; PMCID: PMC3473097.https://doi.org/10.1002/oby.20282

Strobach, T., & Schubert, T. (2017). No evidence for task automatization after dual-task training in younger and older adults. *Psychology and Aging* , 32(1), 28–41.https://doi.org/10.1037/pag0000144

Vallotton, C. D., & Fischer, K. W. (2008). Cognitive development. In M. M. Haith & J. B. Benson (Eds.), *Encyclopedia of infant and early childhood development* (pp. 286–298). Academic Press. ISBN 9780123708779.

Wang, P., Liu, H. H., Zhu, X. T., Meng, T., Li, H. J., & Zuo, X. N. (2016). Action video game training for healthy adults: A meta-analytic study. *Frontiers in Psychology, 7*, Article 907.

Wouters, P., van Nimwegen, C., van Oostendorp, H., & van der Spek, E. D. (2013). A meta-analysis of the cognitive and motivational effects of serious games.*Journal of Educational Psychology*, 105(2), 249–265.https://doi.org/10.1037/a0031311

[illegible]

Wang, [illegible], Liu, H. [illegible], Zhu, X. L., Meng, [illegible] & [illegible] (2016). A[illegible] for healthy adults: A meta-analysis. [illegible] Article 90.

Wouters, P., van Nimwegen, C., van Oostendorp, H., & van der Spek, E. D. (2013). A meta-analysis of the cognitive and motivational effects of serious games. *Journal of Educational Psychology, 105*(2), 249–265. https://doi.org/10.1037/a0031311

Chapter 5
Gaming and Social Development: Implications for Adolescents and Young Adults

Rafael Ângulo Condoretti Barros

Abstract Games, both video and board games, have long been recognized as influential in human development, fostering cognitive, emotional, and social growth. During adolescence and young adulthood, critical periods of neurobiological and social transition games serve as platforms for peer interaction, identity exploration, and social competence development. This chapter examines the role of games in social and cognitive development, drawing insights from developmental psychology, neuroscience, and communication studies. While strategic and cooperative gaming can enhance executive functions, problem-solving, and teamwork, excessive gaming may pose risks such as social withdrawal and impaired face-to-face interactions. Research indicates that moderate gaming can strengthen peer relationships, improve empathy, and serve as a medium for stress relief and emotional resilience. However, unregulated gaming habits, particularly in toxic online environments, may contribute to negative behavioral patterns. The chapter explores both the benefits and challenges of gaming, emphasizing the need for balanced engagement to maximize its positive contributions while mitigating potential risks. Future research should investigate long-term effects and develop strategies to optimize gaming experiences for cognitive and social well-being.

Keywords Social Development · Gaming and Cognitive Skills · Adolescent Psychology · Video Game Effects · Gaming and Mental Health

R. Â. C. Barros (✉)
Mental Health Department, Santa Casa de São Paulo School of Medical Sciences, São Paulo, SP, Brazil

L. Murrins Marques et al. (eds.), *Social and Affective Neuroscience of Gaming*,
https://doi.org/10.1007/978-3-032-00325-6_5

Introduction

Both video games and board games have been known way back as an integral aspect of human development, supporting mental, emotional, and social health. For today's increasing generation of youngsters and young adults, when several neurobiological and psychosocial transitions occur, games play a significant role in interrelation building, self-concept development, and social interaction. Social development in this phase is especially prominent as people shift from family-focused to peer-based interactions, negotiating intricate social hierarchies and norms (Blakemore & Mills, 2014). Games, both cooperative and competitive, offer a guided yet adaptive space for testing and experimenting with these dynamics.

An estimate of about 71% adolescents participate in online gaming (Pew Research Center, 2022). It has also emerged that "digital games" are the typical context in which adolescents might engage either with best friends to form tighter bonds or with strangers to enhance collaborative skills and resolve conflicts (Kaye & Bryce, 2012). The engaging and involving nature of video games provides opportunities for real-time communication, role-play experiences, and context-relevant problem-solving, making them especially powerful in influencing social development (Granic et al., 2014). However, another side of these interactions consists of several difficulties, as the nature of online gaming might also expose players to various risks, such as cyberbullying and social ostracism (Kowert & Quandt, 2016).

The Role of Gaming in Social Development

This chapter outlines the functions of games in the social development of teens and emerging adults. This chapter focuses on a convergence of insights from developmental psychology, neuroscience, and commutation studies on the ways by which games foster social competencies and identity negotiation in peer interaction. We therefore seek, purposively, to offer perspective and understanding into the ways in which it is influential on social development throughout these years by balancing the types of concerns which have been included in the positive and negative views of these activities.

Adolescence is a period of important development in cognitive and social characteristics, marked by significant alterations in both brain architecture and functionality, alongside a growing intricacy in social engagement. The prefrontal cortex is important because it has strong control over executive functions such as decision-making, planning, and impulse regulation; it experiences big maturation in adolescents (Casey et al., 2008). Simultaneously, adolescents become significantly involved in relationships with peers, building social networks that become instrumental in serving their identity and interpersonal competencies (Steinberg & Morris, 2001).

Games, particularly those that require strategic thinking, collaboration, or competition, have the power to enhance cognitive as well as social functioning (Greitemeyer and Osswald, 2010). The cognitive benefits are realized in the development of abilities such as problem-solving, multitasking, and spatial awareness. To illustrate, action video games have shown effectiveness in enhancing attentional control and visual–spatial abilities, which form the foundation for successfully navigating virtual as well as real-world environments (Bavelier et al., 2010). Similarly, strategy-based games like chess and certain video games promote planning and foresight, essential components of executive functioning (Subrahmanyam & Greenfield, 2008).

However, despite the identified benefits, consideration must be given to the associated risks associated with excessive or unregulated gaming. Stronger reliance on virtual interactions may ultimately lower the ability to develop face-to-face social skills, and being exposed to toxic gaming environments may promote negative behavior (Kowert et al., 2014). Thus, the positive impacts on the cognitive and social development of children through gaming are highly dependent on balanced engagement, together with timely, proper parental, or community guidance.

Risks and Challenges of Gaming

One longitudinal study by Przybylski and Weinstein (2017) examined the link between gaming and social adjustment among adolescents. They found moderate gaming, which they defined as up to 3 h a day, to be associated with greater amounts of peer connection and smaller amounts of emotional difficulty. Such may indicate that, in controlled amounts, gaming can serve as a portal to positive social contact and/or emotional regulation, especially in co-op games.

Another recent study, by Azadboni et al. (2024), investigated the impact of serious games designed for prosocial behavior on adolescents' empathetic skills and the ability to take the perspective of others. Results indicated that adolescents participating in such games showed observable differences in the identification of feelings of others and the understanding of different perspectives. Interactively designed stories and role-playing within such games enable rehearsal of social abilities that one can apply in real life.

Another fruitful avenue of research has been the use of MMORPGs as social spaces. In this respect, a study by Cole and Griffiths (2007) demonstrated that MMORPG players form strong interpersonal relationships from within the game context, and for most of the participants, such relationships were meaningful and supportive. The other most important finding of this study was that such games encouraged cooperation and teamwork in that players must discuss and share information about ways of overcoming certain problems in attempts at achieving common goals.

On the other hand, some studies view gaming as posing risks for social development. In one study, Lemola et al. (2015) noted that excessive gaming-more so for those gaming more than 5 h a day-is associated with less face-to-face interaction and increased loneliness among adolescents. Such findings point to the need for balance in gaming habits if it is to have a positive impact on social life.

In young adulthood, gaming continues to influence cognitive and social development ,(Pallavicini et al, 2018) often in more refined and domain-specific ways. In this stage of life, social networks are expanding and demands for higher-order cognitive skills like critical thinking, decision-making, and interpersonal communication increase. Gaming behavior in this period can determine the impact on cognition and social skill sets, especially as regards the type of games one is exposed to and the context within which engagement occurs.

Gaming, especially strategy and role-playing games, enhances cognitive flexibility, working memory, and problem-solving skills. A meta-analysis by (Wang et al. 2017) showed that video gaming enhanced cognitive domains such as attentional control and visuospatial reasoning in young adults, especially in action games. These abilities are very important in real-life situations since they allow multitasking and flexibility when changing rapidly from one task to another.

Furthermore, gaming often requires players to interpret complex scenarios, anticipate outcomes, and devise strategic solutions, which are linked to improvements in higher-order cognitive processes (Bediou et al., 2018). For instance, young adults who frequently engage in games with high strategic demands, such as Starcraft, exhibit superior planning and task-switching abilities compared to non-gamers (Strobach et al., 2012).

Balancing Gaming for Optimal Development

On the social front, gaming environments also offer young adults' opportunities for skill development in collaboration, leadership, and effective communication. Multiplayer online games, especially those with team-based objectives-like league of legends or overwatch-reliably demand that players negotiate real-time strategic plans. These interactions encourage social adaptability and conflict resolution skills that are critical to professional and personal relationships (Barnettt & Coulson, 2010).

Gaming communities can also be the social context for young adults to find their meaningful connections and foster their empathy. Kowert et al. (2015) have shown that online gaming communities can provide a sense of belonging and affective support, especially for such individuals who might experience some difficulties in maintaining such relationships within the offline world. Role-playing games improve perspective-taking by the players of different characters and situations and augment their level of emotional understanding and social perspective (Otani et al., 2024).

However, excessive or uneven gaming may have a negative influence on social skills and cognitive functions. Excessive gaming has been linked to a decrease in offline social interactions, which may result in social isolation or poor emotional management (Lemmens et al., 2011). Similarly, exposure to toxic behaviors, like harassment or hostility in competitive gaming, may undermine positive social skills that gaming is seen to nurture (Kowert & Quandt, 2016).

While gaming is often framed within a discourse of risk in public debate, a growing body of research challenges this by highlighting its positive potential contribution to mental health. For many, it is a source of relaxation, social connection, and emotional resilience. Such outcomes become particularly salient in contexts where players engage in balanced and mindful gaming, using the games as tools for stress release, social bonding, and personal growth.

Game playing, particularly in very engaging or relaxing games, reduces stress levels and manages emotions. The study proves that casual games, like puzzles or even simulation games, can provide immediate relief from stress by engaging a person in non-threatening, enjoyable activities. Russoniello et al. (2009) indicate that such games divert one's attention away from whatever causes stress and create an opportunity for temporary relief by keeping players in the present. The repetitive and predictable nature of some game mechanics lends itself well to soothing effects that are considered salutary, even akin to meditation (Granic et al., 2014).

Multiplayer and cooperative games offer robust opportunities for social interaction, particularly for individuals who might struggle with face-to-face communication. Gaming communities often serve as inclusive spaces where players can find belonging and shared purpose, contributing to reduced feelings of loneliness (Kowert et al., 2014). For instance, massive multiplayer online games (MMOs) frequently facilitate the formation of supportive social networks, providing emotional and social resources to players (Cole & Griffiths, 2007).

Gaming has also been said to enhance psychological resilience by creating opportunities for players to overcome setbacks and mistakes in a low-stakes environment. Many games present challenges that require players to persevere, adjust course, and adopt a variety of problem-solving approaches-skills that easily generalize to real-life coping strategies (Przybylski et al., 2010). The enjoyment of success in meeting in-game goals, after multiple failures, reinforces agency and a growth mindset.

Many of these gamer's report that gaming is a source of pleasure and positive emotion. Indeed, a study by Przybylski et al. in 2021 found that those individuals who can play games for fun and social connectedness show more life satisfaction and wellbeing. Certainly, playing cooperative or narrative-driven games may elicit happy feelings and a sense of satisfaction, especially with friends and family. This is where the power of games-creating immersive narratives or engaging mechanics-leads to a sense of accomplishment, hence a very real benefit to mental health.

Beyond casual gaming, serious games designed for a therapeutic purpose have also demonstrated the mitigation of symptoms of mental health conditions. Games that contain mindfulness, biofeedback, or cognitive-behavioral principles have shown a reduction in symptoms of anxiety and depression (Colder Carras et al., 2018). Other games, such as SPARX, which has been created for adolescents suffering from depression, have been tested in clinical trials to show improvement of mood and a shift in negative thoughts (Merry et al., 2012).

On the other hand, excessive gaming and involvement in toxic atmospheres undermine numerous benefits that gaming has on mental health. In this respect, mindful, balanced gaming practices are important for optimizing its mental health potential. The time used for gaming is a critical factor that determines its effect. According to studies, moderate gaming, which is 1–3 h a day, can be beneficial in terms of mental health, cognitive enhancement, and social connectedness. Przybylski and Weinstein introduced the "Goldilocks Hypothesis" in 2017, which posited that a moderate amount of gaming is related to higher psychological well-being, but too much or too little is associated with reduced benefits.

The World Health Organization (WHO) guidelines for screen time suggest that prolonged periods of sedentary gaming should be avoided, especially in young adults, who require adequate physical activity and sleep for optimal health (World Health Organization, 2020). Breaking gaming sessions into shorter intervals, such as 45–60 min followed by breaks, can help maintain focus and reduce physical strain (Mentzoni et al., 2011).

Excessive gaming may result in addictive tendencies, with adverse effects on physical well-being, academic functioning, and social relationships. The World Health Organization's inclusion of "Gaming Disorder" in its classification indicates the expanding worry regarding the compulsive utilization of games to the detriment of other areas of life (Addo et al., 2021). Extensive gaming periods might also cause physical conditions like repetitive strain injury and posture problems.

Gaming that exceeds recommended durations may interfere with academic responsibilities, leading to poor grades and decreased involvement with academic activities. Socially, excessive gaming can isolate players from real-world interactions, limiting a child's development of face-to-face social skills (Chiu et al., 2004).

Conclusion

Gaming is neither wholly good nor wholly bad; its impact is entirely a matter of approach. Done responsibly, gaming has the potential to enhance cognitive abilities, further social contacts, and yield emotional benefits. On the other hand, excessive or unregulated gaming may raise a variety of physical, psychological, and social problems. Parents, educators, and players themselves can best guarantee that gaming remains a positive, enriching experience by encouraging balanced engagement with games.

References

Addo, P. C., Fang, J., Kulbo, N. B., Gumah, B., Dagadu, J. C., & Li, L. (2021). Violent video games and aggression among young adults: The moderating effects of adverse environmental factors. *Cyberpsychology, Behavior and Social Networking, 24*(1), 17–23. https://doi.org/10.1089/cyber.2020.0018

Azadboni, T. T., Nasiri, S., Khenarinezhad, S., & Sadoughi, F. (2024). Effectiveness of serious games in social skills training to autistic individuals: A systematic review. *Neuroscience & Biobehavioral Reviews, 161*, 105634. https://doi.org/10.1016/j.neubiorev.2024.105634. ISSN 0149-7634.

Barnett, J., & Coulson, M. (2010). Virtually real: A psychological perspective on massively multiplayer online games. *Review of General Psychology, 14*(2), 167–179. https://doi.org/10.1037/a0019442

Bavelier, D., Green, C. S., & Dye, M. W. (2010). Children wired: For better and for worse. *Neuron, 67*(5), 692–701. https://doi.org/10.1016/j.neuron.2010.08.035

Bediou, B., Adams, D. M., Mayer, R. E., Tipton, E., Green, C. S., & Bavelier, D. (2018). Meta-analysis of action video game impact on perceptual, attentional, and cognitive skills [published correction appears in Psychol Bull. 2018 Sep;144(9):978–979. doi: 10.1037/bul0000168]. *Psychological Bulletin, 144*(1), 77–110. https://doi.org/10.1037/bul0000130

Blakemore, S.-J., & Mills, K. L. (2014). Is adolescence a sensitive period for sociocultural processing? *Annual Review of Psychology, 65*, 187–207. https://doi.org/10.1146/annurev-psych-010213-115202

Casey, B. J., Getz, S., & Galvan, A. (2008). The adolescent brain. *Developmental Review, 28*(1), 62–77. https://doi.org/10.1016/j.dr.2007.08.003

Chiu, S. I., Lee, J. Z., & Huang, D. H. (2004). Video game addiction in children and teenagers in Taiwan. *Cyberpsychology & Behavior, 7*(5), 571–581. https://doi.org/10.1089/cpb.2004.7.571

Colder Carras, M., Van Rooij, A. J., Spruijt-Metz, D., Kvedar, J., Griffiths, M. D., Carabas, Y., & Labrique, A. (2018). Commercial video games as therapy: A new research agenda to unlock the potential of a global pastime. *Frontiers in Psychiatry, 8*, 300. https://doi.org/10.3389/fpsyt.2017.003

Cole, H., & Griffiths, M. D. (2007). Social interactions in massively multiplayer online role-playing games. *Cyberpsychology & Behavior, 10*(4), 575–583. https://doi.org/10.1089/cpb.2007.9988

Granic, I., Lobel, A., & Engels, R. C. M. E. (2014). The benefits of playing video games. *American Psychologist, 69*(1), 66–78. https://doi.org/10.1037/a0034857

Greitemeyer, T., & Osswald, S. (2010). Effects of prosocial video games on prosocial behavior. *Journal of Personality and Social Psychology, 98*(2), 211–221. https://doi.org/10.1037/a0016997

Kaye, L. K., & Bryce, J. (2012). Putting the "fun factor" into gaming: The influence of social contexts on experiences of playing videogames. *International Journal of Internet Science, 7*(1), 23–37.

Kowert, R., & Quandt, T. (Eds.). (2016). *The video game debate: Unravelling the physical, social, and psychological effects of video games*. Routledge.

Kowert, R., Vogelgesang, J., Festl, R., & Quandt, T. (2015). Psychosocial causes and consequences of online video game play *Computers in Human Behavior, 45*, 51-58. ISSN0747-5632.https://doi.org/10.1016/j.chb.2014.11.074

Kowert, R., Domahidi, E., & Quandt, T. (2014). The relationship between online video game involvement and gaming-related friendships among emotionally sensitive individuals. *Cyberpsychology, Behavior, and Social Networking, 17*(3), 153–158. https://doi.org/10.1089/cyber.2012.0645

Lemmens, J. S., Valkenburg, P. M., & Peter, J. (2011). Psychosocial causes and consequences of pathological gaming. *Computers in Human Behavior, 27*(1), 144–152. https://doi.org/10.1016/j.chb.2010.07.015

Lemola, S., Perkinson-Gloor, N., Brand, S., Dewald-Kaufmann, J. F., & Grob, A. (2015). Adolescents' electronic media use at night, sleep disturbance, and depressive symptoms in the smartphone age. *Journal of Youth and Adolescence, 44*(2), 405–418. https://doi.org/10.1007/s10964-014-0176-x

Mentzoni, R., Brunborg, G., Molde, H., Myrseth, H., Skouverøe, K., Hetland, J., & Pallesen, S. (2011). Problematic video game use: Estimated prevalence and associations with mental and physical health. *Cyberpsychology, Behavior and Social Networking, 14*, 591–596. https://doi.org/10.1089/cyber.2010.0260

Merry, S. N., Stasiak, K., Shepherd, M., Frampton, C., Fleming, T., Lucassen, M. F. G., et al. (2012). The effectiveness of SPARX, a computerised self help intervention for adolescents seeking help for depression: Randomised controlled non-inferiority trial. *BMJ, 344*, e2598. https://doi.org/10.1136/bmj.e2598

Otani, V. H. O., Novaes, R. A. C. B., Pedron, J., Nabhan, P. C., Rodrigues, T. M., Chiba, R., Guedes, J. V. C., Marques, L. M., & Vissoci, J. R. N. (2024). Framework proposal for role-playing games as mental health intervention: The critical skills methodology. *Frontiers in Psychiatry, 15*, 1297332. https://doi.org/10.3389/fpsyt.2024.1297332

Pallavicini, F., Ferrari, A., & Mantovani, F. (2018). Video games for well-being: A systematic review of the application of computer games for cognitive and emotional training in the adult population. *Frontiers in Psychology, 9*, 2127. https://doi.org/10.3389/fpsyg.2018.02127

Pew Research Center. (2022). *Teens, social media and technology 2022*. Retrieved from https://www.pewresearch.org/

Przybylski, A. K., & Weinstein, N. (2017). A large-scale test of the goldilocks hypothesis: Quantifying the relations between digital-screen use and the mental well-being of adolescents. *Psychological Science, 28*(2), 204–215. https://doi.org/10.1177/0956797616678438

Przybylski, A. K., Rigby, C. S., & Ryan, R. M. (2010). A motivational model of video game engagement. *Review of General Psychology, 14*(2), 154–166. https://doi.org/10.1037/a0019440

Russoniello, C. V., O'Brien, K., & Parks, J. M. (2009). The effectiveness of casual video games in improving mood and decreasing stress. *Journal of Cybertherapy and Rehabilitation, 2*(1), 53–66.

Steinberg, L., & Morris, A. S. (2001). Adolescent development. *Annual Review of Psychology, 52*(1), 83–110. https://doi.org/10.1146/annurev.psych.52.1.83

Strobach, T., Frensch, P. A., & Schubert, T. (2012). Video game practice optimizes executive control skills in dual-task and task switching situations. *Acta Psychologica, 140*(1), 13–24. https://doi.org/10.1016/j.actpsy.2012.02.001. ISSN 0001-6918.

Subrahmanyam, K., & Greenfield, P. (2008). Media use and the development of children and adolescents. *Journal of Developmental and Behavioral Pediatrics, 29*(3), 129–136. https://doi.org/10.1097/DBP.0b013e31816c4586

Wang, P., Zhu, X. T., Liu, H. H., Zhang, Y. W., Hu, Y., Li, H. J., Zuo, X. N. (2017). Age-Related Cognitive Effects of Videogame Playing Across the Adult Life span.*Games Health J, 6* (4):237-248. PMID: 28609152. https://doi.org/10.1089/g4h.2017.0005

World Health Organization. (2020). *Guidelines on physical activity and sedentary behaviour*. Retrieved from WHO.

Chapter 6
Parental Guidance, Regulation, and Support in Gaming

Dayane Antunes dos Santos, Amanda Douat Cardoso, Alexandre Luiz de Oliveira Serpa, and Ana Paula Almeida de Pereira

Abstract The increasing prevalence of digital gaming among children and adolescents necessitates active parental involvement to foster healthy gaming habits and mitigate potential risks. This chapter explores the multifaceted role of parents in regulating, supervising, and supporting gaming behaviors. It examines the cognitive, social, and developmental impacts of gaming, highlighting the benefits of balanced engagement and the risks of excessive use, including Internet gaming disorder (IGD), sleep disruption, and social isolation. Different parental mediation strategies—such as restrictive mediation, active mediation, joint use, and supervised use—are discussed, emphasizing the importance of parental literacy regarding gaming content and practices. Additionally, the chapter provides practical guidelines for setting healthy gaming routines, choosing appropriate age content, and promoting a diversified lifestyle that balances digital and non-digital activities. By fostering open communication, setting clear boundaries, and modeling responsible digital behavior, parents can help children and adolescents maximize the positive aspects of gaming while minimizing adverse outcomes.

Keywords Parental Mediation · Gaming and Child Development · Internet Gaming Disorder · Healthy Gaming Habits · Digital Literacy for Parents

Introduction

The increasing prevalence of video games in childhood and adolescence has sparked significant research interest. Over 90% of children aged two and older engage in video gaming, with an average daily playtime of 1.5–2 h (Alanko, 2023). While

D. A. dos Santos · A. P. A. de Pereira (✉)
Federal University of Paraná, Curitiba, Paraná, Brazil

A. D. Cardoso · A. L. de Oliveira Serpa
Computational Intelligence in Psychometrics and Epidemiology Research, Mackenzie Presbyterian University, São Paulo, Brazil

L. Murrins Marques et al. (eds.), *Social and Affective Neuroscience of Gaming*,
https://doi.org/10.1007/978-3-032-00325-6_6

gaming offers cognitive and motor skill benefits, concerns remain about potential social and psychological risks, such as isolation and aggression. On the positive side, active video games may counteract sedentary behavior and contribute to obesity prevention (Moncada Jiménez & Chacón Araya, 2015). Several factors influence gaming engagement, including early adolescence, cross-gender peer interactions, and parental involvement. Parents play a crucial role in shaping their children's gaming habits (Brooks et al., 2016). Given the increasing integration of digital games into children's lives, researchers emphasize the need for systematic studies on their long-term cognitive and developmental impacts, particularly during middle childhood (Blumberg et al., 2019).

Parental involvement is essential in guiding children's gaming habits and mitigating risks such as Internet gaming disorder (IGD) and academic decline. Clear rules and active supervision help prevent excessive gaming and its negative consequences (Donati et al., 2021; Lloret Irles et al., 2013; Su et al., 2018). Conversely, inconsistent or insufficient monitoring is associated with increased gaming frequency and intensity, potentially leading to academic difficulties and IGD (Lloret Irles et al., 2013). Effective parenting strategies include structured monitoring, appropriate mediation styles, and setting a positive example (Martins et al., 2017; Van Petegem et al., 2019). Research suggests that parents with negative perceptions of gaming often adopt controlling mediation styles, which may inadvertently encourage defiance and problematic gaming behaviors. In contrast, positive and neutral mediation—where parents discuss gaming content and establish reasonable limits—has been found to be more effective (Martins et al., 2017).

The gaming habits of parents themselves also influence their children's behaviors. When parents model excessive gaming, children are more likely to develop problematic gaming tendencies. Conversely, parents who demonstrate balanced gaming habits provide a framework for healthy engagement with digital games (Donati et al., 2021; Wernicke & Montag, 2022). Additionally, many parents lack awareness of the specific games their children play or the time they spend gaming, limiting their ability to intervene effectively. A study by Yalçın et al. (2022) involving 1245 Turkish parents and their children aged 2–5 years found that over a quarter (26.7%) of preschool children played video games, and in nearly a third (29.5%) of households, another family member also played. Gaming prevalence increased with age, yet over half of the parents were unaware of the specific games their children played. This lack of parental awareness underscores the need for active involvement and education in managing children's gaming habits (Funk et al., 1999; Yalçın et al., 2022).

In conclusion, fostering healthy gaming habits among children and adolescents requires proactive parental involvement. Through consistent monitoring, positive mediation, balanced role modeling, and increased awareness of gaming content, parents can help reduce the risks of excessive gaming while promoting a more positive and developmentally beneficial gaming experience.

Impact of Games on Development

Research highlights the cognitive benefits of video gaming, particularly in enhancing executive functions, attention control, and problem-solving skills (Alho et al., 2022; Lieberoth & Fiskaali, 2021). Action and strategy-based games require players to analyze complex scenarios, develop strategies, and adapt to dynamic environments, thereby improving cognitive flexibility and decision-making abilities (Wernicke & Montag, 2022).

From a social perspective, gaming serves as a medium for interaction, collaboration, and communication. Multiplayer games encourage teamwork, negotiation skills, and social competence (Collier et al., 2016). Additionally, gaming communities provide a platform for children and adolescents to form and maintain friendships, some of which extend beyond the virtual realm into real-life interactions (Donati et al., 2021).

Problem-solving is another key skill strengthened through gaming. Puzzle and adventure games challenge players to navigate obstacles and devise creative solutions, fostering critical thinking and adaptability (Granic et al., 2014; Lloret Irles et al., 2013). These cognitive and social benefits indicate that when approached with balance, gaming can be a valuable developmental tool.

Despite its benefits, excessive gaming poses several risks. One concern is social isolation, especially when gaming replaces real-world interactions. Problematic gaming has been linked to diminished face-to-face communication, which can impede social skill development and emotional intelligence (Jang & Ryu, 2016; Lérida-Ayala et al., 2022). However, the extent of this effect depends on gaming habits—some players use online gaming to complement rather than replace offline socialization.

Another risk is sleep disruption. Prolonged screen exposure before bedtime can interfere with sleep patterns, leading to insufficient rest, which negatively affects cognitive function, emotional regulation, and overall well-being (Donati et al., 2021). The combination of blue light exposure and psychological stimulation from gaming can delay sleep onset and reduce sleep quality. These risks are particularly significant when gaming is used as an escape mechanism to avoid real-life stressors, a key characteristic of gaming addiction (Lieberoth & Fiskaali, 2021).

Internet gaming disorder (IGD), recognized by the World Health Organization (World Health Organization, 2019), is characterized by impaired control over gaming, prioritization of gaming over other activities, and continued gaming despite negative consequences. Studies indicate that high neuroticism, low conscientiousness, male gender, poor impulse control, and prolonged screen exposure increase the likelihood of IGD (Macur & Pontes, 2021; Wernicke & Montag, 2022). This disorder can lead to serious consequences, including neglect of responsibilities, reduced participation in offline activities, and strained family relationships.

The debate over gaming and aggression remains ongoing. While early studies suggested a link between violent games and aggression, recent meta-analyses show that this relationship is modest and influenced more by individual temperament and environmental factors than by gaming content alone (Collier et al., 2016). Contextual factors, such as parental discussions about media content, may be more influential than mere exposure to violent games.

Parents play a central role in shaping children's gaming experiences, mitigating risks while fostering benefits (Van Petegem et al., 2019). Active parental engagement, such as discussing gaming experiences and setting reasonable boundaries, promotes healthier gaming habits and reduces problematic behaviors (Lloret Irles et al., 2013). Children are more likely to adopt responsible gaming practices when parents model balanced behaviors themselves (Wernicke & Montag, 2022). Encouraging alternative leisure activities, such as reading and sports, further supports healthy digital engagement and reduces excessive screen time.

Moreover, parental well-being significantly affects children's gaming behaviors. Studies show that parental depression is linked to an increased likelihood of IGD symptoms in children, underscoring the importance of family mental health in gaming regulation (Lam & Cheng, 2022). Furthermore, parental mediation strategies influence broader behavioral outcomes. While restrictive mediation can limit negative effects, harsh criticism without engagement is linked to higher delinquency rates (Martins et al., 2017). In contrast, positive and interactive mediation—where parents actively discuss gaming content and participate in gameplay—proves more effective in fostering responsible gaming habits and strengthening parent–child relationships.

Several myths about gaming persist. A common misconception is that video games directly cause violent behavior, yet research shows that personality traits and environmental influences are stronger predictors of aggression than gaming itself (Collier et al., 2016; Ferguson, 2015). Another myth suggests that gaming harms academic performance, whereas studies indicate that moderate gaming, especially when involving educational or strategic content, enhances cognitive skills (Alho et al., 2022; Lloret Irles et al., 2013; Virgilio, 2024). Finally, the assumption that gaming leads to social withdrawal overlooks the fact that many multiplayer games promote collaboration, teamwork, and social bonding (Donati et al., 2021).

Strategies for Healthy Habits

The Health Behavior in School-aged Children (HBSC) study (2022) examined approximately 280,000 adolescents aged 11–15 across 44 countries and regions in Europe, Central Asia, and Canada. It found that 34% of teenagers played digital games daily, with 22% dedicating at least 4 h to gaming. The study also revealed that 12% of young people were at risk of developing gaming-related problems, with boys exhibiting a higher prevalence (16%) compared to girls (7%). Additionally, research on problematic gaming behavior across five countries indicated correlations with lower life satisfaction, increased psychological distress, and diminished peer support (World Health Organization, 2024).

However, for many teenagers, gaming remains a harmless or even beneficial pastime. What strategies can parents implement to ensure that gaming serves as a meaningful part of their children's lives, providing both entertainment and learning opportunities?

Here we will discuss how the involvement of parents in regulating their children's gaming habits can contribute to promoting a balanced environment in which entertainment and education coexist in a healthy way. We can understand how parents' guide access to games in some ways: restrictive mediation, active mediation, joint use, supervised use, and monitored use (Nikken & Schols, 2015).

Restrictive Mediation

Restrictive mediation involves setting clear rules regarding playtime and game selection to prevent children from exposure to inappropriate content and to ensure gaming does not interfere with essential activities. For instance, parents might establish a rule such as, "You can play for 1 h per day, but no shooting games."

Leading organizations provide guidelines to avoid the potential negative effects of prolonged use of screens. The World Health Organization (WHO) recommends that children aged 1–2 years should have no more than 1 h of screen time per day, preferably less. Similarly, for children aged 3–4, the recommendation remains at 1 h daily, favoring educational content (WHO, 2019). The American Academy of Pediatrics advises a maximum of 1 h per day for children aged 6–12, with consistent limits for teenagers.

To aid control, digital devices and platforms offer tools that allow parents to manage screen time and access to specific content. These features help enforce technical limits, such as time blocks and age restrictions that prevent children from uninterrupted gaming or accessing content that is inappropriate for their age group.

Active Mediation and Joint Use

Active mediation involves parents engaging in conversations with their children about the games they play, encouraging reflection on content and decision-making, and promoting critical thinking and problem-solving skills. For example, during an adventure game, a parent might ask, "Why did you choose to help this character? What do you think will happen next?" This approach fosters critical thinking and ethical reasoning.

Joint use occurs when parents play with their children, making the game a shared activity to strengthen bonds and allow parents and children to learn and have fun together. A cooperative gaming session allows parents to moderate the emotional impact of competitive or challenging games and guide children in handling frustration constructively.

Supervised Use and Monitoring

Another frequent form of mediation is supervised use, which involves parents staying nearby while their children play to ensure the content remains appropriate and safe. Even when not physically present, parents can review gaming history and discuss in-game experiences with their children, identifying any concerns that may require intervention or guidance.

Active Parental Involvement

It is observed that parents who fear the use of media and games tend to limit their use and establish rules, while those who see games as an educational and entertainment opportunity end up using them together with their children. The combination of these strategies to guide habits seems to be the best strategy, varying according to each situation.

In any case, good mediation also depends on the parents' "literacy" in relation to games, that is, how comfortable they feel and understand the content well. It is considered that literate parents can guide usage more effectively, so parental education is something to invest in.

An effective strategy for encouraging responsible gaming habits is to guide game choices based on age rating systems such as PEGI (Pan European Game Information) and ESRB (Entertainment Software Rating Board). These systems help identify whether content is appropriate for the child's age, promoting a safe gaming experience. Games classified as PEGI 3, for example, are suitable for all ages, while PEGI 18 games include more mature and complex content.

The ESRB is the rating system adopted in the United States for video games and applications. It provides ratings that indicate the recommended age and content of games, helping parents choose appropriate products for their children. Rating categories include: "E" (Everyone), "T" (Teen), "M" (Mature), among others, based on the games' content and target audience. These ratings can also be viewed by parents before their child accesses (https://www.esrb.org).

Even though PEGI provides ratings for European countries, the characteristics considered for game ratings can be useful in guiding parents around the world on what to consider when thinking about appropriate games for their child. More information can be found on the official website: https://pegi.info/parental-controls. Beyond ratings, parents should evaluate games based on specific elements, including:

- Representations of violence
- Use of inappropriate language
- Scary images or sounds or horror sequences
- Gambling elements or simulated betting
- Sexual content or suggestive themes

- Depictions of substance use (drugs, alcohol, tobacco)
- Depictions of ethnic, religious, nationalist or other stereotypes that are likely to encourage discrimination and/or hatred
- The inclusion of in-game purchases or microtransactions

Since many online games involve interactions with other players, parental vigilance in setting privacy controls is essential. Children should be advised to engage only with known contacts and avoid sharing personal information. Configuring in-game settings to limit access to chats and external communities reduces the risk of exposure to inappropriate interactions.

Another recommendation is that parents read game summaries and reviews before granting permission to play. This proactive approach enables them to assess a game's narrative, level of intensity, and potential risks, ensuring that the chosen games align with their child's emotional and cognitive maturity. Additionally, allowing children to participate in the decision-making process reinforces the importance of informed choices and responsibility.

Understanding each child's profile is also essential for choosing appropriate games. For instance, children prone to irritability may struggle with highly competitive or fast-paced games, while cooperative and calming games that promote relaxation may be a better fit.

Device selection is another factor influencing healthy gaming habits. Small-screen devices like smartphones may not be ideal for younger children, as they require sustained visual focus and fine motor coordination. Tablets and gaming consoles, which offer more ergonomic controls, larger displays and require larger and more natural movements, are preferable for children aged 3–7.

Regardless of the form of mediation or the content of the game, to promote a healthy gaming experience, it is essential to set regular breaks and avoid very long periods of use. This practice reduces visual, mental, and physical fatigue, preventing adverse effects such as pain or excessive tiredness. Adopting specific and moderate schedules allows children to benefit from the positive aspects of games without the harm of excessive use.

Furthermore, a healthy and balanced life includes a variety of activities outside the digital world, such as sports, reading, and direct social interaction. Encouraging participation in extracurricular activities or hobbies is essential for children to learn to value time outside of screens. This practice diversifies stimuli and provides opportunities for physical, social, and emotional development.

Conclusion

Fostering a healthy relationship with video games requires open communication, well-defined rules, and proactive parental involvement. By setting clear guidelines on screen time, curating appropriate content, and supervising online interactions, parents can maximize the benefits of gaming while mitigating potential risks.

Additionally, parents should lead by example in their use of digital media, demonstrating responsible and conscious engagement with technology. When parent's model balanced screen habits, children are more likely to adopt similar behaviors.

In cases where gaming interferes with daily responsibilities, social interactions, or emotional well-being, seeking professional guidance is recommended. Signs of gaming dependency, difficulties in self-regulation, or declining academic performance may indicate the need for support from psychologists or digital well-being specialists.

A balance between digital and non-digital activities provides a solid foundation for the integral development of children, enabling them to interact safely, consciously, and responsibly with the digital world.

References

Alanko, D. (2023). The health effects of video games in children and adolescents. *Pediatrics in Review, 44*(1), 23–32. https://doi.org/10.1542/pir.2022-005666

Alho, K., Moisala, M., & Salmela-Aro, K. (2022). Effects of media multitasking and video gaming on cognitive functions and their neural bases in adolescents and young adults. *European Psychologist*. https://doi.org/10.1027/1016-9040/a000477

Blumberg, F. C., Deater-Deckard, K., Calvert, S. L., Flynn, R. M., Green, C. S., Arnold, D., & Brooks, P. J. (2019). Digital games as a context for children's cognitive development: Research recommendations and policy considerations. *Social Policy Report, 32*(1), 1–33. https://doi.org/10.1002/sop2.3

Brooks, F. M., Chester, K. L., Smeeton, N. C., & Spencer, N. H. (2016). Video gaming in adolescence: Factors associated with leisure time use. *Journal of Youth Studies, 19*(1), 36–54. https://doi.org/10.1080/13676261.2015.1048200

Collier, K. M., Coyne, S. M., Rasmussen, E. E., Hawkins, A. J., Padilla-Walker, L. M., Erickson, S. E., & Memmott-Elison, M. K. (2016). Does parental mediation of media influence child outcomes? A meta-analysis on media time, aggression, substance use, and sexual behavior. *Developmental Psychology, 52*(5), 798–812. https://doi.org/10.1037/dev0000108

Donati, M. A., Guido, C. A., De Meo, G., Spalice, A., Sanson, F., Beccari, C., & Primi, C. (2021). Gaming among children and adolescents during the COVID-19 lockdown: The role of parents in time spent on video games and gaming disorder symptoms. *International Journal of Environmental Research and Public Health, 18*(12), 6642. https://doi.org/10.3390/ijerph18126642

Ferguson, C. (2015). Do angry birds make for angry children? A meta-analysis of video game influences on children's and adolescents' aggression, mental health, prosocial behavior, and academic performance. *Perspectives on Psychological Science, 10*, 646–666. https://doi.org/10.1177/1745691615592234

Funk, J., Hagan, J., & Schimming, J. (1999). Children and electronic games: A comparison of parents' and children's perceptions of children's habits and preferences in a United States sample. *Psychological Reports, 85*(3), 883–888. https://doi.org/10.2466/pr0.1999.85.3.883

Granic, I., Lobel, A., & Engels, R. C. M. E. (2014). The benefits of playing video games. *American Psychologist, 69*(1), 66–78. https://doi.org/10.1037/a0034857

Jang, Y., & Ryu, S. (2016). The role of parenting behavior in adolescents' problematic mobile game use. *Social Behavior and Personality: An International Journal, 44*(2), 269–282. https://doi.org/10.2224/sbp.2016.44.2.269

Lam, Y.-T., & Cheng, C. (2022). Parental depression and leisure activity engagement on children's gaming disorder: A dyadic study. *International Journal of Environmental Research and Public Health, 19*(10), 5880. https://doi.org/10.3390/ijerph19105880

Lérida-Ayala, V., Aguilar-Parra, J., Collado-Soler, R., Alférez-Pastor, M., Fernández-Campoy, J.-M., & La Rosa, A. L.-D. (2022). Internet and video games: Causes of behavioral disorders in children and teenagers. *Children, 10*. https://doi.org/10.3390/children10010086

Lieberoth, A., & Fiskaali, A. (2021). Can worried parents predict effects of video games on their children? A case-control study of cognitive abilities, addiction indicators and wellbeing. *Frontiers in Psychology, 11*, 586699. https://doi.org/10.3389/fpsyg.2020.586699

Lloret Irles, D., Cabrera Perona, V., & Sanz Baños, Y. (2013). Relationships between video-gaming habits, parental monitoring and school performance. *European Journal of Investigation in Health, Psychology and Education, 3*(3), 237–248. https://doi.org/10.3390/ejihpe3030021

Macur, M., & Pontes, H. M. (2021). Internet Gaming Disorder in adolescence: Investigating profiles and associated risk factors. *BMC Public Health, 21*(1), 1547. https://doi.org/10.1186/s12889-021-11394-4

Martins, N., Matthews, N. L., & Ratan, R. A. (2017). Playing by the rules: Parental mediation of video game play. *Journal of Family Issues, 38*(9), 1215–1238. https://doi.org/10.1177/0192513X15613822

Moncada Jiménez, J., & Chacón Araya, Y. (2015). El efecto de los videojuegos en variables sociales, psicológicas y fisiológicas en niños y adolescentes (The effect of playing videogames on social, psychological and physiological variables in children and adolescents). *Retos, 21*, 43–49. https://doi.org/10.47197/retos.v0i21.34603

Nikken, P., & Schols, M. (2015). How and why parents guide the media use of young children. *Journal of Child and Family Studies, 24*(11), 3423–3435. https://doi.org/10.1007/s10826-015-0144-4

Su, B., Yu, C., Zhang, W., Su, Q., Zhu, J., & Jiang, Y. (2018). Father–child longitudinal relationship: Parental monitoring and internet gaming disorder in chinese adolescents. *Frontiers in Psychology, 9*, 95. https://doi.org/10.3389/fpsyg.2018.00095

Van Petegem, S., De Ferrerre, E., Soenens, B., Van Rooij, A. J., & Van Looy, J. (2019). Parents' degree and style of restrictive mediation of young children's digital gaming: Associations with parental attitudes and perceived child adjustment. *Journal of Child and Family Studies, 28*(5), 1379–1391. https://doi.org/10.1007/s10826-019-01368-x

Virgilio, R. D. L. Á. (2024). The cognition-impairing effect of video games in adolescents. *Science Insights, 44*(3), 1283–1289. https://doi.org/10.15354/si.24.re935

Wernicke, J., & Montag, C. (2022). Linking gaming disorder tendencies in children to their personality and parental gaming behavior. *Frontiers in Psychiatry, 12*, 748195. https://doi.org/10.3389/fpsyt.2021.748195

World Health Organization. (2019). *Guidelines on physical activity, sedentary behaviour, and sleep for children under 5 years of age*. World Health Organization.

World Health Organization. (2024, September 25). *Teens, screens and mental health*. https://www.who.int/europe/news/item/25-09-2024-teens%2D%2Dscreens-and-mental-health

Yalçın, S. S., Çaylan, N., Erat Nergiz, M., Oflu, A., Yıldız, D., Tezol, Ö., Çiçek, Ş., & Yurdakök, K. (2022). Video game playing among preschoolers: Prevalence and home environment in three provinces from Turkey. *International Journal of Environmental Health Research, 32*(10), 2233–2246. https://doi.org/10.1080/09603123.2021.1950653

Part III
Gaming and Mental Health

Chapter 7
Gaming Addiction and Compulsive Gaming Behavior: Psychological Perspectives

Cecilia Cheng and Hao Zhang

Abstract This chapter provides a comprehensive examination of psychological perspectives on gaming disorder and compulsive gaming behavior, beginning with the conceptualization and symptomatology that distinguish the disorder from normative gaming habits. We outline the diagnostic criteria established by the DSM-5 and ICD-11, analyzing how these frameworks have evolved with advancements in clinical understanding and shifting societal attitudes toward gaming. The discussion extends to the adverse impacts of gaming disorder on both mental and physical health, emphasizing associated psychiatric issues such as depression and obsessive-compulsive disorder, as well as broader health concerns such as sleep disturbances and musculoskeletal problems. Furthermore, we identify personal and social risk factors that heighten susceptibility to gaming disorder, including cognitive factors like poor impulse control and socio-environmental influences like family cohesion. Finally, we review a range of treatment approaches, from individual-focused cognitive-behavioral therapy and family-based interventions to emerging digital mindfulness and gamification therapies, evaluating their efficacy and applicability. By synthesizing current research and clinical practices, this chapter offers valuable insights for scholars and mental health professionals seeking to understand and effectively manage gaming disorders across diverse populations.

Keywords Problematic gaming · Pathological gaming · Mental health · Psychiatric issues · Treatment · Therapy · Risk factors

C. Cheng (✉) · H. Zhang
The University of Hong Kong, Pokfulam, Hong Kong
e-mail: ceccheng@hku.hk

L. Murrins Marques et al. (eds.), *Social and Affective Neuroscience of Gaming*,
https://doi.org/10.1007/978-3-032-00325-6_7

Introduction

The rise of digital technologies has transformed how we interact, with video gaming becoming a prevalent global pastime (Zhou et al., 2024). Gaming offers many positive benefits, including improved cognitive skills and social connections (Martinez et al., 2023; Nuyens et al., 2019), and gamification has been utilized as an intervention technique for enhancing physical and mental health (Cheng & Ebrahimi, 2023a, b). However, some individuals exhibit excessive, uncontrollable gaming behavior that impairs their daily functioning, a phenomenon widely referred to as gaming disorder (e.g., Cheng et al., 2018; Şalvarlı & Griffiths, 2022).

"Gaming disorder" encompasses various terms used by researchers and practitioners, including gaming addiction, pathological gaming, and compulsive gaming. For clarity, this chapter will consistently use gaming disorder, which includes both online and offline gaming, as defined in the 11th edition of the International Classification of Diseases (World Health Organization, 2018).

In the present cyber era, the high accessibility of online gaming platforms, mobile gaming devices, and the immersive nature of contemporary games make individuals—particularly adolescents and young adults—vulnerable to developing problematic gaming behaviors (Aziz et al., 2021; Stevens et al., 2021; H.-Y. Wang et al., 2019a). As of 2023, the global prevalence of gaming disorder has risen from 3% to 7% within a few years (Stevens et al., 2021; Zhou et al., 2024), with higher rates in East and Southeast Asia compared to North America and Europe (e.g., Chia et al., 2020; Liao et al., 2022). Demographic differences are also found, with the prevalence rates higher among adolescents and young adults (vs. older adults and the elderly) as well as among males (vs. females; e.g., Gao et al., 2022; Singh et al., 2021).

While gaming disorder is an emerging phenomenon, its adverse impact is not just confined to individuals but rather to society as a whole, especially public mental health. Research increasingly links gaming disorder to significant psychosocial dysfunctions (e.g., impaired cognitive functioning, social isolation, and strained interpersonal relations), which does not only disrupt individuals' lives but also have ripple effects on families and the communities (Kim et al., 2022; Kowalik et al., 2024; Zhuang et al., 2023). Furthermore, co-morbid mental health conditions (e.g., anxiety and depression) are frequently reported among individuals with gaming disorder, emphasizing its broader relevance to mental health outcomes (Zhou et al., 2024).

Taken together, the growing prevalence and societal ramifications of gaming disorder emphasize the urgent need for policy frameworks and public health strategies that address its psychological and societal impacts. A focus on preventive measures, early identification, and tailored interventions could mitigate the public health challenges posed by gaming disorder and its mental health correlates. By situating gaming disorder within the broader context of public mental health, this chapter aims to provide a foundation for understanding its background, diagnostic criteria, symptomology, impacts, risk factors, and treatment approaches.

Background

Over the past decades, the conceptualization of gaming disorder evolved from cultural aspects (e.g., social perceptions, gaming communities), particularly in Asian regions, before transitioning gradually to clinical aspects, including the proposed diagnostic criteria in the DSM-5 and the development of treatment approaches specifically for mitigating symptoms of gaming disorder. In the 1990s and 2000s, the rapid growth of internet cafés and the popularity of massively multiplayer online role-playing games in many Asian societies created an environment that breeds gaming-related problems. For instance, in China and South Korea, gaming became deeply embedded in the social fabric, with a concentration of gaming infrastructure that attracted large numbers of youngsters engaging in gaming excessively (Bax, 2015). This phenomenon was particularly pronounced in East Asia (Chia et al., 2020), where early adoption of online gaming and its socio-cultural significance as a form of entertainment and social interaction were evident.

Researchers in the region were among the first to recognize and investigate excessive gaming as a potential behavioral disorder, driven by the unique context of high engagement with gaming technology and the observable negative impacts on individuals' lives (Zhang & Park, 2023). Early theoretical models of gaming disorder drew from existing frameworks of behavioral disorders, such as substance use and pathological gambling, suggesting that compulsive gaming might exhibit similar symptoms, including tolerance, withdrawal, and significant functional impairments. This marked the onset of broader global recognition of gaming disorder as a mental health issue, ultimately leading to its inclusion in the Fifth edition of the Diagnostic and Statistical Manual of Mental Disorders (DSM-5) as a condition warranting further study.

American Psychiatric Association's Inclusion

A significant turning point came in 2013 with the publication of the Fifth edition of the Diagnostic and Statistical Manual of Mental Disorders (DSM-5) by the American Psychiatric Association. For the first time, the term "Internet Gaming Disorder" appeared in the manual as a potential psychiatric condition warranting further clinical research (American Psychiatric Association, 2013). While internet gaming disorder was not officially classified as a disorder in the main body of the DSM-5, its inclusion in Part III laid a critical foundation for recognizing gaming disorder as a potential issue that compromises mental health. The DSM-5 outlined nine provisional criteria for internet gaming disorder: (1) preoccupation with gaming; (2) withdrawal symptoms following the prohibition of gaming; (3) the development of tolerance symptoms, leading to the need to spend more time gaming; (4) unsuccessful attempts to reduce or stop gaming; (5) loss of interest in previous hobbies and activities as a result of gaming; (6) continued excessive gaming despite recognition of its adverse impact; (7) deception of family members, therapists, or others regarding the amount of gaming; (8) reliance of gaming to evade or relieve unpleasant moods; and (9) jeopardizing or losing a significant relationship, job, or study or work opportunity due to gaming. Overall, a diagnosis of internet gaming disorder is made if an individual exhibits five or more of these nine symptoms over a 12-month period. These criteria emphasize the psychological, emotional, and social impairments resulting from problematic gaming, rather than simply the amount of time spent engaging in gaming (American Psychiatric Association, 2013).

The inclusion of internet gaming disorder in the DSM-5 has major implications for clinical practice. First, it has provided mental health professionals with a standardized set of criteria to guide diagnosis, helping clinicians identify problematic gaming behaviors and their associated psychological impacts. This standardized framework also facilitates communication among healthcare providers, ensuring that individuals experiencing severe gaming-related impairments are consistently assessed and treated across diverse settings.

Second, such inclusion encourages more comprehensive assessments that consider not only the amount of gaming but also its functional and emotional consequences. Clinicians are now better equipped to identify underlying issues such as mood disorders, anxiety, or social isolation that may be contributing to the development of issues related to gaming. This shift towards a more holistic understanding of gaming disorder emphasizes the need for treatment plans that address both the gaming behavior itself and any co-morbid psychiatric conditions.

Third, the recognition of internet gaming disorder as a potential disorder by the DSM-5 has stimulated the development of targeted therapeutic interventions, including cognitive-behavioral therapy and other evidence-based treatments aimed at mitigating gaming-related harms. It has also led to the exploration of preventative strategies, particularly for high-risk populations such as adolescents, who are more vulnerable to developing problematic gaming patterns. Clinicians are now more attuned to the possibility of internet gaming disorder, making it more likely that individuals can receive timely and appropriate care.

Lastly, the inclusion highlights the growing need for public health initiatives that raise awareness of the potential mental health risks associated with excessive gaming. As research attention to this emerging disorder has started to grow, clinicians and policymakers can have a better understanding of the risk and protective factors and mechanisms involved, ultimately leading to more informed approaches to prevention and intervention.

World Health Organization's Formal Inclusion

The World Health Organization (WHO) adopted a more definitive stance in 2018 by officially recognizing gaming disorder in the 11th edition of the International Classification of Diseases (ICD-11; World Health Organization, 2018). This recognition has set a milestone for the formal acknowledgement of categorizing gaming disorder as a mental health condition with a set of specific diagnostic criteria. Unlike the DSM-5's provisional recognition of internet gaming disorder as a condition requiring further study, the ICD-11 classified gaming disorder as an official diagnosis, a decision with far-reaching implications for clinical practice, public health policy, and global health initiatives (Van Rooij et al., 2018). The ICD-11's inclusion marks the first time a global health authority has formally acknowledged the potential harms of excessive gaming in a way that could influence national health policies, prompting a more structured approach to prevention, intervention, and public health awareness worldwide.

According to the ICD-11, GD encompasses both online and offline gaming behavior. Gaming disorder is defined using a set of criteria, focusing more on the functional impairments caused by excessive and maladaptive gaming (World Health Organization, 2018). The ICD-11's diagnostic criteria of gaming disorder included: (1) impaired control over gaming (e.g., onset, frequency, intensity, duration); (2) increasing priority given to gaming over other life interests and daily activities; (3) continuation or escalation of gaming despite the occurrence of undesirable consequences; and (4) significant impairment in personal, family, social, educational, or occupational functioning due to gaming.

Aligned with DSM-5's criteria of internet gaming disorder, one of the most crucial benchmarks in ICD-11's diagnostic criteria is that excessive gaming should significantly disrupt daily life functioning for a considerable period, typically 12 months, to distinguish normal from pathological gaming. This major criterion highlights the disorder's functional impairment as a key diagnostic factor, rather than a sole focus on the experience of behavioral symptoms characterized by gaming disorder (World Health Organization, 2018).

The global impact of the ICD-11's classification is profound, particularly when compared to the more limited influence of the DSM-5. While the DSM-5's inclusion of internet gaming disorder encouraged research and clinical attention in certain regions, the ICD-11's global reach ensures that gaming disorder is now recognized across diverse healthcare systems and public health infrastructures worldwide (e.g., Seok & DaCosta, 2014; Wang & Cheng, 2020; Young, 2009; Zhuang et al., 2023). This shift has catalyzed a global conversation on the mental health implications of gaming, encouraging governments and public health organizations to consider it as part of broader mental health initiatives. For instance, countries such as South Korea, where gaming is deeply embedded in the culture, have already implemented policies aimed at managing gaming behavior, and the WHO's classification provides an official framework for such interventions on an international scale.

However, the formal recognition of gaming disorder has also sparked significant debate. Some critics argue that this inclusion may pathologize normal behaviors, especially in younger populations who engage in gaming as a professional activity, such as e-sports (e.g., Aarseth et al., 2017; Van Rooij et al., 2018). These critics assert that gaming, particularly competitive gaming, is increasingly recognized as a legitimate career, with individuals spending significant hours honing their skills. In this context, what might be considered excessive gaming may, in fact, be a part of a healthy, goal-oriented pursuit for those engaged in e-sports or other gaming-related professions. The concern here is that the ICD-11's broad diagnostic criteria could inadvertently pathologize behaviors that are, for many, entirely normative and even beneficial in fostering skills like teamwork, problem-solving, and cognitive development (Király & Demetrovics, 2017; Na et al., 2017). As such, the challenge lies in differentiating between professional gaming, recreational gaming, and genuinely harmful, maladaptive gaming behaviors.

Despite these critiques, there is strong evidence supporting the necessity of the ICD-11's classification. Research has shown that problematic gaming can lead to significant impairments in mental health, including issues with anxiety, depression, and social isolation (Király et al., 2017; Seok & DaCosta, 2014). The classification is particularly relevant in identifying individuals who engage in gaming as a form of escape from negative emotions or stress, leading to significant disruptions in daily life (World Health Organization, 2018). For many individuals, gaming can become a maladaptive coping mechanism, exacerbating pre-existing mental health issues. Furthermore, the recognition of gaming disorder helps distinguish between those who experience occasional gaming-related issues and those whose gaming behavior results in long-term, life-altering consequences. This distinction ensures that interventions are targeted towards individuals whose gaming behavior genuinely interferes with their ability to function in key areas of life.

The ICD-11's classification does not suggest that all gaming is harmful, but rather advocates for early identification and intervention for those exhibiting problematic gaming patterns. It recognizes that, with proper guidelines, gaming can be a healthy and enriching activity, promoting cognitive and social development, and enhancing well-being (Jo et al., 2019). The ICD-11's classification emphasizes that problematic gaming behaviors are typically characterized by a loss of control and persistent negative outcomes, making early intervention and management crucial to preventing long-term psychological harm.

Overall, the global recognition of gaming disorder in the ICD-11 provides an important framework for understanding and managing problematic gaming behavior on a larger scale. By emphasizing the importance of distinguishing between healthy and harmful gaming practices, the ICD-11 provides a necessary framework for managing this emerging mental health issue. This solid foundation allows for a focused exploration of the symptomatic presentation of gaming disorder, as outlined in the following section.

Symptomology

Gaming disorder is marked by specific behavioral symptoms that compromise an individual's capacity to manage their gaming activities, often resulting in adverse effects on personal, social, and occupational functioning. Recent studies have elaborated on these core symptoms, revealing the interconnected nature of gaming disorder symptoms and their underlying structure, which plays a critical role in determining treatment priorities (Bäcklund et al., 2022; Gundogdu & Eroglu, 2022; Kochuchakkalackal & Reyes, 2020; Tang et al., 2023). The inter-connections among these symptoms reflect that treating one symptom in isolation—such as simply reducing gaming time—may not be sufficient to break the vicious cycles that sustain the disorder. Understanding the mutual influences among these symptoms is thus essential for developing more effective, comprehensive treatment plans.

A primary symptom of gaming disorder is preoccupation with gaming. Individuals with gaming disorder frequently exhibit persistent thoughts about gaming, including planning future gaming sessions or reminiscing about past ones (Mills & Allen, 2020). Such preoccupation often diverts their attention from other important aspects of life, such as job responsibilities or interpersonal relations, creating a significant imbalance in their daily functioning (Bäcklund et al., 2022). This preoccupation is closely linked with the loss of control over gaming, as persistent thoughts about gaming can make it increasingly difficult to resist the urge to play, further exacerbating the disorder.

Another critical symptom is loss of control over gaming behavior (Tang et al., 2023). Individuals may experience a strong, repeated inability to reduce gaming time despite setting personal limits or intentions to do so. Studies highlight that those with gaming disorder often play far beyond their planned duration, demonstrating an impairment in their self-regulation and an escalating need to continue gaming (Xu et al., 2025). This symptom plays a pivotal role in maintaining the disorder, as it creates a self-reinforcing cycle with the preoccupation symptom, where the more time spent gaming, the more it dominates thoughts and priorities. This interconnectedness underscores the need to address both preoccupation and loss of control together, as tackling one without considering the impact of the other may weaken treatment efficacy.

Withdrawal symptoms are also central to gaming disorder, manifesting as psychological discomfort, irritability, or restlessness when gaming is inaccessible. The experience of withdrawal symptoms reinforces the cycle of compulsive gaming by increasing the emotional drive to return to gaming, which in turn exacerbates other symptoms such as preoccupation and loss of control. The presence of these symptoms underscores the addictive nature of the disorder, as individuals seek gaming as an attempt to alleviate their withdrawal symptoms but ironically further reinforce the problematic behavior (Tang et al., 2023).

Overall, the tendency of continued gaming despite adverse consequences illustrates how individuals with gaming disorder prioritize gaming over essential obligations, including family, work, or education, despite being aware of its undesirable and possible long-term impacts. This persistence reflects a degree of psychological attachment to gaming that compromises decision-making and goal-setting abilities in other areas of life (Starzec et al., 2024). The relationship between this symptom and others, such as loss of control or withdrawal, amplifies its adverse impact on daily functioning, making it necessary to focus on the more extensive and interconnected symptom network during treatment.

A recent study utilizes network analysis to examine the centrality and strength of the inter-relationships among the various gaming disorder symptoms, and the results underscore that symptoms such as loss of control and withdrawal are particularly central to the factor structure of gaming disorder, as they exert considerable influences on other symptoms, creating a self-reinforcing vicious cycle (Tang et al., 2023). This empirical evidence is crucial for informing the design of more effective interventions, as addressing these core symptoms (e.g., loss of control, withdrawal) may help disrupt the feedback loops that maintain the disorder, leading to more efficacious treatment outcomes.

In summary, understanding the interconnectedness of symptoms is critical for treatment, as it implies that interventions for gaming disorder should not be confined to a single symptom at a time. Empirical findings indicate that targeting multiple interconnected symptoms simultaneously should exert greater effectiveness in disrupting the cycle of addiction and promote long-term recovery.

Adverse Impacts on Health

As mentioned at the outset, while gaming can be a pleasant and enjoyable leisure activity for many people, compulsive gaming behavior has been associated with a range of health problems. This section reviews recent bodies of studies on the health implications of gaming disorder, including its effects on mental health, psychiatric conditions, and physical wellness.

Mental Health Issues

Anxiety

Anxiety is a prevalent mental health concern that often co-occurs with gaming disorder. The relationship between anxiety and gaming disorder highlights how compulsive gaming behaviors serve as both a cause and consequence of heightened anxiety (Bäcklund et al., 2022; Zhuang et al., 2023). For example, a recent meta-analysis, examining data from 39 studies, found that individuals with gaming disorder consistently exhibit higher levels of generalized anxiety compared to those without this disorder (Zhuang et al., 2023). Anxiety symptoms linked to gaming disorder often include excessive worry, nervousness, and physiological manifestations such as increased heart rate and muscle tension. These symptoms are not only experienced during gaming sessions but may extend to other broader life domains, potentially impacting work, studies, and interpersonal interactions (Gao et al., 2022; Stevens et al., 2021; Wang & Cheng, 2022). Moreover, compulsive gaming behaviors may exacerbate avoidance tendencies, as individuals with gaming disorder often turn to gaming as a coping mechanism to evade real-life stressors and responsibilities (Bäcklund et al., 2022; Király et al., 2017). This reliance can foster a vicious cycle in which the individual feels progressively more isolated and anxious, thereby reinforcing the compulsive gaming behavior as a means of escape (Gao et al., 2022; Wang & Cheng, 2022).

The interaction between gaming disorder and anxiety may also be influenced by the type of games played. For instance, massively multiplayer online games, which often require high levels of social interaction and competition, are linked with increased levels of social anxiety among gamers with gaming disorder (Aziz et al., 2021). Such compulsive gaming behavior can elicit fear of judgment or criticism, leading to a heightened state of social anxiety that compounds with gaming disorder symptoms. Such findings indicate that the nature of the gaming experience itself can contribute to specific forms of anxiety, aggravating the overall mental health impact of gaming disorder.

Depression

Depression is another prevalent mental health issue that is frequently comorbid with gaming disorder. The connection between depression and gaming disorder illustrates the cyclical nature of the relationship, in which problematic gaming behaviors often exacerbate depressive symptoms, and vice versa. Many systematic reviews and meta-analyses have identified depressive symptoms as a common comorbidity among individuals with gaming disorder (Darvesh et al., 2020; Gao et al., 2022; Moore et al., 2022; Noroozi et al., 2021). Depressive symptoms experienced among individuals with gaming disorder manifest as a persistent feeling of sadness, lack of interest in previously enjoyable activities, fatigue, and feelings of worthlessness. In addition, individuals with gaming disorder are found to have a significantly higher likelihood of experiencing depressive symptoms, with severity often increasing in proportion to gaming time and level of engagement with compulsive gaming behaviors (Ostinelli et al., 2021).

Such positive relationships between depression and gaming disorder can be cyclical. Many individuals with depressive tendencies may turn to gaming as a coping strategy for emotional escape, seeking temporary relief from unpleasant feelings (Bäcklund et al., 2022; Stevens et al., 2021; Wang & Cheng, 2022). However, the temporary nature of this relief can lead to a dependence on gaming to maintain pleasant moods, which in turn elicit more depressive symptoms as time spent on gaming increases while more social, occupational, or academic responsibilities are neglected. Moreover, the sedentary lifestyle associated with excessive gaming can exacerbate physical symptoms of depression, such as fatigue and lethargy, further increasing the severity of gaming disorder (Kim et al., 2022; Kowalik et al., 2024; Zhuang et al., 2023).

Longitudinal research on gaming disorder underscores the worsening trajectory of depressive symptoms in individuals with a prolonged experience of gaming disorder symptoms. For instance, studies revealed that over a span of 2 years, individuals with untreated symptoms of gaming disorder experienced a marked increase in depressive symptoms (Kim et al., 2022; Krossbakken et al., 2018; Zhou et al., 2024; Zhuang et al., 2023). Such findings imply that the longer an individual engages in compulsive gaming behaviors, the more intense their experience of depressive symptoms can become, thus highlighting the importance of early intervention for gaming disorder (Cheng et al., 2024).

Psychological Well-being

The psychological well-being of individuals with gaming disorder is significantly impacted by their problematic gaming behaviors. The connection between psychological well-being and gaming disorder reflects how excessive gaming undermines autonomy, positive relations with others, and purpose in life, all of which are key elements of psychological well-being (Ryff & Keyes, 1995). Individuals with gaming disorder experience low levels of psychological well-being, a broad construct encompassing elements such as emotional balance and a sense of purpose (Moore et al., 2022). Studies have shown that gaming disorder can adversely impact psychological well-being by isolating individuals from meaningful social interactions and activities that contribute to a sense of fulfillment and self-worth (Gao et al., 2022; Kavanagh et al., 2023; Lemenager et al., 2020). Individuals with gaming disorder consistently report lower scores on measures of psychological well-being compared to average gamers and non-gamers (Akaroğlu, 2024; Kort-Butler, 2021; Moore et al., 2022; Snodgrass et al., 2022). In addition, their lower levels of psychological well-being often align with the progressive nature of gaming disorder, where their lives become increasingly centered around gaming at the expense of engagement in social and recreational activities.

Research reveals that individuals with gaming disorder frequently report feelings of guilt, lack of control, and frustration related to their gaming behaviors, further deteriorating their psychological well-being (Afriwilda & Mulawarman, 2021; Carragher et al., 2022; Kim et al., 2022; Kowalik et al., 2024). These unpleasant emotions may arise from an awareness of the undesirable consequences of their gaming behaviors, such as failing to meet the requirements of their studies or work. Besides, the lack of real-world social support, combined with the pressure to excel in virtual environments, may intensify feelings of inadequacy and low self-worth (Zhuang et al., 2023). This disconnection from real-world social structures can severely reduce one's sense of purpose and belonging, thus compromising psychological well-being over time (Aziz et al., 2021; Kochuchakkalackal & Reyes, 2020; Noroozi et al., 2021).

Quality of Life

Quality of life—a multidimensional construct comprising facets such as physical functioning, vitality, and social functioning—is notably diminished in individuals having gaming disorder (Akaroğlu, 2024; Kim et al., 2022; Snodgrass et al., 2022). As revealed in a meta-analysis and systematic review, individuals with gaming disorder tend to perceive a lower quality of life across all life domains, with particularly severe impacts on emotional and social aspects (Noroozi et al., 2021). These reductions in quality of life may stem from the disruptions of gaming disorder symptoms on individuals' daily routines, interpersonal relationships, and physical health.

Social isolation, one of the key contributors to decreased quality of life, is prevalent among those with gaming disorder (e.g., Bhagat et al., 2020; Tateno et al., 2019; J.-L. Wang et al., 2019b). Many individuals with gaming disorder withdraw from their significant others, such as family and friends, leading to a reduction in face-to-face social interactions and shrinkage in social support networks (Bhagat et al., 2020). Social isolation not only weakens emotional support but also exacerbates feelings of loneliness, contributing to a lower overall quality of life (e.g., Farrell et al., 2023; J.-L. Wang et al., 2019b).

Studies further indicate that such declines in quality of life may be more pronounced among younger individuals with gaming disorder. Adolescents, for instance, are in a critical period of psychological and social development, and social isolation resulting from gaming disorder can hinder the development of their social skills and impair their ability to establish meaningful social ties (Farrell et al., 2023; Skripkauskaite et al., 2022; Teng et al., 2020). As a result, gaming disorder may limit the formation of social networks that typically enhance individuals' quality of life, leaving those who are affected more susceptible to long-term mental health consequences.

In summary, recent meta-analyses and systematic reviews provide strong evidence of the detrimental impact of gaming disorder on various aspects of mental health, including anxiety, depression, psychological well-being, and quality of life. Individuals with gaming disorder are more likely to experience heightened levels of anxiety and depression, low psychological well-being, and poor quality of life across various life domains. These mental health problems are not confined to the experience of symptoms per se but can also reinforce the severity of gaming disorder in the long term, creating a vicious cycle that requires tremendous effort and time to break. These findings imply that intervention for gaming disorder should be conducted as early as possible, particularly among children and adolescents, before the gaming disorder symptoms develop and are maintained. If such non-clinical mental health issues persist, they may develop into more severe psychiatric problems.

Psychiatric Issues

Gaming disorder, beyond commonly discussed mental health impacts like anxiety and depression, has significant associations with a range of severe psychiatric disorders. Research demonstrates that compulsive gaming behaviors can serve as both a consequence and a contributing factor for various psychiatric conditions, notably attention-deficit/hyperactivity disorder (ADHD), obsessive-compulsive disorder (OCD), and substance use disorder (SUD; e.g., Burleigh et al., 2019; Dullur et al.,

2021; Klugah-Brown et al., 2021; Stevens et al., 2021). The prevalence and nature of these comorbidities tend to vary according to factors such as age and game genre, with different age groups and gaming preferences potentially influencing the severity and manifestation of these psychiatric conditions.

Attention-Deficit/Hyperactivity Disorder

ADHD has been consistently linked to problematic gaming behaviors due to the condition's inherent difficulties with impulse control, sustained attention, and reward processing (e.g., Dullur et al., 2021; Klugah-Brown et al., 2021; Koncz et al., 2023; Mathews et al., 2019; Salerno et al., 2022). Individuals with ADHD may be particularly susceptible to the immediate gratification and continuous engagement offered by gaming (Mathews et al., 2019; Salerno et al., 2022). Moreover, ADHD-related gaming behaviors tend to vary across age groups. For example, adolescents and young adults are more prone to engaging in fast-paced, high-reward games such as action-oriented or competitive multiplayer games, which could exacerbate impulsivity and attention difficulties (Paulus et al., 2018; Tang et al., 2023). In contrast, older individuals with ADHD may play more immersive or solitary games, such as role-playing games, where hyper-focus on in-game tasks may reinforce ADHD-related challenges (Dullur et al., 2021; Koncz et al., 2023). Paulus et al. (2018) observed that individuals with ADHD tend to use gaming as a strategy to cope with the frustration and impulsivity associated with their disorder, often leading to excessive gaming. Longitudinal studies have further indicated that the hyper-focused states induced by gaming can exacerbate ADHD symptoms, creating a reinforcing cycle of dependence on gaming as a maladaptive coping strategy (Dullur et al., 2021; Koncz et al., 2023; Paulus et al., 2018; Singh, 2019).

Obsessive-Compulsive Disorder

OCD is another psychiatric condition frequently comorbid with gaming disorder (Burleigh et al., 2019). For individuals with OCD, gaming may serve as a temporary distraction from their own intrusive thoughts or compulsive urges (Klugah-Brown et al., 2021). The repetitive nature of certain games, such as those that involve completing the same tasks or levels repeatedly—like grinding in role-playing games or repetitive actions in puzzle games—may resonate with patterns observed among individuals with OCD. This can create a feedback loop where compulsive gaming mimics or temporarily alleviates obsessive-compulsive tendencies (Grassi et al., 2020). Age and game genre appear to influence the form of OCD-related gaming. For instance, younger individuals tend to be more drawn to fast-paced, high-stakes games, whereas older individuals tend to engage in games that require repetitive actions or tasks, thereby reinforcing the compulsive aspects of their OCD (Stevens et al., 2021). Over time, however, this form of escapism can in turn reinforce compulsive behavior, making it more challenging for individuals to disengage and potentially intensifying the severity of their OCD symptoms (Stevens et al., 2021).

Substance Use Disorder

Gaming disorder and SUD are often found to share common behavioral and neurobiological underpinnings (Burleigh et al., 2019; Stevens et al., 2021). The reward pathways implicated in SUD are similarly activated by gaming, which provides instant rewards, social connection, and achievement. Some studies highlight that individuals who engage in compulsive gaming behaviors are more likely to develop or struggle with issues pertaining to SUD, as both types of problematic behaviors may fulfill similar dopamine-driven reward needs (Brand et al., 2019). Other evidence also indicates that gaming can serve as an alternative or supplementary form of dependency for those with SUD, exacerbating their overall psychiatric burden (Ko et al., 2012). This association may vary with age and game genre, as younger individuals, particularly adolescents, are more likely to engage in online gaming or gambling, which may overlap with risk factors for substance abuse (Stevens et al., 2021). Older individuals, however, generally turn to gaming as a means of coping with stressors related to life transitions, such as work or family issues, potentially contributing to a greater likelihood of dual dependency on both gaming and substances.

Impacts on Pre-Existing Psychiatric Conditions

For individuals with pre-existing psychiatric disorders, gaming disorder may serve as a double-edged sword. Specifically, gaming may initially relieve these individuals from the experience of their distressing symptoms but ultimately contribute to the worsening of their psychiatric conditions. Gaming as an avoidant coping mechanism may lead to more profound isolation, reduced social functioning, and increased psychological distress over time (Kowalik et al., 2024). This cycle is further consolidated by findings from systematic reviews, which indicate that compulsive gaming often reinforces negative behavioral patterns, making it more challenging for individuals with psychiatric conditions to manage their symptoms without relying on gaming (Estupiñá Puig et al., 2024; Starzec et al., 2024; Wang et al., 2023). Moreover, comorbidities such as ADHD and OCD may become more pronounced or manifest differently depending on the age of the gamer and the game genre they engage in, thus affecting the severity of the overall psychiatric impact (Kowalik et al., 2024).

To conclude, the psychiatric impacts of gaming disorder are broad and severe. Gaming disorder not only occurs together with ADHD, OCD, and SUD but also contributes to the maintenance and exacerbation of these psychiatric conditions over time. A comprehensive understanding of the role of gaming disorder in psychiatric conditions, and how this role varies by age and gaming genre, may inform more effective clinical interventions and preventative measures targeted at clients at heightened risks due to pre-existing psychiatric vulnerabilities.

Physical Health Problems

Beyond its psychological implications, gaming disorder has also been found to be associated with a range of physical health issues, including eye strain, musculoskeletal pain, and sleep disturbances (Aziz et al., 2021; Noroozi et al., 2021; Starzec et al., 2024; Tholl et al., 2022). These health consequences often result from prolonged gaming sessions, lack of physical activities, and exposure to screen light, which, when unaddressed, can worsen physical health (Alimoradi et al., 2019; Kaur et al., 2022; Kristensen et al., 2021). To mitigate these physical health issues, preventive measures such as adopting ergonomic gaming setups, limiting screen time, and establishing proper sleep hygiene practices are crucial (e.g., Martha et al., 2022; Rahmillah et al., 2023).

Eye Strain

One of the most common physical health complaints linked to excessive gaming is eye strain, frequently referred to as "computer vision syndrome." Prolonged screen exposure without adequate breaks has been shown to cause symptoms such as blurred vision, dry eyes, headaches, and eye discomfort. Systematic reviews indicate that extensive screen time is a significant contributor to visual fatigue, with potential long-term effects on vision health (Kaur et al., 2022; Mylona et al., 2020). Both cross-sectional and longitudinal studies involving adolescent gamers revealed a higher prevalence of eye strain among those who spent multiple hours gaming consecutively (Kaur et al., 2022). These vision-related issues are particularly exacerbated in environments with improper lighting or when users fail to implement strategies to mitigate eye strain (Johnson & Rosenfield, 2023; Turgut, 2018). To prevent eye strain, individuals should adopt the "20-20-20 rule," which involves taking a 20-s break every 20 min by focusing on something 20 feet away (Talens-Estarelles et al., 2023). Besides, it is essential to create a well-lit gaming environment and avoid glare on screens. Using blue light filters or glasses designed for screen use can help minimize strain (Coles-brennan et al., 2019).

Back Pain and Musculoskeletal Issues

Back pain and other musculoskeletal problems are also widespread among individuals engaged in excessive gaming (Ahmed et al., 2024; Tholl et al., 2022). Gaming often requires prolonged periods of sitting in a single position, which, if combined with poor posture, can lead to muscle strain and chronic pain in areas such as the lower back, shoulders, and neck. Longitudinal studies have highlighted that the physical demands of extended gaming sessions, especially without ergonomic support, increase the risk of developing chronic back pain and related musculoskeletal disorders over time (Cankurtaran et al., 2022; Nasir et al., 2024; Tholl et al., 2022). A systematic review shows that sedentary gaming behaviors correlate strongly with physical discomfort and musculoskeletal complaints among young adults, emphasizing the need for preventive ergonomic practices while gaming (Tholl et al., 2022). To prevent musculoskeletal issues, individuals should invest in ergonomic furniture, such as chairs with lumbar support, and ensure that their screen is at eye level to avoid neck strain (Gonçalves et al., 2024). Taking frequent breaks, practicing stretching exercises, and adjusting the gaming setup to allow for a more neutral body position can also help alleviate back pain and improve posture.

Sleep Disturbance

Sleep disturbance is another critical health concern associated with gaming disorder. Excessive gaming, particularly during late hours, disrupts normal sleep patterns, frequently leading to issues such as delayed sleep onset and insomnia. Studies indicate that the stimulating content of games, combined with screen light exposure, interferes with the production of melatonin, thus disturbing the body's circadian rhythms (Gundogdu & Eroglu, 2022; Wong et al., 2020; Zaman et al., 2022). A meta-analysis revealed that individuals with gaming disorder tend to report poorer sleep quality and higher incidences of sleep disorders compared to those without this disorder (Alimoradi et al., 2019). The association between gaming and poor sleep quality has been documented in both cross-sectional and longitudinal studies, demonstrating that consistent sleep disruption can lead to cumulative physical and cognitive health deficits, including decreased mental focus, impaired immune function, and increased fatigue (Kristensen et al., 2021). To promote better sleep, individuals may limit gaming time, especially during the evening hours, and establish a consistent sleep schedule. Implementing a "screen curfew" at least an hour before bedtime can help reduce the negative effects of screen light on sleep (Perrault et al., 2019). Using blue light filters on devices or employing apps that reduce screen brightness at night can mitigate the impact on melatonin production (Bossi & Hopker, 2017).

Overall, the physical health problems associated with gaming disorder—eye strain, back pain, and sleep disturbances—highlight the need for awareness and preventive strategies for physical health enhancements among heavy gamers. As gaming continues to be a popular leisure activity, addressing these physical health risks through ergonomic interventions, screen time management, and sleep hygiene practices is essential to mitigating the adverse effects on individuals' overall well-being.

Risk Factors

Understanding the risk factors for gaming disorder is essential for creating targeted prevention and intervention strategies for mitigating symptoms of gaming disorder and the health issues related to this problem. Although gaming disorder can compromise mental and physical health among individuals from diverse backgrounds, certain psychosocial factors can increase such risks. This section examines two primary categories of risk factors: cognitive and familial factors.

Cognitive Risk Factors

Cognitive risk factors such as gaming self-esteem, maladaptive gaming rules, and poor impulse control, play central roles in the development and maintenance of gaming disorder (Gao et al., 2022; Kochuchakkalackal & Reyes, 2020; Tang et al., 2023; Zhuang et al., 2023). Findings on gaming self-esteem—one of the most central cognitive risk factors—indicate that individuals with gaming disorder often use gaming as a means to fulfill unmet esteem needs, reflecting a reliance on gaming to experience validation and accomplishment absent in other life areas (Tang et al., 2023). This finding aligns with self-determination theory, which explains how individuals with gaming disorder are drawn to gaming to satisfy fundamental psychological needs (Bäcklund et al., 2022; Mills & Allen, 2020). Specifically, individuals with low self-esteem were more likely to engage in prolonged gaming sessions, often seeking the structured and goal-oriented nature of games to experience a sense of accomplishment and validation that are lacking in their daily lives (Kavanagh et al., 2023). For example, teenagers who struggle academically and socially may turn to multiplayer online games where they can earn greater achievements or high rankings, gaining recognition and praise from fellow players, which they may not experience in school or at home. Similarly, employees who have not received much appreciation at work may immerse themselves in a game where they can lead teams or conquer challenges, providing a sense of value and self-worth that are largely absent from their daily routine.

Maladaptive gaming rules—rigid, often irrational beliefs about gaming—also act as key cognitive risk factors by linking cognitive distortions to gaming disorder (Tang et al., 2023). These maladaptive cognitions create distinct symptom profiles, where specific beliefs intensify related gaming disorder symptoms, contributing to an enduring and self-perpetuating disorder pattern (Kowalik et al., 2024; Zhou et al., 2024). Impulse control deficits further exacerbate gaming disorder by reducing an individual's ability to regulate gaming behaviors, even when aware of adverse consequences (Darvesh et al., 2020; Wartberg et al., 2019). This lack of self-regulation, particularly pronounced in adolescents, often leads to compulsive gaming behaviors and avoidance of real-life responsibilities (Gao et al., 2022).

Maladaptive coping mechanisms also contribute to increasing the risk of the development of gaming disorder. For individuals who lack effective coping strategies to handle stress or emotional challenges, gaming becomes a primary strategy for them to evade those adverse events (Melodia et al., 2020). Recent systematic reviews document that individuals who rely on gaming for emotional regulation often face increased risks of gaming disorder, as they turn to games to avoid confronting stressful events, thereby reinforcing their dependency on gaming as a major coping strategy for handling life stress (Estupiñá et al., 2024; Estupiñá Puig et al., 2024; Marchica et al., 2019). Together, these cognitive risk factors underscore how unmet psychological needs and maladaptive cognitive patterns maintain and intensify gaming disorder (Tang et al., 2023).

Family Risk Factors

In addition to the influences of cognitive risk factors at the individual level, the social environment also plays an influential role in the development and maintenance of gaming disorder. Family dynamics—including parenting style, family cohesion, and parental involvement in gaming—can exert influences on elevating individual's likelihood of developing gaming disorder (Nielsen et al., 2020). Research consistently shows that individuals from families with low cohesion and limited emotional support are at a higher risk of excessive gaming (Bonnaire & Phan, 2017; Nielsen et al., 2020; Schneider et al., 2017). This is because these individuals may use gaming as a substitute for their connection with family members and as a coping strategy to mitigate feelings of loneliness derived from emotional neglect by family members.

The familial factor of parenting style is particularly impactful. Supportive and balanced parenting styles, which combine warmth and structure (e.g., Tehrani et al., 2024), have been shown to provide protective effects against gaming disorder (Li et al., 2018). Authoritarian parenting—characterized by strict rules without emotional support—has been associated with higher incidences of gaming disorder. A systematic review revealed that adolescents with parents exhibiting an authoritarian style were more likely to seek autonomy and emotional fulfillment through gaming, heightening the risks of gaming disorder (Nielsen et al., 2020). However, it is noteworthy that permissive parenting, which involves few restrictions or limits on gaming time, can also foster problematic gaming behaviors. Studies have documented that permissive parenting often fails to provide children with the structure needed to develop self-discipline while gaming (Akaroğlu, 2024; Bonnaire & Phan, 2017). In contrast, a supportive and balanced approach—setting clear expectations, providing emotional support, and encouraging open communication—has been related to a lower likelihood of children turning to gaming to cope with unmet needs (Li et al., 2018).

In addition to parenting style, parental involvement in gaming habits can either mitigate or exacerbate the risks of gaming disorder. Parents who actively monitor their child's gaming behavior and encourage non-gaming leisure activities can prevent the development of compulsive gaming behaviors (Bonnaire & Phan, 2017; Lam & Cheng, 2022; Li et al., 2018). Specific strategies, such as establishing daily or weekly time limits for gaming, ensuring that gaming does not interfere with essential activities like schoolwork or socializing, and engaging in regular discussions about the content and impact of games, can significantly reduce the risk of gaming disorder. However, a lack of parental oversight or permissiveness around gaming time may unintentionally increase the risk of addiction. These findings show that children are more prone to problematic gaming behaviors if their parents are unaware of their gaming habits or do not place any limits on children's gaming time, such that the children are left to self-regulate without parental guidance (Nielsen et al., 2020).

Taken together, cognitive and familial factors can interact in complex ways to have conjoint influences on the risk of developing gaming disorder. Cognitive factors such as low self-esteem, poor impulse control, and reliance on gaming for coping can predispose individuals to compulsive gaming behaviors, while family dynamics—including parenting style and parental involvement—further intensify such risks. By understanding these risk factors, preventive efforts can be better tailored to address the unique vulnerabilities that lead to gaming disorder.

Treatment Approaches

With the growing scholarly and public concerns of gaming disorder as a legitimate mental health disorder, a growing body of research effort has been devoted to identifying effective treatment approaches to mitigating symptoms of gaming disorder. This section reviews recent meta-analyses and systematic reviews, focusing on cognitive-behavioral therapy (CBT), pharmacological interventions, and family-based therapies, all of which are the most widely adopted approaches for treating gaming disorder (King et al., 2017; Xu et al., 2021; Zajac et al., 2017, 2020). Emerging digital interventions, such as visual mindfulness-based therapies, will also be discussed in the following section. Table 7.1 summarizes the key benefits and limitations of the major treatment approaches available for gaming disorder.

Cognitive-Behavioral Therapy

Cognitive-behavioral therapy (CBT) is widely regarded as the gold standard treatment for gaming disorder. It targets the maladaptive thoughts and behaviors that perpetuate compulsive gaming behaviors, with the overarching goals of modifying cognitive distortions and behavioral patterns that reinforce compulsive gaming behaviors. Recent meta-analyses underscore the efficacy of CBT in relieving gaming disorder symptoms, improving emotional regulation, and promoting long-term behavior change among those with gaming disorder (Stevens et al., 2019; Wang et al., 2023; Zajac et al., 2017). A systematic review with a specific focus on randomized controlled trials found that CBT led to substantial reductions in gaming time and improvements in psychological well-being across heterogenous samples

Table 7.1 Key benefits and caveats of major treatments for gaming disorder

Treatment	Key benefits	Caveats
Cognitive-behavioral therapy	Addresses both cognitive and behavioral aspects of gaming disorder	Requires significant time commitment and regular engagement
	Effective in reducing gaming time and improving emotional regulation	Some clients may struggle with self-reflection or cognitive restructuring
	Provides relapse prevention strategies for long-term recovery	
Pharmacotherapy	Helps manage co-occurring symptoms, such as anxiety or depression	Not directly address underlying cognitive or behavioral factors of gaming disorder
	Alleviates mood dysregulation and impulsivity in some clients	Dependency risks (especially with stimulant medications), potentially leading to abuse
		Inconsistent efficacy
Family-based therapy	Engages family members to address dynamics contributing to gaming disorder	Less effective for adult clients or those from families with poor cohesion or ongoing conflict
	Effective for younger clients by improving family communication and support	Families with low cohesion or dysfunctional communication structures struggle to engage fully
	Promotes healthy family dynamics, setting boundaries and encouraging healthier habits	Not provide sufficient support for adult clients who no longer have parental guidance
Digital mindfulness-based therapy	Flexible, accessible options for clients unable to attend in-person therapy	Require client motivation and self-discipline to remain engaged
	Apps to help regulate gaming behaviors, manage stress, and improve focus	The efficacy needs further validation, especially for long-term or severe cases
		Limited personal interaction may be a barrier for individuals who prefer face-to-face or more intensive therapies
Gamification therapy	Fosters a structured, engaging therapeutic environment that encourages active participation in treatment	More effective in relieving gaming disorder symptoms for normal and at-risk clients, but less so for those with gaming disorder who need more intensive treatment
	Builds an enjoyable atmosphere that foster effective skill acquisition	Over-reliance on gamified therapy elements may reinforce problematic gaming behaviors and hinder recovery

of clients with gaming disorder (Stevens et al., 2019). All these findings reveal CBT as an efficacious approach for treating gaming disorder.

Recent research illustrates not only the efficacy of CBT in addressing gaming disorder but also highlights the structured techniques that contribute to its effectiveness. The psychological benefits of CBT are partly attributed to its structured approach, which involves specific techniques such as cognitive restructuring, behavioral activation, and skills training. Cognitive restructuring helps clients identify and challenge distorted thoughts related to gaming, such as overestimating the rewards of gaming or perceiving gaming as an escape from real-life problems. For instance, a client may learn to replace biased thoughts such as "I need to game to feel good about myself" with more balanced thoughts such as "I can find enjoyment in a variety of activities." Moreover, behavioral activation encourages clients to engage in rewarding, non-gaming activities to enhance a pleasant mood and reduce reliance on gaming as a coping strategy. For example, a client may set small goals to engage in social or physical activities that foster a sense of accomplishment, such as exercising, engaging in hobbies, or spending time with family members. Lastly, skills training empowers clients to acquire practical strategies for managing stress, enhancing impulse control, and effectively addressing gaming triggers.

A systematic review further specifies that CBT improves clients' self-control and helps them build more active coping mechanisms to manage stress, addressing poor impulse control and maladaptive coping mechanisms that constitute core cognitive risks of gaming disorder (Zajac et al., 2017). Furthermore, CBT interventions often include relapse prevention strategies to maintain therapeutic gains over time, an essential aspect for long-term recovery from gaming disorder (de Sá et al., 2023; Kumari & Dhiksha, 2022; Stevens et al., 2019).

Pharmacological Interventions

Pharmacological interventions for gaming disorder, while less prevalent, have shown some potential in alleviating some of the associated symptoms, particularly when used together with psychotherapy. The primary medications prescribed for treating symptoms of gaming disorder include antidepressants, such as selective serotonin reuptake inhibitors, and methylphenidate, that have been commonly used to treat ADHD. These medications alleviate underlying psychiatric conditions—such as depression, anxiety, and ADHD—that frequently co-occur with gaming disorder (de Sá et al., 2023; Leeman & Potenza, 2012; Seo et al., 2021).

Meta-analytic studies yield mixed findings on the effectiveness of pharmacological treatments for gaming disorder, often highlighting that while these medications can alleviate specific symptoms (e.g., mood dysregulation, attention deficits), they are generally insufficient to be adopted as standalone treatments (Moore et al., 2022; Seo et al., 2021; Wölfling et al., 2019; Zajac et al., 2020). In a systematic review of pharmacological interventions (de Sá et al., 2023), limited evidence supporting the long-term efficacy of these treatments when used alone, but pharmacological approaches were most effective only when applied in combination with CBT. This combined approach should help clients manage the biological underpinnings of their gaming behavior, such as impulse control issues, while simultaneously addressing their cognitive and behavioral deficits through the application of CBT techniques.

While pharmacotherapy shows some promise in mitigating certain symptoms of gaming disorder, it should be used as a supplementary approach. For example, research on selective serotonin reuptake inhibitors, such as fluoxetine, indicates some psychological benefits in mitigating anxiety symptoms that may drive compulsive gaming behaviors; yet there are major concerns regarding the inconsistent empirical evidence and the potential side effects of medications, which range from gastrointestinal issues to sexual dysfunction, making long-term adherence challenging for many clients (Bonnaire et al., 2019; King & Delfabbro, 2019; King et al., 2017; Wölfling et al., 2019). Furthermore, the use of medications (e.g., methylphenidate for clients with ADHD) presents additional challenges. Although such medications have been found to reduce impulsive behavior in clients with both gaming disorder and ADHD, no evidence has shown that these medications directly decrease compulsive gaming behaviors in those with gaming disorder but without ADHD (Wang et al., 2023). Such empirical inconsistencies in the efficacy of pharmacotherapy raise concerns about its broader applicability in treating gaming disorder alone.

It is noteworthy that the dependency risks associated with certain pharmacological treatments, such as stimulant medications for ADHD, may further complicate their efficacy (e.g., Cunill et al., 2014; Judge & Mellen, 2019). Stimulants have the potential for abuse and can lead to dependency in some clients, especially when used without proper monitoring, thus limiting their feasibility as a primary treatment for gaming disorder.

Family-Based Therapy

Apart from individual-oriented therapy, family-based therapy has gained prominence, especially for treating gaming disorder in children and adolescents. This approach involves actively engaging family members in the treatment process to address dynamics within the home environment that may contribute to or exacerbate gaming disorder symptoms. Family-based therapies are often tailored to improve communication among members, establish boundaries around gaming, and create a supportive family environment that encourages healthier leisure habits (Apisitwasana et al., 2018; Li et al., 2019; Wang et al., 2024).

Recent findings indicate that family-based therapies are particularly effective for younger individuals, as they allow for direct intervention in the clients' daily environment (Apisitwasana et al., 2018; Harshini et al., 2023; Wang et al., 2024). A latest meta-analysis indicates that family-based interventions can alleviate gaming disorder symptoms and improve family cohesion, implying that family engagement is critical for sustained recovery in adolescent clients (Wang et al., 2024). This is particularly relevant given that adolescents are more susceptible to family influence and may rely heavily on parental guidance in behavioral regulation (Kumari & Dhiksha, 2022; Paulus et al., 2018).

Family-based interventions, such as multidimensional family therapy and behavioral family systems therapy, have demonstrated desirable outcomes in fostering better parent-child communication, setting clear behavioral expectations, and addressing parental concerns about gaming (Wang et al., 2024). By involving parents in the treatment process, these approaches help ensure that adolescents receive the necessary support in their social environment to modify their gaming behavior. Besides, family-based therapies address common risk factors for gaming disorder, such as low family cohesion and permissive parenting (Akaroğlu, 2024; Harshini et al., 2023). The therapies seek to mitigate these risk factors, which are often associated with increased severity of gaming disorder.

It is important to highlight that the effectiveness of family-based therapy may be limited for adult clients and families characterized by poor communication or low cohesion (e.g., Asen, 2002; Lloyd et al., 2023). Specifically, for adult clients, the involvement of family members can be less impactful due to established patterns of behavior, a lack of parental authority, or a diminished role of family influence in adulthood. In addition, for low-cohesion or disengaged families, it may be difficult to foster the trust and cooperation needed to make family-based interventions work effectively. In such cases, family-based therapy may not provide the necessary support structure, and clients may be left without the positive reinforcement that typically comes from a well-functioning family environment. These limitations should be addressed by modifying this approach to accommodate the unique needs of adults or families with low cohesion or by adopting other treatment approaches (Asen, 2002).

Some Emerging Therapies

In addition to the aforementioned more traditional common approaches, emerging treatment approaches, including digital mindfulness-based therapies and gamified therapies, are gaining popularity for their accessibility and novel approach to the treatment of gaming disorder. Mindfulness-based therapies emphasize developing awareness and acceptance of urges without acting on them, which is especially beneficial for individuals with a gaming disorder. These approaches foster self-regulation skills that can diminish compulsive behaviors. A recent meta-analysis demonstrated that mindfulness interventions effectively alleviate stress and anxiety symptoms in clients with gaming disorder, enabling them to gain better control over their impulses (Fendel et al., 2024). In the current cyber age, these interventions are delivered online, providing more flexible and accessible options for those clients who are unable to attend face-to-face therapy sessions (Fendel et al., 2024; Mettler et al., 2020; Sharma et al., 2021). Studies document that such digital interventions can achieve therapeutic outcomes comparable to face-to-face CBT (Lei & Zhang, 2020; Ni et al., 2024), making them a viable option in remote or underserved geographic regions.

Often, these therapies are combined with cognitive behavioral therapy to address the cognitive and behavioral aspects of gaming disorder while promoting emotional resilience. Notable mindfulness programs, such as "Headspace" and "Calm," provide guided meditation sessions aimed at reducing stress and enhancing focus, both of which assist individuals in managing their impulses more effectively (Collins et al., 2019; O'Daffer et al., 2022). These apps offer a variety of mindfulness exercises, ranging from brief breathing techniques to longer meditation practices, which can easily be integrated into daily routines to improve emotional regulation and boost self-control.

Recent studies highlight the potential of gamification therapies in addressing the problems of gaming disorder, particularly among youth. Gamification techniques elicit enjoyment and relaxation, establishing a fun environment and enjoyable atmosphere that encourages clients to acquire new strategies in a low-pressure setting (Cheng & Ebrahimi, 2023a; Fleming et al., 2023). Moreover, gamification fosters the creation of a structured, engaging environment that promotes participation and adherence to treatment regimens (Cheng & Ebrahimi, 2023b). Specifically, gamification incorporates motivational affordances that cater to users' psychological needs, encouraging them to actively engage in interventions. This active engagement can lead to prominent psychological and physiological changes, such as reduced anxiety and improved self-regulation. Such changes can be further translated into real-world outcomes, including symptom relief and improved quality of life. Overall, gamification not only makes interventions more enjoyable but also facilitates deeper engagement and meaningful behavioral transformations.

In recent years, prevention programs, such as Wise IT-use (WIT) and the Digital Netizen Alliance (D.N.A.), have been developed (Chau et al., 2019; Cheng et al., 2024). These novel programs leverage gamification techniques to enhance awareness and strengthen self-regulation related to gaming disorder. Evaluations of these initiatives have shown that program participation alleviates GD symptoms and improves self-control and resilience among primary school students. In addition, augmented-reality exercise games have been utilized as a therapeutic approach for adolescents diagnosed with gaming disorder (Lee & Huh, 2024). The program evaluation findings reveal that this gamified program not only promotes physical health but also enhances mental well-being. Overall, these studies indicate that gamified interventions can effectively reduce gaming disorder symptoms while fostering mental and physical health in young individuals. While gamification has been found effective for children and adolescents, their efficacy in mitigating gaming disorder symptoms among adults remains unexamined. Also, gamification interventions are found to effectively relieve symptoms for non-clinical and at-risk clients but may be less adequate for those already diagnosed with gaming disorder, who often need more intensive treatment. Furthermore, an over-reliance on gaming elements may reinforce gaming behaviors, complicating recovery for vulnerable individuals.

Comparative Efficacy and Limitations of Treatment Approaches

A common finding across meta-analyses is the comparative efficacy of CBT over other treatment approaches, with family-based interventions also demonstrating substantial effectiveness, particularly for adolescents. Pharmacological approaches are recognized for their supportive role, mainly addressing co-occurring psychiatric symptoms rather than the gaming disorder itself. A systematic review further consolidates these findings, emphasizing that the most effective therapeutic outcomes typically result from a multimodal approach that combines CBT with family therapy or pharmacological support whenever necessary (Stevens et al., 2019).

It is worth noting that limitations remain in the current treatment landscape. CBT, while effective, requires sustained engagement and may not be accessible to all clients, especially those in settings with scant resources. Family-based therapies, though impactful for adolescents, may be less effective for adult clients or those from families with limited parental involvement or family cohesion. Pharmacological treatments are not widely recommended as first-line treatments due to their limitations in directly reducing problematic gaming behaviors.

In this light, effective treatment of gaming disorder should involve a multifaceted approach, with CBT as the core because it is the most widely validated and applied intervention approach. Family-based interventions hold particular promise for younger individuals by actively involving the home environment in the treatment process, while pharmacological approaches provide supplementary support for managing co-occurring psychiatric symptoms. Emerging treatments, including gamified and mindfulness-based interventions, offer flexible alternatives and additional tools for self-regulation, though further evaluation research is needed to establish their efficacy across diverse populations. Future studies should focus on identifying optimal combinations of these treatments and refining digital interventions to improve accessibility and effectiveness.

Conclusions

This chapter highlights that gaming disorder is a complex disorder with major implications for mental and physical health, shaped by various cognitive and familial risk factors. Effective treatment requires a multimodal approach tailored to the individual's needs, incorporating evidence-based therapies and supportive family dynamics. However, to fully understand the long-term effects of these treatments for gaming disorder, longitudinal research should be conducted to track the far-reaching impact of various types of intervention over time. Continued research on tailored interventions and the long-term treatment outcomes should be valuable for advancing the field, improving clinical care, and addressing public concerns for the emerging problem of gaming disorder. This methodology provides a more sophisticated assessment of the effectiveness and sustainability of interventions, helping clinicians refine strategies that reduce relapse and foster recovery in the long run.

Acknowledgements GPV was financed in part by the Coordenação de Aperfeiçoamento de Pessoal de Nível Superior - Brasil (CAPES) - Finance Code 001

References

Aarseth, E., Edman, J., Helmersson Bergmark, K., & van Rooij, A. J. (2017). Scholars' open debate paper on the World Health Organization ICD-11 Gaming Disorder proposal. *Journal of Behavioral Addictions, 6*(3), 267. https://doi.org/10.1556/2006.5.2016.088

Afriwilda, M. T., & Mulawarman, M. (2021). The effectiveness of motivational interviewing counseling to improve psychological well-being on students with online game addiction tendency. *Islamic Guidance and Counseling Journal, 4*(1), 106–115. https://doi.org/10.25217/igcj.v4i1.1235

Ahmed, S., Samuel, A. J., Mishra, A., Rahman, M. S., Islam, M. A., Rashaduzzaman, M., Roy, S. K., Akter, R., & Islam, M. J. (2024). Mobile game addiction and its association with musculoskeletal pain among students: A cross-sectional study. *PLoS One, 19*(8), e0308674. https://doi.org/10.1371/journal.pone.0308674

Akaroğlu, G. (2024). Parental attitudes and social emotional well-being predict digital game addiction in Turkish children. *The American Journal of Family Therapy, 52*(3), 311–329. https://doi.org/10.1080/01926187.2022.2117249

Alimoradi, Z., Lin, C.-Y., Broström, A., Bülow, P. H., Bajalan, Z., Griffiths, M. D., Ohayon, M. M., & Pakpour, A. H. (2019). Internet addiction and sleep problems: A systematic review and meta-analysis. *Sleep Medicine Reviews, 47*, 51–61. https://doi.org/10.1016/j.smrv.2019.06.004

American Psychiatric Association. (2013). *Diagnostic and statistical manual of mental disorders* (5th ed.). American Psychiatric Association.

Apisitwasana, N., Perngparn, U., & Cottler, L. B. (2018). Effectiveness of school-and family-based interventions to prevent gaming addiction among grades 4–5 students in Bangkok, Thailand. *Psychology Research and Behavior Management, 11*, 103–115. https://doi.org/10.2147/prbm.s145868

Asen, E. (2002). Outcome research in family therapy. *Advances in Psychiatric Treatment, 8*(3), 230–238. https://doi.org/10.1192/apt.8.3.230

Aziz, N., Nordin, M. J., Abdulkadir, S. J., & Salih, M. M. M. (2021). Digital addiction: Systematic review of computer game addiction impact on adolescent physical health. *Electronics (Basel), 10*(9), 996. https://doi.org/10.3390/electronics10090996

Bäcklund, C., Elbe, P., Gavelin, H. M., Eriksson Sörman, D., & Ljungberg, J. K. (2022). Gaming motivations and gaming disorder symptoms: A systematic review and meta-analysis. *Journal of Behavioral Addictions, 11*(3), 667. https://doi.org/10.1556/2006.2022.00053

Bax, T. (2015). "Internet gaming disorder" in China: Biomedical sickness or sociological badness? *Games and Culture, 11*(3), 233–255. https://doi.org/10.1177/1555412014568188

Bhagat, S., Jeong, E. J., & Kim, D. J. (2020). The role of individuals' need for online social interactions and interpersonal incompetence in digital game addiction. *International Journal of Human–Computer Interaction, 36*(5), 449–463. https://doi.org/10.1080/10447318.2019.1654696

Bonnaire, C., & Phan, O. (2017). Relationships between parental attitudes, family functioning and internet gaming disorder in adolescents attending school. *Psychiatry Research, 255*, 104–110. https://doi.org/10.1016/j.psychres.2017.05.030

Bonnaire, C., Liddle, H. A., Har, A., Nielsen, P., & Phan, O. (2019). Why and how to include parents in the treatment of adolescents presenting internet gaming disorder? *Journal of Behavioral Addictions, 8*(2), 201–212. https://doi.org/10.1556/2006.8.2019.27

Bossi, A. H., & Hopker, J. (2017). Twilight: Filter the blue light of your device and sleep better. *British Journal of Sports Medicine, 51*(14), 1103–1104. https://doi.org/10.1136/bjsports-2017-098062

Brand, M., Wegmann, E., Stark, R., Müller, A., Wölfling, K., Robbins, T. W., & Potenza, M. N. (2019). The Interaction of Person-Affect-Cognition-Execution (I-PACE) model for addictive behaviors: Update, generalization to addictive behaviors beyond internet-use disorders, and specification of the process character of addictive behaviors. *Neuroscience and Biobehavioral Reviews, 104*, 1–10. https://doi.org/10.1016/j.neubiorev.2019.06.032

Burleigh, T. L., Griffiths, M. D., Sumich, A., Stavropoulos, V., & Kuss, D. J. (2019). A systematic review of the co-occurrence of gaming disorder and other potentially addictive behaviors. *Current Addiction Reports, 6*(4), 383–401. https://doi.org/10.1007/s40429-019-00279-7
Cankurtaran, F., Menevşe, Ö., Namlı, A., Kızıltoprak, H., Altay, S., Duran, M., Demir, E., Şahan, A., & Ekşi, C. (2022). The impact of digital game addiction on musculoskeletal system of secondary school children. *Nigerian Journal of Clinical Practice, 25*(2), 153–159. https://doi.org/10.4103/njcp.njcp_177_20
Carragher, N., Billieux, J., Bowden-Jones, H., Achab, S., Potenza, M. N., Rumpf, H. J., Long, J., Demetrovics, Z., Gentile, D., Hodgins, D., Aricak, O. T., Baigent, M., Gandin, C., Rahimi-Movaghar, A., Scafato, E., Assanangkornchai, S., Siste, K., Hao, W., King, D. L., et al. (2022). Brief overview of the WHO collaborative project on the development of new international screening and diagnostic instruments for gaming disorder and gambling disorder. *Addiction (Abingdon, England), 117*(7), 2119–2121. https://doi.org/10.1111/add.15780
Chau, C.-L., Tsui, Y. Y.-Y., & Cheng, C. (2019). Gamification for internet gaming disorder prevention: Evaluation of a Wise IT-use (WIT) program for Hong Kong primary students. *Frontiers in Psychology, 10*, 2468. https://doi.org/10.3389/fpsyg.2019.02468
Cheng, C., & Ebrahimi, O. V. (2023a). Gamification: A novel approach to mental health promotion. *Current Psychiatry Reports, 25*(11), 577–586. https://doi.org/10.1007/s11920-023-01453-5
Cheng, C., & Ebrahimi, O. V. (2023b). A meta-analytic review of gamified interventions in mental health enhancement. *Computers in Human Behavior, 141*, 107621. https://doi.org/10.1016/j.chb.2022.107621
Cheng, C., Cheung, M. W. L., & Wang, H.-Y. (2018). Multinational comparison of internet gaming disorder and psychosocial problems versus well-being: Meta-analysis of 20 countries. *Computers in Human Behavior, 88*, 153–167. https://doi.org/10.1016/j.chb.2018.06.033
Cheng, C., Li, S., & Chen, S. (2024). Evaluation of a multicomponent positive psychology program to prevent gaming disorder and enhance mental wellness in primary pupils: A randomized controlled trial. *Journal of Behavioral Addictions, 13*(3), 871–883. https://doi.org/10.1556/2006.2024.00052
Chia, D. X. Y., Ng, C. W. L., Kandasami, G., Seow, M. Y. L., Choo, C. C., Chew, P. K. H., Lee, C., & Zhang, M. W. B. (2020). Prevalence of internet addiction and gaming disorders in Southeast Asia: A meta-analysis. *International Journal of Environmental Research and Public Health, 17*(7), 2582. https://doi.org/10.3390/ijerph17072582
Coles-brennan, C., Sulley, A., & Young, G. (2019). Management of digital eye strain. *Clinical and Experimental Optometry, 102*(1), 18–29. https://doi.org/10.1111/cxo.12798
Collins, E., Cox, A., Wilcock, C., & Sethu-Jones, G. (2019). Digital games and mindfulness apps: Comparison of effects on post work recovery. *JMIR Mental Health, 6*(7), e12853. https://doi.org/10.2196/12853
Cunill, R., Castells, X., Tobias, A., & Capellà, D. (2014). Pharmacological treatment of attention deficit hyperactivity disorder with co-morbid drug dependence. *Journal of Psychopharmacology, 29*(1), 15–23. https://doi.org/10.1177/0269881114544777
Darvesh, N., Radhakrishnan, A., Lachance, C. C., Nincic, V., Sharpe, J. P., Ghassemi, M., Straus, S. E., & Tricco, A. C. (2020). Exploring the prevalence of gaming disorder and internet gaming disorder: A rapid scoping review. *Systematic Reviews, 9*(1), 68–68. https://doi.org/10.1186/s13643-020-01329-2
de Sá, R. R. C., Coelho, S., Parmar, P. K., Johnstone, S., Kim, H. S., & Tavares, H. (2023). A systematic review of pharmacological treatments for internet gaming disorder. *Psychiatry Investigation, 20*(8), 696. https://doi.org/10.30773/pi.2022.0297
Dullur, P., Krishnan, V., & Diaz, A. M. (2021). A systematic review on the intersection of attention-deficit hyperactivity disorder and gaming disorder. *Journal of Psychiatric Research, 133*, 212–222. https://doi.org/10.1016/j.jpsychires.2020.12.026
Estupiñá Puig, F. J., Bernaldo De Quirós Aragón, M., Vallejo Achón, M., Fernández Arias, I. G., & Labrador Encinas, F. J. (2024). Emotional regulation in gaming disorder: A systematic review. *The American Journal on Addictions, 33*, 605–620. https://doi.org/10.1111/ajad.13621

Estupiñá, F. J., Bernaldo-de-Quirós, M., Vallejo-Achón, M., Fernández-Arias, I., & Labrador, F. (2024). Review manuscript: Emotional regulation in gaming disorder: A systematic review. *The American Journal on Addictions, 33*, 1–16.

Farrell, A. H., Vitoroulis, I., Eriksson, M., & Vaillancourt, T. (2023). Loneliness and well-being in children and adolescents during the COVID-19 pandemic: A systematic review. *Children, 10*(2), 279. https://doi.org/10.3390/children10020279

Fendel, J. C., Vogt, A., Brandtner, A., & Schmidt, S. (2024). Mindfulness programs for problematic usage of the internet: A systematic review and meta-analysis. *Journal of Behavioral Addictions, 13*, 327–353. https://doi.org/10.1556/2006.2024.00024

Fleming, T., Poppelaars, M., & Thabrew, H. (2023). The role of gamification in digital mental health. *World Psychiatry, 22*(1), 46–47. https://doi.org/10.1002/wps.21041

Gao, Y.-X., Wang, J.-Y., & Dong, G.-H. (2022). The prevalence and possible risk factors of internet gaming disorder among adolescents and young adults: Systematic reviews and meta-analyses. *Journal of Psychiatric Research, 154*, 35–43. https://doi.org/10.1016/j.jpsychires.2022.06.049

Gonçalves, H., Sereno, P., Santos, R., & Rodrigues, M. A. (2024). Lumbar postural responses during gaming activity: A study with semi-professional and amateur gamers. In P. M. Arezes, R. B. Melo, P. Carneiro, J. Castelo Branco, A. Colim, N. Costa, S. Costa, J. Duarte, J. C. Guedes, G. Perestrelo, & J. S. Baptista (Eds.), *Occupational and environmental safety and health V* (pp. 151–161). Springer Nature Switzerland. https://doi.org/10.1007/978-3-031-38277-2_13

Grassi, G., Makris, N., & Pallanti, S. (2020). Addicted to compulsion: Assessing three core dimensions of addiction across obsessive-compulsive disorder and gambling disorder. *CNS Spectrums, 25*(3), 392–401. https://doi.org/10.1017/s1092852919000993

Gundogdu, U., & Eroglu, M. (2022). The relationship between dissociation symptoms, sleep disturbances, problematic internet use and online gaming in adolescents. *Psychology, Health & Medicine, 27*(3), 686–697. https://doi.org/10.1080/13548506.2021.1984542

Harshini, V. S., Abraham, M., Prabhu, S., Shi, J., & Dsouza, D. D. (2023). Mapping the theories, content, and outcomes of family-based interventions for children and young people with gaming disorder: A scoping review protocol [version 1; peer review: Awaiting peer review]. *F1000 Research, 12*, 1178–1178. https://doi.org/10.12688/f1000research.134800.1

Jo, Y. S., Bhang, S. Y., Choi, J. S., Lee, H. K., Lee, S. Y., & Kweon, Y.-S. (2019). Clinical characteristics of diagnosis for internet gaming disorder: Comparison of DSM-5 IGD and ICD-11 GD diagnosis. *Journal of Clinical Medicine, 8*(7), 945. https://doi.org/10.3390/jcm8070945

Johnson, S., & Rosenfield, M. (2023). 20-20-20 Rule: Are these numbers justified? *Optometry and Vision Science, 100*(1), 52–56. https://doi.org/10.1097/OPX.0000000000001971

Judge, D., & Mellen, L. (2019). Pharmacological interventions. In I. Crome & R. Williams (Eds.), *Substance misuse and young people* (pp. 415–440). CRC Press.

Kaur, K., Gurnani, B., Nayak, S., Deori, N., Kaur, S., Jethani, J., Singh, D., Agarkar, S., Hussaindeen, J. R., & Sukhija, J. (2022). Digital eye strain-a comprehensive review. *Ophthalmology and therapy, 11*(5), 1655–1680. https://doi.org/10.1007/s40123-022-00540-9

Kavanagh, M., Brett, C., & Brignell, C. (2023). What is the reported relationship between self-esteem and gaming disorder? A systematic review and meta-analysis. *Computers in Human Behavior, 145*, 107776. https://doi.org/10.1016/j.chb.2023.107776

Kim, H. S., Son, G., Roh, E.-B., Ahn, W.-Y., Kim, J., Shin, S.-H., Chey, J., & Choi, K.-H. (2022). Prevalence of gaming disorder: A meta-analysis. *Addictive Behaviors, 126*, 107183–107183. https://doi.org/10.1016/j.addbeh.2021.107183

King, D. L., & Delfabbro, P. H. (2019). *Internet gaming disorder: Theory, assessment, treatment, and prevention*. Academic Press, an imprint of Elsevier. https://doi.org/10.1016/C2016-0-04107-4

King, D. L., Delfabbro, P. H., Wu, A. M., Doh, Y. Y., Kuss, D. J., Pallesen, S., Mentzoni, R., Carragher, N., & Sakuma, H. (2017). Treatment of internet gaming disorder: An international systematic review and CONSORT evaluation. *Clinical Psychology Review, 54*, 123–133. https://doi.org/10.1016/j.cpr.2017.04.002

Király, O., & Demetrovics, Z. (2017). Inclusion of gaming disorder in ICD has more advantages than disadvantages: Commentary on: Scholars' open debate paper on the World Health Organization ICD-11 gaming disorder proposal (Aarseth et al.). *Journal of Behavioral Addictions, 6*(3), 280–284. https://doi.org/10.1556/2006.6.2017.046

Király, O., Tóth, D., Urbán, R., Demetrovics, Z., Maraz, A., & Petry, N. M. (2017). Intense video gaming is not essentially problematic. *Psychology of Addictive Behaviors, 31*(7), 807–817. https://doi.org/10.1037/adb0000316

Klugah-Brown, B., Zhou, X., Pradhan, B. K., Zweerings, J., Mathiak, K., Biswal, B., & Becker, B. (2021). Common neurofunctional dysregulations characterize obsessive–compulsive, substance use, and gaming disorders—An activation likelihood meta-analysis of functional imaging studies. *Addiction Biology, 26*(4), e12997. https://doi.org/10.1111/adb.12997

Ko, C. H., Yen, J. Y., Yen, C. F., Chen, C. S., & Chen, C. C. (2012). The association between internet addiction and psychiatric disorder: A review of the literature. *European Psychiatry, 27*(1), 1–8. https://doi.org/10.1016/j.eurpsy.2010.04.011

Kochuchakkalackal, G. K., & Reyes, M. E. S. (2020). An emerging mental health concern: Risk factors, symptoms, and impact of internet gaming disorder. *Journal of Technology in Behavioral Science, 5*(1), 70–78. https://doi.org/10.1007/s41347-019-00117-7

Koncz, P., Demetrovics, Z., Takacs, Z. K., Griffiths, M. D., Nagy, T., & Király, O. (2023). The emerging evidence on the association between symptoms of ADHD and gaming disorder: A systematic review and meta-analysis. *Clinical Psychology Review, 106*, 102343. https://doi.org/10.1016/j.cpr.2023.102343

Kort-Butler, L. A. (2021). The well-being of gamers, video game players, and non-players. *The Social Science Journal*, 1–17. https://doi.org/10.1080/03623319.2021.1910774

Kowalik, B. A., Delfabbro, P. H., & King, D. L. (2024). Impaired control and gaming-related harm in relation to gaming disorder. *Addictive Behaviors, 151*, 107926. https://doi.org/10.1016/j.addbeh.2023.107926

Kristensen, J. H., Pallesen, S., King, D. L., Hysing, M., & Erevik, E. K. (2021). Problematic gaming and sleep: A systematic review and meta-analysis. *Frontiers in Psychiatry, 12*, 675237. https://doi.org/10.3389/fpsyt.2021.675237

Krossbakken, E., Pallesen, S., Mentzoni, R. A., King, D. L., Molde, H., Finserås, T. R., & Torsheim, T. (2018). A cross-lagged study of developmental trajectories of video game engagement, addiction, and mental health. *Frontiers in Psychology, 9*, 2239–2239. https://doi.org/10.3389/fpsyg.2018.02239

Kumari, S., & Dhiksha, J. (2022). Effect of intervention for gaming addiction among adolescents: A systematic review. *International Journal of Health Sciences, 6*(III), 431219. https://doi.org/10.53730/ijhs.v6nS3.8718

Lam, Y., & Cheng, C. (2022). Parental depression and leisure activity engagement on children's gaming disorder: A dyadic study. *International Journal of Environmental Research and Public Health, 19*(10), 5880. https://doi.org/10.3390/ijerph19105880

Lee, Y. S., & Huh, J. H. (2024). Enabling physical activity with augmented reality gamification for reducing internet gaming disorder. *Applied Sciences, 14*(1), 121.

Leeman, R. F., & Potenza, M. N. (2012). Similarities and differences between pathological gambling and substance use disorders: A focus on impulsivity and compulsivity. *Psychopharmacology, 219*(2), 469–490. https://doi.org/10.1007/s00213-011-2550-7

Lei, Y., & Zhang, E. (2020). Intervention effect of mindfulness based cognitive therapy on college students dependent on online games in the internet era. In *Data processing techniques and applications for cyber-physical systems (DPTA 2019)* (pp. 1613–1619). Springer.

Lemenager, T., Neissner, M., Sabo, T., Mann, K., & Kiefer, F. (2020). "Who am I" and "how should I be": A systematic review on self-concept and avatar identification in gaming disorder. *Current Addiction Reports, 7*, 166–193. https://doi.org/10.1007/s40429-020-00307-x

Li, A. Y., Lo, B. C. Y., & Cheng, C. (2018). It is the family context that matters: Concurrent and predictive effects of aspects of parent-child interaction on video gaming-related problems.

Cyberpsychology, Behavior, and Social Networking, 21(6), 374–380. https://doi.org/10.1089/cyber.2017.0566

Li, A. Y., Chau, C., & Cheng, C. (2019). Development and validation of a parent-based program for preventing gaming disorder: The game over intervention. *International Journal of Environmental Research and Public Health, 16*(11), 1984. https://doi.org/10.3390/ijerph16111984

Liao, Z., Chen, X., Huang, Q., & Shen, H. (2022). Prevalence of gaming disorder in East Asia: A comprehensive meta-analysis. *Journal of Behavioral Addictions, 11*(3), 727–738. https://doi.org/10.1556/2006.2022.00050

Lloyd, A., Broadbent, A., Brooks, E., Bulsara, K., Donoghue, K., Saijaf, R., Sampson, K. N., Thomson, A., Fearon, P., & Lawrence, P. J. (2023). The impact of family interventions on communication in the context of anxiety and depression in those aged 14–24 years: Systematic review of randomised control trials. *British Journal of Psychiatry Open, 9*(5), e161. https://doi.org/10.1192/bjo.2023.545

Marchica, L. A., Mills, D. J., Derevensky, J. L., & Montreuil, T. C. (2019). The role of emotion regulation in video gaming and gambling disorder: A systematic review. *Canadian Journal of Addiction, 10*(4), 19–29. https://doi.org/10.1097/CXA.0000000000000070

Martha, G. I. R., Budaya, I. G. B. A., Agustino, D. P., Harsemadi, I. G., & Pande, I. M. S. A. (2022). An empirical analysis of ergonomic gaming peripherals improving gaming performance. *Journal of Games, Game Art, and Gamification, 7*(1), 15–21. https://doi.org/10.21512/jggag.v7i1.8258

Martinez, L., Gimenes, M., & Lambert, E. (2023). Video games and board games: Effects of playing practice on cognition. *PLoS One, 18*(3), e0283654. https://doi.org/10.1371/journal.pone.0283654

Mathews, C. L., Morrell, H. E., & Molle, J. E. (2019). Video game addiction, ADHD symptomatology, and video game reinforcement. *The American Journal of Drug and Alcohol Abuse, 45*(1), 67–76. https://doi.org/10.1080/00952990.2018.1472269

Melodia, F., Canale, N., & Griffiths, M. D. (2020). The role of avoidance coping and escape motives in problematic online gaming: A systematic literature review. *International Journal of Mental Health and Addiction, 20*, 1–27. https://doi.org/10.1007/s11469-020-00422-w

Mettler, J., Mills, D. J., & Heath, N. L. (2020). Problematic gaming and subjective well-being: How does mindfulness play a role? *International Journal of Mental Health and Addiction, 18*, 720–736. https://doi.org/10.1007/s11469-018-9978-5

Mills, D. J., & Allen, J. J. (2020). Self-determination theory, internet gaming disorder, and the mediating role of self-control. *Computers in Human Behavior, 105*, 106209. https://doi.org/10.1016/j.chb.2019.106209

Moore, S., Satel, J., & Pontes, H. M. (2022). Investigating the role of health factors and psychological well-being in gaming disorder. *Cyberpsychology, Behavior and Social Networking, 25*(2), 94–100. https://doi.org/10.1089/cyber.2021.0050

Mylona, I., Deres, E. S., Dere, G.-D. S., Tsinopoulos, I., & Glynatsis, M. (2020). The impact of internet and videogaming addiction on adolescent vision: A review of the literature. *Frontiers in Public Health, 8*, 63. https://doi.org/10.3389/fpubh.2020.00063

Na, E., Lee, H., Choi, I., & Kim, D. J. (2017). Comorbidity of internet gaming disorder and alcohol use disorder: A focus on clinical characteristics and gaming patterns. *The American Journal on Addictions, 26*(4), 326–334. https://doi.org/10.1111/ajad.12528

Nasir, E., Arif, F., Ali, Z., Zafar, M., Rasool, A., & Idrees, K. (2024). Relationship between internet gaming disorder severity and musculoskeletal complaints among university students: Internet gaming disorder and musculoskeletal complaints. *Journal of Health and Rehabilitation Research, 4*(3), 1563. https://doi.org/10.61919/jhrr.v4i3.1563

Ni, H., Wang, H., Ma, X., Li, S., Liu, C., Song, X., Potenza, M. N., & Dong, G.-H. (2024). Efficacy and neural mechanisms of mindfulness meditation among adults with internet gaming disorder: A randomized clinical trial. *JAMA Network Open, 7*(6), e2416684. https://doi.org/10.1001/jamanetworkopen.2024.16684

Nielsen, P., Favez, N., & Rigter, H. (2020). Parental and family factors associated with problematic gaming and problematic internet use in adolescents: A systematic literature review. *Current Addiction Reports, 7*, 365–386. https://doi.org/10.1007/s40429-020-00320-0

Noroozi, F., Hassanipour, S., Eftekharian, F., Eisapareh, K., Kaveh, M. H., Raggi, A., & Alberto, R. (2021). Internet addiction effect on quality of life: A systematic review and meta-analysis. *The Scientific World Journal, 2021*, 2556679–2556612. https://doi.org/10.1155/2021/2556679

Nuyens, F. M., Kuss, D. J., Lopez-Fernandez, O., & Griffiths, M. D. (2019). The empirical analysis of non-problematic video gaming and cognitive skills: A systematic review. *International Journal of Mental Health and Addiction, 17*(2), 389–414. https://doi.org/10.1007/s11469-018-9946-0

O'Daffer, A., Colt, S. F., Wasil, A. R., & Lau, N. (2022). Efficacy and conflicts of interest in randomized controlled trials evaluating Headspace and Calm apps: Systematic review. *JMIR Mental Health, 9*(9), e40924. https://doi.org/10.2196/40924

Ostinelli, E. G., Zangani, C., Giordano, B., Maestri, D., Gambini, O., D'Agostino, A., Furukawa, T. A., & Purgato, M. (2021). Depressive symptoms and depression in individuals with internet gaming disorder: A systematic review and meta-analysis. *Journal of Affective Disorders, 284*, 136–142. https://doi.org/10.1016/j.jad.2021.02.014

Paulus, F. W., Ohmann, S., von Gontard, A., & Popow, C. (2018). Internet gaming disorder in children and adolescents: A systematic review. *Developmental Medicine and Child Neurology, 60*(7), 645–659. https://doi.org/10.1111/dmcn.13754

Perrault, A. A., Bayer, L., Peuvrier, M., Afyouni, A., Ghisletta, P., Brockmann, C., Spiridon, M., Hulo Vesely, S., Haller, D. M., Pichon, S., Perrig, S., Schwartz, S., & Sterpenich, V. (2019). Reducing the use of screen electronic devices in the evening is associated with improved sleep and daytime vigilance in adolescents. *Sleep, 42*(9), zsz125. https://doi.org/10.1093/sleep/zsz125

Rahmillah, F. I., Tariq, A., King, M., & Oviedo-Trespalacios, O. (2023). Evaluating the effectiveness of apps designed to reduce mobile phone use and prevent maladaptive mobile phone use: Multimethod study. *Journal of Medical Internet Research, 25*, e42541. https://doi.org/10.2196/42541

Ryff, C. D., & Keyes, C. L. M. (1995). The structure of psychological well-being revisited. *Journal of Personality and Social Psychology, 69*, 719–727. https://doi.org/10.1037/0022-3514.69.4.719

Salerno, L., Becheri, L., & Pallanti, S. (2022). ADHD-gaming disorder comorbidity in children and adolescents: A narrative review. *Children, 9*(10), 1528. https://doi.org/10.3390/children9101528

Şalvarlı, Ş. İ., & Griffiths, M. D. (2022). The association between internet gaming disorder and impulsivity: A systematic review of literature. *International Journal of Mental Health and Addiction, 20*(1), 92–118. https://doi.org/10.1007/s11469-019-00126-w

Schneider, L. A., King, D. L., & Delfabbro, P. H. (2017). Family factors in adolescent problematic internet gaming: A systematic review. *Journal of Behavioral Addictions, 6*(3), 321–333. https://doi.org/10.1556/2006.6.2017.035

Seo, E. H., Yang, H.-J., Kim, S.-G., Park, S.-C., Lee, S.-K., & Yoon, H.-J. (2021). A literature review on the efficacy and related neural effects of pharmacological and psychosocial treatments in individuals with internet gaming disorder. *Psychiatry Investigation, 18*(12), 1149. https://doi.org/10.30773/pi.2021.0207

Seok, S., & DaCosta, B. (2014). Distinguishing addiction from high engagement: An investigation into the social lives of adolescent and young adult massively multiplayer online game players. *Games and Culture, 9*(4), 227–254. https://doi.org/10.1177/1555412014538811

Sharma, M. K., Bhargav, H., Kumar, A., Digambhar, V., & Mani, T. A. (2021). Mindfulness-based interventions: Potentials for management of internet gaming disorder. *International Journal of Yoga, 14*(3), 244–247. https://doi.org/10.4103/ijoy.ijoy_46_21

Singh, M. (2019). Compulsive digital gaming: An emerging mental health disorder in children. *The Indian Journal of Pediatrics, 86*(2), 171–173. https://doi.org/10.1007/s12098-018-2785-y

Singh, Y. M., Prakash, J., Chatterjee, K., Khadka, B., Shah, A., & Chauhan, V. S. (2021). Prevalence and risk factors associated with internet gaming disorder: A cross-sectional study. *Industrial Psychiatry Journal, 30*(Suppl 1), S172–S177. https://doi.org/10.4103/0972-6748.328810

Skripkauskaite, S., Fazel, M., & Team, O. S. (2022). Time spent gaming, device type, addiction scores, and well-being of adolescent english gamers in the 2021 OxWell survey: Latent profile analysis. *JMIR Pediatrics and Parenting, 5*(4), e41480. https://doi.org/10.2196/41480

Snodgrass, J. G., Lacy, M. G., & Cole, S. W. (2022). Internet gaming, embodied distress, and psychosocial well-being: A syndemic-syndaimonic continuum. *Social Science & Medicine, 295*, 112728. https://doi.org/10.1016/j.socscimed.2019.112728

Starzec, S., Starosta, J., Zajas, A., Kiszka, P., Śpiewak, S., & Strojny, P. (2024). Scoping review of withdrawal's role in contemporary gaming disorder research: Conceptualizations and operationalizations. *Clinical Psychology Review, 114*, 102478. https://doi.org/10.1016/j.cpr.2024.102478

Stevens, M. W. R., King, D. L., Dorstyn, D., & Delfabbro, P. H. (2019). Cognitive–behavioral therapy for internet gaming disorder: A systematic review and meta-analysis. *Clinical Psychology & Psychotherapy, 26*(2), 191–203. https://doi.org/10.1002/cpp.2341

Stevens, M. W. R., Dorstyn, D., Delfabbro, P. H., & King, D. L. (2021). Global prevalence of gaming disorder: A systematic review and meta-analysis. *Australian and New Zealand Journal of Psychiatry, 55*(6), 553–568. https://doi.org/10.1177/0004867420962851

Talens-Estarelles, C., Cerviño, A., García-Lázaro, S., Fogelton, A., Sheppard, A., & Wolffsohn, J. S. (2023). The effects of breaks on digital eye strain, dry eye and binocular vision: Testing the 20-20-20 rule. *Contact Lens and Anterior Eye, 46*(2), 101744. https://doi.org/10.1016/j.clae.2022.101744

Tang, M. C., Ebrahimi, O. V., & Cheng, C. (2023). Symptomatology, risk, and protective factors of gaming disorder: A network analytical approach. *Computers in Human Behavior, 148*, 107899. https://doi.org/10.1016/j.chb.2023.107899

Tateno, M., Teo, A. R., Ukai, W., Kanazawa, J., Katsuki, R., Kubo, H., & Kato, T. A. (2019). Internet addiction, smartphone addiction, and Hikikomori trait in Japanese young adult: Social isolation and social network. *Frontiers in Psychiatry, 10*, 455. https://doi.org/10.3389/fpsyt.2019.00455

Tehrani, H. D., Yamini, S., & Vazsonyi, A. T. (2024). Parenting styles and Big Five personality traits among adolescents: A meta-analysis. *Personality and Individual Differences, 216*, 112421. https://doi.org/10.1016/j.paid.2023.112421

Teng, Z., Pontes, H. M., Nie, Q., Xiang, G., Griffiths, M. D., & Guo, C. (2020). Internet gaming disorder and psychosocial well-being: A longitudinal study of older-aged adolescents and emerging adults. *Addictive Behaviors, 110*, 106530. https://doi.org/10.1016/j.addbeh.2020.106530

Tholl, C., Bickmann, P., Wechsler, K., Froböse, I., & Grieben, C. (2022). Musculoskeletal disorders in video gamers – A systematic review. *BMC Musculoskeletal Disorders, 23*(1), 678. https://doi.org/10.1186/s12891-022-05614-0

Turgut, B. (2018). Ocular ergonomics for the computer vision syndrome. *Journal of Eye and Vision, 1*(1–2), 1–9. https://www.researchgate.net/profile/Burak-Turgut-2/publication/324112318_Ocular_Ergonomics_for_the_Computer_Vision_Syndrome/links/5abe6e990f7e9bfc0459a311/Ocular-Ergonomics-for-the-Computer-Vision-Syndrome.pdf

Van Rooij, A. J., Ferguson, C. J., Colder Carras, M., Kardefelt-Winther, D., Shi, J., Aarseth, E., Bean, A. M., Helmersson Bergmark, K., Brus, A., Coulson, M., Deleuze, J., Dullur, P., Dunkels, E., Edman, J., Elson, M., Etchells, P. J., Fiskaali, A., Granic, I., Jansz, J., et al. (2018). A weak scientific basis for gaming disorder: Let us err on the side of caution. *Journal of Behavioral Addictions, 7*(1), 1. https://doi.org/10.1556/2006.7.2018.19

Wang, H.-Y., & Cheng, C. (2020). Psychometric evaluation and comparison of two gaming disorder measures derived from the DSM-5 and ICD-11 frameworks. *Frontiers in Psychiatry, 11*, 577366. https://doi.org/10.3389/fpsyt.2020.577366

Wang, H.-Y., & Cheng, C. (2022). The associations between gaming motivation and internet gaming disorder: Systematic review and meta-analysis. *JMIR Mental Health, 9*(2), e23700. https://doi.org/10.2196/23700

Wang, H.-Y., Sigerson, L., & Cheng, C. (2019a). Digital nativity and information technology addiction: Age cohort versus individual difference approaches. *Computers in Human Behavior, 90*, 1–9. https://doi.org/10.1016/j.chb.2018.08.031

Wang, J.-L., Sheng, J.-R., & Wang, H.-Z. (2019b). The association between mobile game addiction and depression, social anxiety, and loneliness. *Frontiers in Public Health, 7*, 247. https://doi.org/10.3389/fpubh.2019.00247

Wang, X., Zhang, Y., Lin, J., Wong, A. C. W., Chan, K. K.-Y., Wong, S. Y.-S., & Yang, X. (2023). Treatments of internet gaming disorder and comorbid mental disorders: A systematic review and meta-analysis. *Computers in Human Behavior, 149*, 107947. https://doi.org/10.1016/j.chb.2023.107947

Wang, H., Li, X., Lok, G. K. I., Meng, C., Tan, Y., Lee, U. M., Lei, L. S. M., & Chan, N. (2024). Family-based therapy for internet addiction among adolescents and young adults: A meta-analysis. *Journal of Behavioral Addictions, 13*, 295–312. https://doi.org/10.1556/2006.2024.00015

Wartberg, L., Kriston, L., Zieglmeier, M., Lincoln, T., & Kammerl, R. (2019). A longitudinal study on psychosocial causes and consequences of internet gaming disorder in adolescence. *Psychological Medicine, 49*(2), 287–294. https://doi.org/10.1017/s003329171800082x

Wölfling, K., Müller, K. W., Dreier, M., Ruckes, C., Deuster, O., Batra, A., Mann, K., Musalek, M., Schuster, A., & Lemenager, T. (2019). Efficacy of short-term treatment of internet and computer game addiction: A randomized clinical trial. *JAMA Psychiatry, 76*(10), 1018–1025. https://doi.org/10.1001/jamapsychiatry.2019.1676

Wong, H. Y., Mo, H. Y., Potenza, M. N., Chan, M. N. M., Lau, W. M., Chui, T. K., Pakpour, A. H., & Lin, C.-Y. (2020). Relationships between severity of internet gaming disorder, severity of problematic social media use, sleep quality and psychological distress. *International Journal of Environmental Research and Public Health, 17*(6), 1879. https://doi.org/10.3390/ijerph17061879

World Health Organization. (2018). *The 11th revision of the international classification of diseases (ICD-11)*. World Health Organization.

Xu, L.-X., Wu, L.-L., Geng, X.-M., Wang, Z.-L., Guo, X.-Y., Song, K.-R., Liu, G.-Q., Deng, L.-Y., Zhang, J.-T., & Potenza, M. N. (2021). A review of psychological interventions for internet addiction. *Psychiatry Research, 302*, 114016. https://doi.org/10.1016/j.psychres.2021.114016

Xu, X., Cao, M., Chen, Q., Xu, F., & Zhou, Z. (2025). Escape the uncertainty by internet gaming: A serial mediation model of intolerance of uncertainty and internet gaming disorder. *Personality and Individual Differences, 233*, 112917. https://doi.org/10.1016/j.paid.2024.112917

Young, K. (2009). Understanding online gaming addiction and treatment issues for adolescents. *The American Journal of Family Therapy, 37*(5), 355–372. https://doi.org/10.1080/01926180902942191

Zajac, K., Ginley, M. K., Chang, R., & Petry, N. M. (2017). Treatments for internet gaming disorder and internet addiction: A systematic review. *Psychology of Addictive Behaviors, 31*(8), 979–994. https://doi.org/10.1037/adb0000315

Zajac, K., Ginley, M. K., & Chang, R. (2020). Treatments of internet gaming disorder: A systematic review of the evidence. *Expert Review of Neurotherapeutics, 20*(1), 85–93. https://doi.org/10.1080/14737175.2020.1671824

Zaman, M., Babar, M. S., Babar, M., Sabir, F., Ashraf, F., Tahir, M. J., Ullah, I., Griffiths, M. D., Lin, C.-Y., & Pakpour, A. H. (2022). Prevalence of gaming addiction and its impact on sleep quality: A cross-sectional study from Pakistan. *Annals of Medicine and Surgery, 78*, 103641. https://doi.org/10.1016/j.amsu.2022.103641

Zhang, M. W. B., & Park, S. Y. (2023). A bibliometric analysis of research into internet gaming disorders in Korea. *International Journal of Environmental Research and Public Health, 20*(5), 3786.

Zhou, R., Morita, N., Ogai, Y., Saito, T., Zhang, X., Yang, W., & Yang, F. (2024). Meta-analysis of internet gaming disorder prevalence: Assessing the impacts of DSM-5 and ICD-11 diagnostic criteria. *International Journal of Environmental Research and Public Health, 21*(6), 700. https://www.mdpi.com/1660-4601/21/6/700

Zhuang, X., Zhang, Y., Tang, X., Ng, T. K., Lin, J., & Yang, X. (2023). Longitudinal modifiable risk and protective factors of internet gaming disorder: A systematic review and meta-analysis. *Journal of Behavioral Addictions, 12*(2), 375–392. https://doi.org/10.1556/2006.2023.00017

Chapter 8
Internet Gaming Disorder: Diagnosis, Models, Assessment, and Management

Matheus Figueiredo, Gian Luigi Martins Menegazzo, Thomás Garboggini Moretti, Gabriel Branquinho Pedone, Felipe Almeida Picon, and Ricardo Riyoiti Uchida

Abstract Internet gaming disorder (IGD) has emerged as a complex and multifactorial behavioral condition that affects individuals worldwide, particularly adolescents and young adults. This chapter provides a comprehensive overview of IGD, including its diagnostic criteria, epidemiology, psychological mechanisms, risk and protective factors, and evidence-based prevention and treatment strategies. We examine key theoretical models—such as cognitive–behavioral and motivational frameworks—as well as game design features that contribute to excessive gaming behaviors. The role of individual traits (e.g., impulsivity, emotional dysregulation), family environment, and social context is discussed through an integrative biopsychosocial model. The chapter also outlines validated screening tools and highlights the need for nuanced assessment to differentiate between high engagement and clinical impairment. Finally, we present preventive approaches and policy recommendations, emphasizing the importance of multisectoral collaboration to promote digital well-being.

Keywords Internet gaming disorder · Behavioral addiction · Risk and protective factors · Game design · Digital mental health

M. Figueiredo (✉) · G. L. M. Menegazzo · T. G. Moretti · G. B. Pedone
Mental Health Department, Santa Casa de São Paulo School of Medical Sciences, São Paulo, Brazil
e-mail: matheus.fidelis@fcmsantacasasp.edu.br

F. A. Picon
Universidade Federal Do Rio Grande Do Sul (UFRGS), Porto Alegre, Brazil

ADHD Outpatient Program, Clinical Research Center, Department of Psychiatry, Hospital de Clínicas de Porto Alegre, Universidade Federal do Rio Grande do Sul, Porto Alegre, Brazil

R. R. Uchida
Mental Health Department, Santa Casa de São Paulo School of Medical Sciences, São Paulo, Brazil

Artificial Intelligence Research Division, Infinity Doctor's Inc., Dover, DE, USA

L. Murrins Marques et al. (eds.), *Social and Affective Neuroscience of Gaming*,
https://doi.org/10.1007/978-3-032-00325-6_8

Introduction

Electronic games have become part of the daily lives of billions of people worldwide since their creation and popularization. According to the Global Games Market Report, it is estimated that there are currently 3.42 billion gamers globally, of whom 2.8 billion are exclusively mobile gamers, primarily using smartphones—an increase of 4.5% compared to the previous year (Newzoo, 2024). This continuous growth has drawn attention not only from markets involved in game production but also from the fields of scientific research and healthcare.

The first time risks associated with electronic games were described in medical literature was in 1981, in an article published in the *New England Journal of Medicine*, which discussed repetitive strain injury due to prolonged gaming sessions (McCowan, 1981). Shortly thereafter, the *Journal of the American Medical Association* published the first article raising concerns about mental health related to excessive video game engagement (Ross et al., 1982). Since then, interest in the potential harms associated with video gaming has only deepened, eventually leading to the proposal of "Internet Gaming Disorder" (IGD) as a new diagnostic category in the DSM-5. In this framework, the disorder is understood as a behavioral addiction, similar to gambling and problematic pornography use, given the clinical similarities it shares with previously described substance use disorders.

In this chapter, we will explore this disorder in depth, covering its definition, diagnostic criteria, treatment, and pathophysiology.

Definition and Diagnosis

According to the DSM-5, a diagnosis of Internet gaming disorder can be made when five or more of the following criteria are present:

- Preoccupation and excessive planning related to electronic games
- Irritability, sadness, or anxiety when gaming is interrupted
- Increasing need to play to achieve satisfaction or pleasure
- Unsuccessful attempts to reduce or stop gaming
- Decreased or lost interest in activities other than gaming
- Continued gaming despite psychosocial harm
- Minimizing the time or significance of gaming in daily life
- Gaming to forget problems or escape reality
- Significant losses in family, occupational, or emotional domains due to gaming

These symptoms must lead to clinically significant distress or impairment. Notably, these diagnostic criteria closely mirror those for other substance-related and behavioral addictions included in the manual.

The creation of this diagnostic category and its inclusion in the DSM-5—and later in the ICD-11—has not been free of controversy. One point of debate lies in the category name: the inclusion of "Internet" could suggest that only online games are capable of fostering addictive behavior, even though the clinical intent was to encompass both online and offline digital games. Furthermore, although the criteria are similar to other diagnostic categories, the pathologization of gaming behavior can be inappropriate in many contexts: gamers often discuss strategies and game mechanics as part of normal engagement, and professional E-sports athletes may spend extended periods training and competing without necessarily exhibiting pathological behavior (Kuss et al., 2017).

Comparing the DSM-5 and ICD-11 criteria reveals that the latter emphasizes longer symptom duration and more significant functional impairment. A South Korean study with 188 participants found that 38.8% would qualify for IGD under DSM-5 criteria, but only 6.4% met ICD-11 diagnostic thresholds (Jo et al., 2019).

Epidemiology

The prevalence of IGD varies depending on country and research methodology, ranging from 1% to 9% in study samples (Gentile et al., 2017). The disorder is more common among adolescents and young adults (Gao et al., 2022) and is also more prevalent in males than in females (Wartberg et al., 2017). A UK-based study found that most individuals diagnosed with IGD identified as white (Piper et al., 2024), whereas ethnic minorities in China were associated with higher prevalence rates (Meng et al., 2024). These findings suggest that IGD is influenced not only by neurobiological factors but also by social and cultural contexts that shape access to and engagement with digital games.

Common psychiatric comorbidities associated with IGD include depression (Liu et al., 2018), social phobia, generalized anxiety disorder, obsessive-compulsive disorder, and attention-deficit/hyperactivity disorder (Pearcy et al., 2017). The strong correlations between IGD and these comorbidities highlight the importance of identifying underlying risk factors and suggest that appropriate treatment of comorbid conditions may help reduce the harms associated with gaming addiction.

Differential Diagnosis

Patients with IGD often present with symptoms that are relatively common and nonspecific, making differential diagnosis essential to avoid mislabeling high gaming engagement as pathological. Crucially, a distinguishing feature between individuals with IGD and the general gaming population is impaired impulse control, as demonstrated in studies using neuroimaging and behavioral analysis (Zha et al., 2022). Clinical distress must be directly related to gaming behavior and must reach a level that justifies clinical intervention. This requirement must be carefully evaluated against the DSM-5 diagnostic criteria and the empirical evidence supporting their use.

Some criteria are more predictive of IGD than others—for example, withdrawal symptoms and diminished behavioral repertoire are particularly relevant (Ballou & Zendle, 2022). Finally, gaming may sometimes function as a coping mechanism for symptoms of other underlying disorders, particularly those mentioned earlier as common comorbidities. In such cases, gaming behavior may obscure the primary psychiatric condition and hinder clinical investigation (Jo et al., 2019).

Psychological Mechanisms in Gaming Disorder

Gaming Disorder (GD), as recognized by the ICD-11 of the World Health Organization, is characterized by a persistent and recurrent pattern of behavior related to digital gaming, often associated with significant functional impairment. Understanding the psychological mechanisms underlying this behavior is essential for the development of effective clinical interventions and public policies.

Cognitive-Behavioral Models

Cognitive-behavioral models have emerged as central explanatory frameworks for GD. These approaches propose that the disorder is maintained by dysfunctional thought patterns, such as distorted beliefs about control, gaming-related self-efficacy, and the catastrophizing of failures in real life (King et al., 2010).

One of the most widely known models is Young's model (2011), adapted from the treatment of internet addiction (CBT-IA), which postulates three main components: behavioral modification, cognitive restructuring, and the development of coping strategies. Studies suggest that restructuring maladaptive beliefs—such as the notion that the virtual world is more rewarding than the real one—is associated with symptom improvement (Wölfling et al., 2019).

Moreover, negative reinforcement—i.e., using games to avoid negative emotions or stressful situations—is identified as a key factor in maintaining the disorder (Kuss & Griffiths, 2015). These findings are consistent with broader behavioral addiction models, which describe the repetitive use of an activity as a dysfunctional emotional regulation strategy.

Motivational Theories

Motivational theories provide a complementary understanding to the cognitive-behavioral approach by exploring the reasons why individuals engage in—and continue engaging in—gaming.

Self-Determination Theory (Ryan & Deci, 2000) has been widely applied to the context of GD. According to this theory, games satisfy basic psychological needs for competence, autonomy, and social relatedness. When these needs are not fulfilled in real life, games may become a powerful motivational substitute, potentially leading to excessive use and, in more severe cases, to disorder (Przybylski et al., 2010).

Empirical models, such as Yee's Three-Factor Model (2006), identify three core motivations for gaming involvement: achievement (e.g., progression and competition), socialization (interactions with other players), and immersion (fantasy and escapism). Longitudinal studies indicate that high scores in the immersion and escapism dimensions are consistently associated with a higher risk of developing GD (Billieux et al., 2013).

More recently, the use of specific inventories, such as the Gaming Motivation Inventory (GMI), has enabled the identification of distinct motivational profiles, offering potential markers of vulnerability to GD (López-Fernández et al., 2020).

The Role of Game Design

Game design is often an overlooked but highly relevant factor in understanding GD. Many commercial games are developed based on principles of intermittent reinforcement, a behavioral mechanism from operant conditioning theory, which is directly related to compulsivity (Skinner, 1965). Features such as loot boxes, variable reward systems, leaderboards, and limited-time events act as extrinsic reinforcers that extend gameplay and promote continued engagement (King & Delfabbro, 2010).

In addition, social elements—such as the ability to form groups or compete online—act as social reinforcers, increasing the salience and subjective value of the game (Griffiths et al., 2012). These mechanisms, when interacting with individual cognitive and motivational factors, may facilitate the progression from recreational gaming to disorder.

Research also points to the intentionality behind these design mechanisms, which often exploit psychological vulnerabilities such as impulsivity and the need for validation. This discussion has gained prominence in regulatory policy debates, particularly regarding the prohibition of loot boxes in games targeted at minors (Zendle & Cairns, 2019).

Loot boxes are virtual reward mechanisms offering random in-game items—such as weapons, skins, characters, or upgrades—functioning like a "chest" or "mystery box." The content is unpredictable and luck-based. They share characteristics with gambling and have been linked to problematic behaviors, especially among adolescents and young people vulnerable to compulsive gaming. Key characteristics of loot boxes:

- Variable and unpredictable rewards: players never know what they will receive, encouraging repeated behavior (intermittent reinforcement, according to behavioral theory).
- Monetization: often involves real-money payments to purchase or open boxes.
- Social and aesthetic pressure: some items are rare or visually desirable, generating competition and a sense of belonging.
- Gamification of risk: simulate the thrill and uncertainty of gambling through animations, sounds, flashing lights, and time-limited offers.

Model Integration and Clinical Implications

An integrated understanding of cognitive, motivational, and design-related factors allows for a more robust clinical formulation of GD. For instance, an individual with high impulsivity (a psychological trait) may engage excessively in games with variable rewards (a design factor) while maintaining a belief that gaming is the only space where they feel valued (a dysfunctional cognition).

These findings suggest that effective interventions should consider broad psychosocial aspects, addressing everything from motivational context to the digital environment the patient inhabits. In this regard, transdiagnostic approaches that combine psychoeducation on game design, cognitive restructuring, and emotional regulation strategies appear particularly promising (Torres-Rodríguez et al., 2018).

Risk and Protective Factors in Gaming Disorder

The etiology of gaming disorder (GD) involves the complex interaction of multiple domains—individual, familial, social, and environmental. Understanding risk and protective factors is crucial not only for therapeutic planning and prevention but also for the development of more robust explanatory models.

Individual Factors: Personality Traits and Coping Strategies

Longitudinal studies and meta-analyses indicate that certain personality traits are strongly associated with the development of GD. The key traits implicated include:

- High impulsivity and low emotional self-regulation

- Elevated neuroticism, often linked to avoidance-based coping strategies
- Low conscientiousness and low agreeableness, according to the Big Five personality model

These traits directly influence the coping styles adopted by individuals. Dysfunctional strategies—such as experiential avoidance, cognitive escape, or rumination—have been significantly associated with problematic gaming as a means of psychological escape.

On the other hand, adaptive (or positive) coping—such as positive reappraisal and seeking social support—appears to act as a protective factor, reducing the impact of environmental stressors on the predisposition to GD. These strategies help mitigate stress, promote mental health, enhance problem-solving capacity, and support psychological growth and maturation.

Coping refers to the set of thoughts and behaviors used to manage internal or external demands perceived as stressful or overwhelming.

Family and Environmental Factors

The family environment is a powerful mediator in the manifestation and clinical course of gaming disorder. Research indicates that low parental monitoring, dysfunctional home environments, and parental modeling of excessive technology use are associated with increased risk, while an authoritative parenting style—marked by a balance of affection and clear boundaries—is linked to lower prevalence of the disorder in adolescents.

Perceived loneliness, low family cohesion, and domestic conflict further contribute as indirect risk factors, often leading individuals to seek immersion in digital environments as a substitute for real-life social engagement. Beyond the family, external factors such as dysfunctional school contexts, bullying, social isolation, and weak community support networks also elevate the risk, particularly during adolescence.

Game-Related Factors

The characteristics of games themselves also function as risk or protective factors. Risk-associated mechanisms include:

- Presence of unpredictable rewards (loot boxes, random achievements)
- Compulsive social components, such as competitive rankings and daily login requirements
- Massively multiplayer online role-playing games (MMORPGs), are more strongly associated with compulsive use than casual games

- Conversely, games that support self-determination, cognitive skill development, and structured, time-limited use tend to have a more adaptive profile.

Psychological Resilience and Adaptive Use of Games

Not all intense gaming use is pathological. The concept of resilience has been applied to explain why certain individuals, even when exposed to risk factors, do not develop the disorder.

Identified protective factors against gaming disorder include high self-efficacy and mature emotional regulation, which enhance individuals' capacity to manage stress and resist maladaptive gaming behaviors. Engagement with games for purposes such as learning, socialization, and creativity also contributes to a healthier interaction with digital environments. Additionally, metacognitive abilities—such as awareness of one's own gaming behavior and the capacity for self-regulation—serve as critical safeguards. When integrated into balanced and structured routines, games can be used adaptively to provide emotional and cognitive benefits, including mood improvement, stress reduction, and the development of strategic thinking skills.

Integration of Factors and Proposed Model

The interaction among the discussed factors can be represented through a biopsychosocial model, in which:

- Individual vulnerabilities (e.g., impulsivity, neuroticism) → increase sensitivity to the compulsive design of certain games.
- Negative family and social factors → reinforce the use of games as an escape mechanism.
- Lack of adaptive coping strategies → maintains dysfunctional gaming behavior.
- Protective factors, such as resilience and social support, can modulate this trajectory and prevent chronicity of the condition.

This model may be useful in formulating personalized interventions, as well as in developing public policies aimed at prevention.

Risk and Protective Factors

Internet gaming disorder (IGD) results from a complex interplay of individual, familial, and environmental influences. Intrapersonal traits such as impulsivity, neuroticism, aggressiveness, and low conscientiousness are consistently linked to IGD, with impulsivity emerging as a particularly strong predictor due to its impact on self-regulation (Gervasi et al., 2017; Müller et al., 2014; Warburton et al., 2022). Emotional difficulties—loneliness, anxiety, depression, and attention problems—also increase vulnerability, often making gaming a form of escapism. Low self-esteem contributes further by encouraging validation-seeking in virtual settings (Warburton et al., 2022).

Conversely, high self-control, conscientiousness, resilience, and self-esteem serve as protective traits, enhancing emotional regulation and decreasing reliance on gaming (Zhuang et al., 2023; Müller et al., 2014). Family dynamics play a critical role: risk increases in contexts of low cohesion, poor communication, violence, and excessive academic pressure (Gao et al., 2022; Park et al., 2008; Schneider et al., 2017), particularly among maltreated adolescents. Protective factors include emotional warmth, flexible rules, and parental monitoring (Bonnaire & Phan, 2017; Schneider et al., 2017).

At the school level, bullying and disengagement elevate risk, while peer support and academic involvement are protective (Gao et al., 2022; Ji et al., 2022). Community-level strategies—such as promoting physical activity and adaptive coping skills—further reduce vulnerability (Nowotarski et al., 2024). Gaming-related factors like extended playtime and reinforcing reward systems also contribute by encouraging prolonged use and reinforcing online social validation (Zhuang et al., 2023; Kochuchakkalackal Kuriala & Reyes, 2020). However, when practiced mindfully, gaming may offer psychological benefits such as stress relief and enhanced self-efficacy (Warburton et al., 2022).

Assessment and Screening Tools

Screening Tools

Several validated instruments are used to assess IGD based on DSM-5 criteria. The Internet Gaming Disorder Scale—Short Form (IGDS9-SF) (Pontes & Griffiths, 2015) and the Ten-Item Internet Gaming Disorder Test (IGDT-10) (Király et al., 2017) are brief, reliable, and culturally validated, though their coverage of ICD-11 criteria is limited. The Game Addiction Scale—7 Items (GAS-7) and the Lemmens IGD-9 also demonstrate strong psychometric properties but similarly fall short of capturing ICD-11 criteria in full (Lemmens et al., 2009, 2015; King et al., 2020). All tools are useful for initial screening but require clinical follow-up for diagnostic confirmation.

Structured Clinical Interviews

Structured interviews offer standardized diagnostic assessment. The Diagnostic Interview for Internet Addiction (DIA) is DSM-5-based, child- and caregiver-inclusive, and includes a craving item, offering detailed symptom evaluation (Ryu et al., 2019; Lee et al., 2025). The Structured Clinical Interview for Internet Gaming Disorder (SCI-IGD) targets adolescents and includes both demographic and diagnostic sections, with strong evidence of reliability and validity (Koo et al., 2017).

Challenges in Early Identification

Early detection of IGD faces multiple barriers. Diagnostic criteria remain debated, with concerns over false positives due to overlap with normative gaming (Ko et al., 2020). There is no universally validated diagnostic tool, and variation among scales (e.g., IGD-20, IGDS9-SF) leads to inconsistent findings (Kuss et al., 2014; Gao et al., 2022). Adolescents may underreport symptoms due to stigma or normalization of gaming (Jeong et al., 2017). Comorbid conditions like ADHD, depression, and anxiety complicate differential diagnosis (Ko et al., 2020). Additionally, cultural norms, limited awareness among adults, and a lack of longitudinal studies hinder the identification of early risk markers and delay intervention (Gao et al., 2022).

Prevention and Policy Strategies

Internet gaming disorder (IGD) is increasingly recognized as a global public health concern, prompting the development of comprehensive prevention strategies and policy responses across various countries. While prevalence rates vary, the consistent identification of gaming-related harms has led to international efforts focused on early intervention, health education, and multi-sectoral collaboration (King et al., 2018; King & Delfabbro, 2017).

Prevention Strategies

School-Based Programs

School-based prevention initiatives are among the most commonly implemented strategies for reducing the risk of disordered gaming. These programs often aim to promote digital literacy, critical thinking about game use, and emotional self-regulation among adolescents. Studies have shown that educational content delivered in school contexts can increase awareness of gaming-related risks and foster healthier attitudes toward play (Stevens et al., 2021). However, evaluations of program effectiveness remain limited, and long-term behavioral outcomes are still being assessed (King et al., 2018).

Digital Interventions

Digital tools such as online psychoeducation modules, app-based tracking systems, and interactive e-health programs have shown potential in promoting self-monitoring and reducing excessive gaming behaviors. These interventions are particularly scalable and cost-effective, making them attractive options for reaching younger populations. Nevertheless, their success often depends on sustained engagement and personalization of content (King & Delfabbro, 2017).

Stepped Care Approaches

A growing number of researchers advocate for stepped-care frameworks that provide tiered support according to the severity of the user's needs. These models can range from low-intensity interventions (e.g., automated feedback tools) to moderate support (e.g., group counseling) and more intensive clinical services when necessary. Early trials suggest that such flexible, multi-level approaches may be effective in optimizing resource allocation and treatment matching (King et al., 2018).

Policy and Public Health Approaches

Formal Recognition of IGD

The inclusion of IGD in diagnostic classification systems, such as the DSM-5 (as a condition for further study) and the ICD-11 (as Gaming Disorder), has catalyzed national-level prevention planning. In countries such as South Korea and China, formal recognition has underpinned the development of legal frameworks and funding mechanisms to address gaming-related harms (King & Delfabbro, 2017).

Government-Led Initiatives

South Korea offers a widely cited example of a coordinated public health response to IGD, with universal, selective, and indicated prevention strategies deployed across educational, clinical, and community settings (King et al., 2018). In contrast, Western nations typically rely more on non-governmental and industry-supported efforts, which can lack uniformity and central oversight (Stevens et al., 2021).

Regulation and Harm Reduction

Policymakers have explored a variety of harm reduction strategies, including age-based restrictions on gaming hours, regulation of in-game purchases (e.g., loot boxes), and requirements for real-name user registration. These approaches seek to mitigate the structural features of games that may foster problematic use. However, such measures can be controversial and raise concerns regarding personal freedoms and enforcement feasibility (King & Delfabbro, 2017).

Risk and Protective Factors

The etiology of IGD appears to be multifactorial, involving a complex interplay of psychological, neurobiological, and environmental factors. Key risk factors include maladaptive cognitions (e.g., escapism, gaming-related self-worth), comorbid psychopathology (e.g., depression, anxiety), and poor impulse control (King et al., 2018). Excessive use of social networks and isolation from offline peers may also exacerbate vulnerability (see Table 8.1). On the other hand, protective factors such as high self-esteem, strong social support, and effective emotional regulation skills have been linked to a lower risk of IGD (Stevens et al., 2021).

Table 8.1 Summary of key risk and protective factors for Internet gaming disorder (IGD), along with their associated evidence strength based on recent literature

Factor type	Examples	Risk/ protection	Evidence strength
Psychological	Maladaptive cognitions, gaming motivations	↑ Risk	Strong
Psychopathological	Depression, anxiety, impulsivity	↑ Risk	Moderate
Social/ environmental	Excessive social network use	↑ Risk	Moderate
Protective	Self-control, self-esteem, education	↓ Protection	Modest

Stakeholder Engagement

Industry Involvement

Recent scholarships have emphasized the importance of engaging the gaming industry in prevention efforts. This can include voluntary participation in responsible gaming initiatives, transparency in data sharing, and design modifications to reduce addictive features. While some companies have begun funding public awareness campaigns or integrating time-limit features, broader adoption remains inconsistent (King et al., 2018).

Multisectoral Collaboration

Effective prevention of IGD requires coordinated action among diverse stakeholders, including policymakers, educators, mental health professionals, researchers, families, and the gaming industry. Stevens et al. (2021) stress that community-based dialogue and cross-sector partnerships are essential for designing interventions that are both contextually relevant and ethically grounded.

Conclusion

The formal recognition of Internet gaming disorder (IGD) as a clinical condition underscores the imperative for precise diagnostic frameworks, high-quality assessment instruments, and effective prevention and intervention strategies. Contemporary meta-analytical evidence highlights the necessity of identifying clinically significant distress and functional impairment to distinguish pathological gaming from normative or highly engaged behavior (Ballou & Zendle, 2022; Gao et al., 2022). Although the DSM-5 and ICD-11 offer foundational diagnostic criteria, ongoing research emphasizes the need for refinement to improve diagnostic validity and reliability, particularly with respect to the sensitivity and specificity of available screening tools (Jo et al., 2019; Király et al., 2017; Pontes et al., 2021). The development and validation of structured clinical interviews and psychometrically robust self-report instruments constitute essential advancements facilitating early detection and accurate diagnosis (Koo et al., 2017; Lee et al., 2025).

Prevention and early identification approaches must adopt a multifactorial perspective that encompasses individual personality traits, familial and environmental influences, and cultural context (Bonnaire & Phan, 2017; Ji et al., 2022; Schneider et al., 2017). Empirically supported public health strategies include school- and family-based educational programs, media literacy initiatives, and policy interventions addressing game design features and regulated playtime (King & Delfabbro,

2017; Stevens et al., 2021). A risk-resilience framework further highlights the protective role of adaptive coping mechanisms and social support in mitigating vulnerability to IGD (Warburton et al., 2022; Meng et al., 2024). Given the rapid evolution of digital gaming technologies, including virtual reality and metaverse platforms, prevention strategies must remain adaptive and culturally sensitive (King et al., 2018; Zhuang et al., 2023).

Future research priorities include longitudinal studies to elucidate etiological mechanisms, identification of neurobiological biomarkers, and continued validation of assessment instruments that capture the complex clinical presentation of IGD (Liu et al., 2018; Zha et al., 2022). Advancements in understanding the neurocognitive and motivational underpinnings of IGD will inform the development of tailored therapeutic interventions, including cognitive-behavioral approaches, digital detoxification protocols, and pharmacotherapy options, thereby improving clinical outcomes (Ko et al., 2020; Piper et al., 2024). Moreover, interdisciplinary collaboration among clinicians, researchers, policymakers, and industry stakeholders is critical to balance the beneficial aspects of gaming with the mitigation of associated harms, while addressing ethical considerations related to medicalization and stigma (Griffiths et al., 2012; King & Delfabbro, 2010).

In sum, Internet gaming disorder represents a complex clinical entity that demands rigorous diagnostic clarity, comprehensive prevention efforts, and continued empirical inquiry. Through the integration of scientific evidence and clinical expertise, the field can advance a balanced, evidence-based approach that supports individuals affected by IGD and promotes healthy digital gaming engagement.

References

Ballou, N., & Zendle, D. (2022). "Clinically significant distress" in internet gaming disorder: An individual participant meta-analysis. *Computers in Human Behavior, 129*, 107140. https://doi.org/10.1016/j.chb.2021.107140

Billieux, J., Van der Linden, M., Achab, S., Khazaal, Y., Paraskevopoulos, L., Zullino, D., & Thorens, G. (2013). Why do you play World of Warcraft? An in-depthexploration of self-reported motivations to play online and in-game behaviours in the virtual world of Azeroth. *Computers in Human Behavior*, 29(1), 103-109

Bonnaire, C., & Phan, O. (2017). Relationships between parental attitudes, family functioning and internet gaming disorder in adolescents attending school. *Psychiatry Research, 255*, 104–110.

Gao, Y. X., Wang, J. Y., & Dong, G. H. (2022). The prevalence and possible risk factors of internet gaming disorder among adolescents and young adults: Systematic reviews and meta-analyses. *Journal of Psychiatric Research, 154*, 35–43. https://doi.org/10.1016/j.jpsychires.2022.06.049

Gentile, D., Bailey, K., Bavelier, D., Brockmyer, J., Cash, H., Coyne, S., Doan, A., Grant, D., Green, C., Griffiths, M., Markle, T., Petry, N., Prot, S., Rae, C., Rehbein, F., Rich, M., Sullivan, D., Woolley, E., & Young, K. (2017). Internet gaming disorder in children and adolescents. *Pediatrics, 140*(S2), S81–S85. https://doi.org/10.1542/peds.2016-1758H

Gervasi, A. M., La Marca, L., Costanzo, A., Pace, U., Guglielmucci, F., & Schimmenti, A. (2017). Personality and internet gaming disorder: A systematic review of recent literature. *Current Addiction Reports, 4*(3), 293–307

Jeong, H., Yim, H. W., Lee, S. Y., Lee, H. K., Potenza, M. N., Jo, S. J., et al. (2017). A partial mediation effect of father-child attachment and self-esteem between parental marital conflict and subsequent features of internet gaming disorder in children: A 12-month follow-up study. *BMC Public Health, 20*(1), 1–10

Ji, Y., Yin, M. X. C., Zhang, A. Y., & Wong, D. F. K. (2022). Risk and protective factors of internet gaming disorder among Chinese people: A meta-analysis. *Australian and New Zealand Journal of Psychiatry, 56*(4), 332–346

Jo, Y., Bhang, S., Choi, J., Lee, H., Lee, S., & Kweon, Y. (2019). Clinical characteristics of diagnosis for internet gaming disorder: Comparison of DSM-5 IGD and ICD-11 GD diagnosis. *Journal of Clinical Medicine, 8*(7), 945. https://doi.org/10.3390/jcm8070945

King, D., & Delfabbro, P. (2017). Prevention and policy related to internet gaming disorder. *Current Addiction Reports, 4*, 284–292. https://doi.org/10.1007/s40429-017-0157-8

King, D., & Delfabbro, P. (2010). The convergence of gambling and digital gaming: Implications for gambling in young people Mechanisms of influence. *Journal of gambling studies*, 26(2)75-187. https://doi.org/10.1007/s10899-009-9153-9

King, D., Delfabbro, P., Doh, Y., Wu, A., Kuss, D., Pallesen, S., Mentzoni, R., Carragher, N., & Sakuma, H. (2018). Policy and prevention approaches for disordered and hazardous gaming and internet use: An international perspective. *Prevention Science, 19*, 233–249. https://doi.org/10.1007/s11121-017-0813-1

King D, Delfabbro P, Griffiths M. (2010). The convergence of gambling and digital media: implications for gambling in young people. *J Gambl Stud*, 26(2):175-87. https://doi.org/10.1007/s10899-009-9153-9. Epub 2009 Sep 16. PMID: 19756979

King, D. L., Chamberlain, S. R., Carragher, N., Billieux, J., Kardefelt-Winther, D., Khazaal, Y., Lee, H. K., Potenza, M. N., Saunders, J. B., Starcevic, V., Stein, D. J., & Yücel, M. (2020). Screening and assessment tools for gaming disorder: A comprehensive systematic review. *Clinical Psychology Review, 77*, 101831. https://doi.org/10.1016/j.cpr.2020.101831

Király, O., Sleczka, P., Pontes, H. M., Urbán, R., Griffiths, M. D., & Demetrovics, Z. (2017). Validation of the ten-item internet gaming disorder test (IGDT-10) and evaluation of the nine DSM-5 internet gaming disorder criteria. *Addictive Behaviors, 64*, 253–260. https://doi.org/10.1016/j.addbeh.2015.11.005

Ko, C. H., Király, O., Demetrovics, Z., Chang, Y. M., & Yen, J. Y. (2020). Identifying individuals in need of help for their uncontrolled gaming: A narrative review of concerns and comments regarding gaming disorder diagnostic criteria. *Journal of Behavioral Addictions, 9*(3), 572–588.

Kochuchakkalackal Kuriala, G., & Reyes, M. E. S. (2020). An emerging mental health concern: Risk factors, symptoms, and impact of internet gaming disorder. *Journal of Technology in Behavioral Science, 5*(1), 70–78

Koo, H. J., Han, D. H., Park, S.-Y., & Kwon, J.-H. (2017). The structured clinical interview for DSM-5 internet gaming disorder: Development and validation for diagnosing IGD in adolescents. *Psychiatry Investigation, 14*(1), 21–29

Kuss, D. J., Griffiths, M. D., Karila, L., & Billieux, J. (2014). Internet addiction: A systematic review of epidemiological research for the last decade. *Current Pharmaceutical Design, 20*(25), 4026–4052.

Kuss DJ, Griffiths MD. (2012). Online gaming addiction in children and adolescents: A review of empirical research. *J Behav Addict*, 1(1):3-22. https://doi.org/10.1556/JBA.1.2012.1.1. PMID: 26166826

Kuss, D.J., Griffiths, M.D. (2015). Internet Addiction in Psychotherapy. Palgrave Studies in Cyberpsychology. *Palgrave Pivot, London*https://doi.org/10.1057/9781137465078_3

Kuss, D. J., Griffiths, M. D., & Pontes, H. M. (2017). DSM-5 diagnosis of Internet Gaming Disorder: Some ways forward in overcoming issues and concerns in the gaming studies field. *Journal of behavioral addictions*, 6(2), 133–141.https://doi.org/10.1556/2006.6.2017.032

Lee, M. S., Choi, J. S., Kweon, Y. S., & Bhang, S. Y. (2025). Structural validity of the Diagnostic Interview for Internet Addiction Scale for clinical samples in Korean children and adolescents: Exploratory and confirmatory factor analysis. *Psychiatry Investigation, 22*(2), 167–174

Lemmens, J. S., Valkenburg, P. M., & Peter, J. (2009). Development and validation of a game addiction scale for adolescents. *Media Psychology*, 12(1), 77–95. https://doi.org/10.1080/15213260802669458

Lemmens, J. S., Valkenburg PM, Gentile DA. (2015). The Internet Gaming Disorder Scale. *Psychol Assess*27(2):567-82. Epub 2015 Jan5. PMID: 25558970. https://doi.org/10.1037/pas0000062

Liu, L., Yao, Y., Li, C., Zhang, J., Xia, C., Lan, J., Zhou, N., & Fang, X. (2018). The comorbidity between internet gaming disorder and depression: Interrelationship and neural mechanisms. *Frontiers in Psychiatry, 9*, 154. https://doi.org/10.3389/fpsyt.2018.00154

López-Fernández FJ, Mezquita L, Griffiths MD, Ortet G, Ibáñez MI. (2020). The development and validation of the Videogaming Motives Questionnaire (VMQ). *PLoS One* Oct 23;15(10):e0240726. https://doi.org/10.1371/journal.pone.0240726. Erratum in: PLoS One. 2022 Dec 30;17(12):e0280007. https://doi.org/10.1371/journal.pone.0280007 PMID: 33095762; PMCID: PMC7584249

McCowan, T. C. (1981). Space-invaders wrist. *New England Journal of Medicine, 304*(22), 1368.

Meng, C., Li, Q., Cui, K., Xie, Y., Liu, W., & Xu, Y. (2024). Low self-control, perceived social support and internet gaming addiction: Findings from an ethnicity minority region in China. *Frontiers in Psychiatry, 15*, 1458626. https://doi.org/10.3389/fpsyt.2024.1458626

Müller, K. W., Beutel, M. E., Egloff, B., & Wölfling, K. (2014). Investigating risk factors for internet gaming disorder: A comparison of patients with addictive gaming, pathological gamblers and healthy controls regarding the Big Five personality traits. *European Addiction Research, 20*(3), 129–136.

Newzoo. (2024). Global Games Market Report and Forecasts. Amsterdam: *Newzoo*, 127(6):3066-3088.https://resources.newzoo.com/hubfs/Free%20Reports/Games%20Market%20Report%20and%20Forecasts/2024_Newzoo_Free_Global_Games_Market_Report.pdf

Nowotarski, A., Rothen, S., Kasina, F., Zullino, D., & Thorens, G. (2024). Physical activity as a predictor of internet gaming disorder in a Swiss male cohort (C-SURF). *The Canadian Journal of Psychiatry, 69*(12), 855–860

Park, S. K., Kim, J. Y., & Cho, C. B. (2008). Prevalence of internet addiction and correlations with family factors among South Korean adolescents. *Adolescence, 43*(172), 895–909.

Pearcy, B., McEvoy, P., & Roberts, L. (2017). Internet gaming disorder explains unique variance in psychological distress and disability after controlling for comorbid depression, OCD, ADHD, and anxiety. *Cyberpsychology, Behavior, and Social Networking, 20*(2), 126–132. https://doi.org/10.1089/cyber.2016.0304

Piper, T., Da Silva Youngs, M., Tomy, H., Lockwood, R., & Bowden-Jones, H. (2024). The National Centre for Gaming Disorders—The demographic profile and clinical characteristics of individuals accessing our service. *British Journal of Psychiatry Open, 10*, S203. https://doi.org/10.1192/bjo.2024.507

Pontes, H. M., & Griffiths, M. D. (2015). Measuring DSM-5 internet gaming disorder: Development and validation of a short psychometric scale. *Computers in human behavior* , 45, 137-143. https://doi.org/10.1016/j.chb.2014.12.006

Pontes, H. M., Schivinski, B., Sindermann, C., Li, M., Becker, B., Zhou, M., Montag, C., & Spector, P. E. (2021). Measurement and conceptualization of gaming disorder according to the World Health Organization framework: The development of the gaming disorder test. *International Journal of Mental Health and Addiction, 19*, 508–528. https://doi.org/10.1007/s11469-019-00088-z

Przybylski, A. K., Rigby, C. S., & Ryan, R. M. (2010). A Motivational Model of Video Game Engagement. *Review of General Psychology*, 14(2), 154-166. https://doi.org/10.1037/a0019440

Ross, D. R., Finestone, D. H., & Lavin, G. K. (1982). Space invaders obsession. *JAMA, 248*(10), 1177.

Ryan, R. M., & Deci, E. L. (2000). Self-determination theory and the facilitation of intrinsic motivation, social development, and well-being. *American Psychologist, 55*(1), 68–78.
Ryu, H., Lee, J. Y., Choi, A. R., Chung, S. J., Park, M., Bhang, S.-Y., Kwon, J.-G., Kweon, Y.-S., & Choi, J.-S. (2019). Application of diagnostic interview for internet addiction (DIA) in clinical practice for Korean adolescents. *Journal of Clinical Medicine, 8*(2), 202.
Schneider, L. A., King, D. L., & Delfabbro, P. H. (2017). Family factors in adolescent problematic internet gaming: A systematic review. *Journal of Behavioral Addictions, 6*(3), 321–333.
Skinner, B. F. (1965). *Science and human behavior* (No. 92904). Simon and Schuster
Stevens, M., Delfabbro, P., & King, D. (2021). Prevention approaches to problem gaming: A large-scale qualitative investigation. *Computers in Human Behavior, 115*, 106611. https://doi.org/10.1016/j.chb.2020.106611
Torres-Rodríguez A, Griffiths MD, Carbonell X. (2018). The Treatment of Internet Gaming Disorder: a Brief Overview of the PIPATIC Program. *Int J Ment Health Addict.* Oct 23;15(10):e0240726. https://doi.org/10.1007/s11469-017-9825-0. Epub 2017 Nov 13. PMID: 30147635; PMCID: PMC6096606
Warburton, W., Parkes, S., & Sweller, N. (2022). Internet gaming disorder: Evidence for a risk and resilience approach. *International Journal of Environmental Research and Public Health, 19*(9), 5587
Wartberg, L., Kriston, L., & Thomasius, R. (2017). The prevalence and psychosocial correlates of internet gaming disorder. *Deutsches Ärzteblatt International, 114*(25), 419–424. https://doi.org/10.3238/arztebl.2017.0419
Young, K. S. (2011). CBT-IA: the first treatment model for Internet addiction. *Journal of Cognitive Psychotherapy*, 25(4)
Yee N. Motivations for play in online games. (2006). *Cyberpsychol Behav* 9(6):772-5. https://doi.org/10.1089/cpb.2006.9.772. PMID: 17201605
Wölfling, K., Müller, K. W., Dreier, M., Ruckes, C., Deuster, O., Batra, A., Mann, K., Musalek, M., Schuster, A., Lemenager, T., Hanke, S., & Beutel, M. E. (2019). Efficacy of Short-term Treatment of Internet and Computer Game Addiction: A Randomized Clinical Trial. *JAMA psychiatry*, 76(10), 1018–1025. https://doi.org/10.1001/jamapsychiatry.2019.1676
Zendle, D., & Cairns, P. (2019). Loot boxes are again linked to problem gambling: Results of a replication study. *PLoS One, 14*(3), e0213194. https://doi.org/10.1371/journal.pone.0213194
Zha, R., Tao, R., Kong, Q., Li, H., Liu, Y., Huang, R., Wei, Z., Hong, W., Wang, Y., Zhang, D., Fallgatter, A., Yang, Y., Zhang, X., Liang, P., & Rao, H. (2022). Impulse control differentiates internet gaming disorder from non-disordered but heavy internet gaming use: Evidence from multiple behavioral and multimodal neuroimaging data. *Computers in Human Behavior, 130*, 107184. https://doi.org/10.1016/j.chb.2022.107184
Zhuang, X., Zhang, Y., Tang, X., Ng, T. K., Lin, J., & Yang, X. (2023). Longitudinal modifiable risk and protective factors of internet gaming disorder: A systematic review and meta-analysis. *Journal of Behavioral Addictions, 12*(2), 375–392

Chapter 9
Emotional and Affective Responses to Gaming

Ana Luiza Urbini, Carolina Ribeiro Ferraz, Gabriel Kadri, Raphael Santos, Pedro Makoto Uchida, Rafael Ângulo Condoretti Barros, and Lucas Murrins Marques

Abstract This chapter explores how social interaction and emotional connectivity emerge and unfold within gaming communities, highlighting the complex interplay between individual experience, collective dynamics, and technological design. Video games are not only platforms for entertainment but also spaces where players cultivate relationships, develop identities, and express emotions—often in deeply personal and socially meaningful ways. Drawing on recent advances in affective neuroscience, psychology, and game studies, we examine how games trigger emotional responses through narrative immersion, multisensory design, and the reward system, while also serving as tools for coping, stress relief, and mental health support. We further analyze how players form emotional bonds with non-playable characters (NPCs), explore identity through avatars, and experience social inclusion—or exclusion—in multiplayer and online environments. Finally, the chapter discusses the ethical responsibilities of the gaming industry in shaping emotional experiences and fostering healthy, inclusive digital ecosystems. By integrating neurobiological, psychological, and cultural perspectives, we offer a comprehensive understanding of how gaming fosters emotional resonance, social bonding, and affective transformation.

Keywords Emotional engagement · Gaming communities · Social connectivity · Affective neuroscience · Player–NPC bonding

---

A. L. Urbini · C. R. Ferraz · G. Kadri · R. Santos · R. Â. C. Barros
Mental Health Department, Santa Casa de São Paulo School of Medical Sciences, São Paulo, Brazil

P. M. Uchida
Pontifícia Universidade Católica de Campinas, Campinas, São Paulo, Brazil

L. M. Marques (✉)
Mental Health Department, Santa Casa de São Paulo School of Medical Sciences, São Paulo, Brazil

Artificial Intelligence Research Division, Infinity Doctor's Inc., Dover, DE, USA

L. Murrins Marques et al. (eds.), *Social and Affective Neuroscience of Gaming*, https://doi.org/10.1007/978-3-032-00325-6_9

Introduction

Video games have long been a subject of discussion regarding their influence on human emotions. Whether evoking excitement, frustration, joy, or even fear, gaming experiences are deeply intertwined with affective and emotional responses. The immersive nature of games—where players interact with dynamic narratives, rich audiovisual stimuli, and complex decision-making scenarios—positions them as powerful tools for emotional engagement. These emotional reactions are not only momentary but can also have long-term psychological and social effects, shaping players' moods, behaviors, and even cognitive functions.

At a fundamental level, games can elicit both primary emotions, such as happiness, anger, or surprise, and complex emotions, such as nostalgia, empathy, and moral dilemmas. The ways in which these emotions emerge depend on multiple factors, including the game's design, the player's personality, and the broader social context in which the game is played. Additionally, different gaming genres, ranging from single-player narrative-driven experiences to competitive multiplayer environments, can lead to varied emotional outcomes, influencing stress levels, motivation, and overall well-being.

Beyond individual emotional responses, video games also play a role in social and affective interactions. Multiplayer games create spaces for socialization, cooperation, and even competition, allowing players to develop relationships and experience collective emotions. The incorporation of advanced artificial intelligence and virtual characters has further expanded the emotional depth of games, fostering bonds between players and non-playable characters (NPCs). These interactions have implications for mental health, stress management, and even therapeutic applications, positioning video games as an increasingly relevant medium for emotional regulation.

As gaming technology continues to evolve, so does our understanding of its psychological and neurological impact. By exploring the mechanisms behind emotional engagement in gaming—ranging from neurochemical processes to multisensory synchronization—this chapter aims to provide a comprehensive overview of how games shape emotional experiences. The discussion will examine key elements such as dopamine-driven reward systems, the role of aesthetics and sound design in emotional modulation, and the diverse emotional effects of different game genres. Ultimately, understanding the emotional and affective dimensions of gaming is crucial for leveraging its benefits while mitigating potential risks, offering valuable insights for both researchers and game developers.

How Games Influence Emotions

In the context of the current scenario, in which video games emerged in their role as immersive and multifaceted experiences, it has become increasingly important to understand the complexity of the emotions they induce (Frome, 2007). Emotions are not merely individual experiences, but deeply rooted phenomena shaped by social, cultural, and political contexts. They emerge from the interplay between body, mind, and language, all of which are influenced by power structures (Burkitt, 2002). Furthermore, emotions encompass more than an affective component, also include cognitive, behavioral, and physiological dimensions, making them central to the human experience (Panksepp, 2004).

Human emotions are understood at different levels of complexity, ranging from primary processes to more intricate experiences. Primary emotions, such as seeking, fear, rage, care, panic, and play, are associated with basic emotional brain systems and exist within instinctual circuits. Such responses, modulated by limbic and subcortical structures, serve as the foundation for the development of more complex emotions. While primary emotions, such as fear and joy, are universal and innate, complex emotions arise from the interaction between biological mechanisms, linguistic, and sociocultural frameworks. They include guilt, shame, envy, and nostalgia, and are deeply embedded in human social dynamics and power structures (Panksepp, 2004), manifesting both in real-life interactions and within virtual environments.

A deeper analysis of emotional responses during gameplay requires looking beyond conventional assessments of joy, fear, or frustration. Frome (2007) proposes a model in which players' emotional responses emerge from their interaction with the game's structure—whether as observer-participants or actor-participants. According to this model, emotions are influenced by various factors, including in-game performance, narrative construction, and artifactual expression, highlighting often overlooked emotional dimensions such as ecological and artifactual emotions. As the boundaries between real and virtual environments continue to blur, recognizing these emotional layers is essential to fully grasp the impact of video games (Frome, 2007).

Human behavior in virtual environments is driven by intrinsic motivations, leading players to seek pleasurable cognitive states such as "flow" as they immerse themselves and remain engaged in games for long periods. When we analyze the reasons why players stay in this state, we can understand the underlying mechanism of positive and comforting emotions that drive them to play and continue playing, as well as the distancing from negative emotions outside that environment. Games facilitate mood regulation, providing relief from boredom and stress, promoting relaxation, and offering an escape from daily pressures. As they are highly immersive and demand significant attention, they can act as a powerful tool for stress distraction (Caro & Popovac, 2020). On the other hand, video games can also become a maladaptive coping strategy to avoid stressors, potentially leading to excessive dependence on this defense mechanism. This ambivalent dynamic shows that playing for long periods in a state of "flow" has the potential to aid in post-trauma recovery, promote positive emotions, and serve as a means to channel distress and aggression, improving mood. At the same time, it can represent a flawed defense mechanism, making games a complex phenomenon with both benefits and risks (Caro & Popovac, 2020).

Beyond the simplistic mind-screen relationship, video games have a profound impact on both the body and mind of players. A clear example of this is Exergames, which seamlessly combine physical activity with digital interactivity. In the systematic review "The Impact of Exergames on Emotional Experience," Lucas Murrins proposes to evaluate the impact of physical training with Exergames on emotional experience, finding that there is a significant effect on various experiences, such as depressive and anxious symptoms, self-esteem, relaxation, behavioral control, and motivation. Here, we can understand the use of Exergames as a modality of both physical activity entertainment and, promoting not only physical well-being but also emotional regulation. When analyzing physical exercise in gamification, it is possible to observe that exercise practice becomes intertwined with game characteristics, such as the sense of challenge, competition, and playfulness, making the experience of exercising more enjoyable and desirable. The gamified format enhances engagement in healthy behaviors, increases the time dedicated to these exercises, strengthens the establishment of a routine, and, by transcending the simplistic experience of a game, when applied to the medical reality of health promotion, it can potentially increase adherence to medications, the number of medical appointments, and engagement in treatments (Marques et al., 2023).

In this context, emotions in the gaming world can correlate with emotions in the sports world, with a particular emphasis on motivation, which influences both practice and the mindset of players. Indeed, motivation is needed to start something new, to maintain consistency and fluency in practice. Engagement requires motivation, which can be influenced by positive emotions such as enthusiasm or "demotivated" by feelings of frustration and boredom. Interestingly, in the Marques et al. (2023) literature review, one of the emotions that keeps gamers engaged in games is the escape from the stress and boredom of the real world outside the screens. These sensations negatively impact motivation, preventing individuals from adhering to treatment plans or engaging in social life projects. In this sense, gamification offsets such sensations by promoting excitement and encouraging users to set goals and build self-confidence, while engaging dopamine-driven reward systems. Even though these goals may seem confined to the gaming environment, the experience with Exergames extends beyond the screens, generating real-world outcomes, such as increased physical activity and positive effects on mental health, particularly in areas such as adherence to exercise routines, satisfaction with activities, and motivation to continue exercising.

When considering community dynamics, video games, through their multiplayer functionality, promote social support and strengthen the sense of community, connection, and belonging, crucial elements for psychological well-being and motivation development. The reviewed studies analyzed a variety of emotional states, including anxiety, depression, appreciation, happiness, mood, motivation, optimism, perceived behavioral control, perceived energy, relaxation, self-efficacy, self-esteem, stress, subjective well-being, and vigor. Gaming can serve as an effective alternative for interventions that aim not only to improve patients' physical condition but also to provide emotional and psychological benefits.

Key Systems Influencing Emotional Processes

Video games have become a prevalent form of entertainment, engaging individuals across various age groups. Beyond their recreational value, video games have been the subject of scientific investigations to understand their effects on the brain's dopaminergic system, associated neural structures, behavior, and mood. This section explores how video gaming influences dopaminergic synapses, alters brain regions, and subsequently affects behavior and mood.

The dopaminergic system plays a crucial role in reward processing, motivation, and learning. Dopamine, a key neurotransmitter, is released in response to rewarding stimuli, reinforcing behaviors essential for survival and well-being. Video games, designed to provide rewarding experiences, can significantly influence this system (Koepp et al., 1998). Engaging in video gaming has been shown to stimulate dopamine release in the striatum, a brain region integral to reward and motivation pathways. A study using positron emission tomography (PET) scans demonstrated that playing video games leads to dopamine release in the human striatum, highlighting the impact of gaming on the brain's circuitry (Brilliant et al., 2019).

The ventral striatum, a key structure in the brain's reward network, is involved not just in reward anticipation but also in processing emotional responses to these rewards. Research indicates that the ventral striatum's activation during gaming is associated with the pleasure derived from accomplishment and goal completion (e.g., winning a match or unlocking an achievement). However, the same brain region is also implicated in processing emotional responses, particularly those related to positive emotions and motivation.

The emotional experience during gaming, especially in games with immersive storylines or intense competitive environments, can also influence how rewards are perceived. Granic et al. (2014) demonstrated that emotionally charged experiences within games—whether thrilling victories or the frustration of setbacks—can shape emotional regulation. These emotional experiences are further modulated by game mechanics such as progress bars, in-game rewards, and feedback loops, which continuously trigger the brain's reward centers.

Interestingly, games that provide social or collaborative components—like multiplayer games—can further enhance these emotional responses by tapping into the emotional bonds players form with others. For example, achieving a team victory or supporting a teammate can induce feelings of pride, camaraderie, or shared excitement, all of which activate the ventral striatum and other emotion-related regions. This heightened emotional engagement can make the experience of playing more rewarding, further drawing players into the game world.

Khun et al. (2013) describe how video game players show changes in areas related to social interaction and reward processing (such as the ventral striatum), which are activated during cooperative or competitive play. Games that require skill-building or continuous learning (such as puzzle games or certain strategy games) can lead to enhanced neuroplasticity, which allows the brain to reorganize itself by forming new neural connections. Oei and Patterson (2013) found that playing brain-training games can lead to improved cognitive skills in non-game tasks as well (Oei & Patterson, 2013; Bavelier et al., 2012).

The dopamine surge experienced during gaming reinforces the behavior, potentially leading to prolonged gaming sessions. This mechanism is similar to that observed in other forms of behavioral reinforcement, where the anticipation of reward motivates continued engagement (Bavelier et al., 2015). Long-term video gaming has been associated with changes in brain structure. For instance, studies have indicated that excessive gaming can lead to alterations in brain regions responsible for various cognitive functions such as executive functions, attention, and emotional regulation (Brilliant et al., 2019).

Beyond the reward system, video games also engage the emotional centers of the brain responsible for empathy and emotional regulation. Games with narrative depth and complex characters often require players to navigate moral choices or manage emotional challenges, both of which engage the prefrontal cortex—the area of the brain involved in higher-order cognitive functions like decision-making and emotional regulation. For instance, games such as The Last of Us or Red Dead Redemption 2 put players in situations where they must make choices that affect both their character's fate and the emotional experiences of others within the story.

Pichon et al. (2021) noted that games that involve emotionally charged narratives or that require players to consider the perspectives and emotions of other characters can enhance emotional intelligence and empathy. These games encourage players to understand and respond to emotional cues, which can contribute to better emotional regulation both within and outside the game. Furthermore, the ability to navigate complex emotional situations in games has been shown to transfer to real-world emotional processing, helping players to better understand their own feelings and those of others.

While the reward system can enhance emotional and cognitive functions, it can also lead to less desirable effects. Overstimulation of the reward system through prolonged gaming can create patterns of behavior similar to addiction. Just as with substances or gambling, the constant gratification from achieving goals within a game can lead to compulsive behavior. Kuss et al. (2013) found that individuals with video game addiction exhibit changes in the brain's reward processing regions, including diminished activity in the prefrontal cortex—an area crucial for impulse control. This suggests that excessive gaming can impair the brain's ability to regulate emotional responses, leading to negative consequences both in terms of mental health and social functioning.

The overstimulation of the reward system through video games can also lead to emotional dysregulation. When players fail to achieve the rewards they seek, they may experience frustration, anxiety, or depression. The emotional "high" that gaming provides can quickly turn into a low when the cycle is interrupted, contributing to emotional volatility.

Emotional Design and Multisensory Integration

In his article, Frome discusses the concept of artifactual emotions—which come into play in the gaming experience through the analysis of aesthetic design and mechanics. Much like in other artistic experiences, such as museums or movies, video games provoke a response as an artifact or an artistic object. However, they also include nuances derived from the medium's particularities, such as impatience from a particularly short story, frustration from a challenging interface, or admiration for the graphic details rendered on screen (Frome, 2007).

When considering the modern gaming experience, it is essential to go beyond the analysis of mechanics and plot and to consider the emotional interaction between a player and the game. Interactive emotional design is an innovative approach that explores the affective connections that amplify immersion into a game. This concept seeks to introduce new layers of emotion to the gaming experience, promoting recognition, identification, and emotional response to the user during gameplay. An example of this is the introduction of increasingly more complex characters, with detailed personalities and rich emotional expressions, that create a higher sense of closeness and immersion in a story. To achieve this, the interface becomes a crucial aspect of the gaming experience. Colors, layouts, detailed sceneries, and emotional soundtracks are carefully planned to stimulate further emotional ties and a feeling of belonging to a game. Therefore, this is not limited to aesthetics but expands into creating a sensorial atmosphere that encompasses a range of emotions and enriches the experience.

Interactive emotional design transforms a game into a space for creative and emotional expression, in which a player feels each challenge and accomplishment more authentically. This way, the visual and mechanical construction of a game aligns intimately with the experience proposed for each phase or, in other words, the narrative strengthens the aesthetic and the aesthetic strengthens the narrative. This creates a state of emotional resonance with gripping plots, filled with conflicts and twists. This approach is often associated with open-world, role-playing games. In these games, players are forced to explore the different regions of a map to advance in a story, and choices based on ethics and beliefs affect the progression of a plot. The results can cause a range of feelings, from accomplishment to frustration, joy to sadness, or pride to embarrassment (Weng, 2024).

Emotional Impact of Gaming

Common sense often associates video games with negative effects on well-being, such as addiction, social isolation, and even aggressive behavior. However, recent studies show that when used in moderation and with appropriate games, video games can have positive impacts, including improved mood, increased well-being, socialization, and even the encouragement of prosocial behaviors.

Improvement in Well-Being

The act of playing video games can be considered a modern and technological form of leisure, which, at the same time, connects to something inherent to human nature and essential for healthy development: playing. Indeed, the practice of video games can provide a moment of relaxation and enjoyment, promoting an improvement in mood, even if temporary. This positive effect is supported by the study conducted by Matti Vuorre et al. (2024), which highlights how playing contributes to emotional well-being.

Moreover, the research (Jones et al., 2014) points out that moderate gaming can be an excellent tool for promoting positive emotions, serving as an effective means of stress reduction. By creating deep immersion, many games induce meditative states, allowing players to disconnect from the stresses of everyday life. This immersion is even more significant in games that promote the development of social relationships and offer a broader sense of purpose, connecting players to larger communities. The feeling of accomplishment and overcoming challenges also plays an important role, boosting self-esteem and perceived competence. Furthermore, the study addresses the fact that the complete absence of video games may be associated with negative consequences, such as an increased risk of emotional disorders, especially in male teenagers. On the other hand, spending excessive time playing video games can bring harm both physically and socially. It can lead to obesity, body aches, poor posture, as well as anxiety, depression, social isolation, and neglect of real-life responsibilities, ultimately damaging relationships with family and friends. Therefore, balance is crucial, with moderate gaming being beneficial for mental well-being, provided that factors such as playtime, game type, and individual characteristics are considered.

The study conducted by Rachel Kowert et al. (2015) supports the finding that moderate video game play can enhance life satisfaction in adolescents more than not playing at all. The study observed that adolescent online gamers (ages 14–18) reported higher life satisfaction compared to those who did not engage in online gaming.

Finally, the relationship between stress and video games also deserves attention. Although racing and shooting games can temporarily increase stress levels, in many cases, video games serve as a break from daily obligations, offering an escape that contributes to stress reduction and increased satisfaction. This effect is closely related to the sense of control and autonomy that games provide, allowing players to feel more balanced and satisfied with their experience (Wang et al., 2023).

Violent Video Games and Aggression Versus Cooperative Video Games and Prosocial Behaviors

Research on the relationship between gaming and aggression is extensive and often contradictory. Several studies indicate that violent games can increase aggressive behavior and desensitization to violence (Anderson et al., 2010). As a counterpoint, other research shows that cooperative games can encourage prosocial behaviors (Gentile et al., 2009).

Given this scenario, the most relevant discussion should not be limited to the issue of violence in games but should also consider the quality of the content and the messages conveyed to children through their stories and narratives. Many game narratives still perpetuate sexist and homophobic stereotypes, which can negatively impact the development of values.

Another crucial point is the role of frustration. More challenging games can cause a temporary increase in aggression—not necessarily due to violence itself, but because of the difficulty imposed on the player. Studies suggest that even non-violent games, when extremely difficult, can lead to destructive reactions—such as streamers breaking their equipment after losing a match (Ferguson et al., 2022).

Psychosocial Impact

Socialization in the virtual world, especially in online video games, has been the subject of various scientific studies that seek to understand its impact on social interactions and the cognitive development of players. Studies indicate that electronic games can promote socialization, attention, concentration, imagination, creativity, persistence, rule acceptance, and morality—skills necessary for future situations (Cotonhoto & Rossetti, 2016).

Furthermore, the practice of video games is associated with the development of cognitive skills, such as attention, perception, and executive functions. A systematic review on the impact of video game use on these cognitive functions suggests that games can have positive effects, although the results are still not consensual in the literature (Rivero et al., 2012).

However, it is important to consider the potential negative effects associated with the excessive use of video games. Research highlights that socialization is a condition for the development of moral thinking, promoting a sense of responsibility, respect for rules, and interaction with peers. These assumptions may not be possible in games where the player plays alone, which can impact moral development (Fernandes et al., 2017).

In summary, socialization in the virtual world through online games represents a significant dimension of human interactions in the digital age. These environments provide opportunities for connection, learning, and entertainment, while also presenting challenges that require attention and balance. Understanding this duality is essential to maximizing the benefits of online games while mitigating potential negative impacts.

Artificial Intelligence (AI)

Interactions in the virtual world between humans and artificial intelligence (AI) have become an area of growing academic interest, especially regarding the social and psychological dynamics emerging from these relationships. One relevant theory in this context is the embodied social presence theory (ESP), which proposes that bodily representation in virtual spaces influences users' perceptions and cognitive engagement (Mennecke et al., 2010). This theory suggests that avatars and virtual agents can serve as extensions of the self, affecting how individuals communicate and interact in the digital environment.

Another pertinent phenomenon is the proteus effect, which describes how avatar characteristics can shape user behavior both in the virtual and real worlds (Yee & Bailenson, 2007). For instance, individuals using more attractive avatars tend to exhibit greater confidence and assertiveness in subsequent interactions. This effect highlights the influence of digital self-representation on technology-mediated interpersonal relationships.

Additionally, the ELIZA effect, initially identified in the context of computer programs like ELIZA, reveals the human tendency to attribute human-like qualities to machines, especially when they display linguistic behaviors similar to humans. This anthropomorphization can lead users to develop emotional bonds with AI agents, even when they are aware of their artificial nature.

In the contemporary context, the concept of the metaverse has gained prominence as an evolution of the internet, integrating the virtual and physical in a continuous digital space. Experts claim that the metaverse represents a new form of social and economic interaction, where digital and physical identities coexist, expanding the possibilities for relationships between humans and AIs.

Enterprises like "Friend," a startup offering digital companions through AI chatbots, exemplify the practical application of these interactions. These digital companions are designed to mitigate loneliness by offering complex emotional narratives that seek to engage users on deep levels, challenging the boundaries between human and artificial interaction.

These initiatives raise ethical questions about the depth of relationships that humans can or should develop with artificial entities. The ease with which users attribute emotions and intentions to AI agents underscores the need for a critical understanding of these interactions and their associated social and psychological implications.

Character Bonding and NPC Engagement

The interactions between players and non-playable characters (NPCs) in video games play a significant role in the gaming experience, contributing to immersion and enriching the narrative. These characters, although controlled by artificial intelligence, are designed to simulate human behaviors, allowing players to form emotional and social bonds with them. The construction of NPCs' identities directly influences player engagement, as elements such as narrative, personality, and aesthetics of the characters are developed to facilitate empathy and emotional connection.

The development of characters in video games has evolved to incorporate elements that facilitate player identification with NPCs. In games like League of Legends, characters were developed with a narrative depth that resonates with players, promoting a more engaging and emotionally meaningful gaming experience. This relationship with NPCs can lead to an emotional involvement comparable to that established between humans, especially when the character presents moral dilemmas and complex stories.

Additionally, the presence of dialogue trees in games allows players to interact more meaningfully with NPCs. These conversation structures offer multiple response options, allowing players to influence the course of the narrative and deepen their relationships with virtual characters. Studies on narratives in games show that the degree of interactivity in dialogue choices can generate different levels of emotional involvement and identification with the NPC (Fontoura, 2022a, b).

The representation of minorities in video games has also been an important factor in forming bonds between players and NPCs. The inclusion of diverse characters allows a broader audience to identify and connect emotionally with the game's narrative (Fontoura, 2022a, b). The presence of NPCs reflecting different cultures, ethnicities, and orientations can generate greater engagement and promote a more inclusive virtual environment.

The bonds established between players and NPCs in video games function as a form of socialization, allowing players to experience complex social interactions in virtual environments. These interactions not only enrich the gaming experience but also offer insights into the social and emotional dynamics that can be applied outside of the gaming context.

Coping in Virtual Worlds

The world of video games offers a unique platform for overcoming adversity, serving as an accessible and innovative tool for managing emotional and social challenges. Gaming can play a crucial role in situations of social isolation, such as during the COVID-19 pandemic, as well as provide a viable alternative for those with limited access to support or in-person treatment. In this section, we will explore examples that illustrate how video games can contribute to coping with difficulties and enhancing players' well-being.

Video games can be highly beneficial for training social skills in individuals with autism spectrum disorder, as they provide a controlled environment where users can manage interactions, allowing them to practice socialization safely. The study conducted by J Autism Dev Disord (Beaumont et al., 2021) evaluated the effectiveness of the Secret Agent Society (SAS) program, a video game-based social skills training for children on the autism spectrum. The results showed a significant improvement in social skills, along with a reduction in problematic behaviors, as reported by parents and teachers. Additionally, the study indicated a decrease in children's anxiety levels. Another study also highlighted benefits, such as improvements in metaphor recognition, interaction and social initiation rates, and the enhancement of communication skills (Walsh et al., 2025).

Another community that can benefit from the gaming environment is the transgender community, which finds in avatars a powerful tool for safely and immersively exploring and expressing their gender identity. The ability to customize characters allows these players to experiment with different gender expressions, test names, pronouns, and appearances, aiding in the process of self-discovery even before making changes in the real world. Additionally, the acceptance and interaction within games can provide a sense of belonging and validation, contributing to emotional and psychological well-being. According to the study conducted by Helen Morgan et al. (2020), avatar customization plays a therapeutic role, allowing these young people to experiment with their gender identity before coming out publicly. For many, creating an avatar that aligned with their experienced gender was a crucial first step in recognizing their identity. Additionally, playing as an avatar that reflected their gender identity generated positive emotions and boosted self-confidence, making it an essential outlet before social transition. Thus, gaming serves as a powerful tool for self-acceptance and gender identity development in a safe space.

Another interesting use of video games is as a complementary therapy in the treatment of adolescents with depression. Gaming stands out as an innovative approach, capable of overcoming geographical barriers and minimizing the stigma associated with seeking psychological help. In this way, it facilitates access to therapeutic support for those who might otherwise not seek treatment. The study found that adolescents who played SPARX—a game designed to help children and adolescents with depression—had higher remission rates and lower depressive symptoms compared to those on a waiting list. The game's results were comparable to those of in-person counseling treatment (Halldorsson et al., 2021).

Considering the therapeutic uses of video games, another notable example is the use of Tetris to help patients overcome trauma. The study conducted by Oisin Butler et al. (2020) provided evidence that integrating visuospatial games, specifically Tetris, into the treatment of post-traumatic stress disorder (PTSD) can bring significant benefits, including an increase in hippocampal volume, a brain structure essential for memory and emotional regulation. Tetris acts as a cognitive interference tool, helping to disrupt the consolidation of traumatic memories while strengthening new memories formed during therapy. By engaging patients in a challenging visuospatial task, the game prevents intrusive memories from becoming fully ingrained, reducing their recurrence. Additionally, the study suggests that sustained focus on the game may stimulate neurogenesis, promoting structural brain changes that, in the long term, contribute to the reduction of PTSD symptoms. Beyond its neurological effects, Tetris stands out as an accessible, low-cost, and portable intervention that can be self-administered without the risk of adverse effects. Its simplicity and engaging nature make it a promising tool to complement traditional therapeutic approaches in trauma treatment.

Empathy Development

The development of empathy through video games has been the subject of various scientific studies, indicating that this skill can be enhanced through interactive experiences. Simulation games, for example, allow players to make decisions and face consequences in safe environments, promoting the use of intelligence, skill development, and decision-making (Herrera et al., 2018). Furthermore, these games make learning more dynamic, encouraging greater interaction among participants. By integrating motor, mental, emotional, and social aspects, simulation games also stimulate teamwork, as each decision affects not only the individual player but also other participants (Gentile et al., 2004). This collaborative dynamic contributes to the development of empathy and social values.

In addition, games such as "Life Is Strange: True Colors" and "Journey" exemplify how interactive narratives can evoke emotions and promote empathy. In "Life Is Strange: True Colors," players follow the story of Alex Chen, a young woman with the ability to experience the emotions of others, which encourages players to put themselves in the characters' shoes and understand their perspectives (Ferguson et al., 2013). In "Journey," the multiplayer experience was designed to facilitate cooperation among players without forcing it, allowing participants to feel an emotional connection with others even without verbal communication (Greitemeyer & Osswald, 2011).

Evidence suggests that it is possible to train and enhance empathy through video games, especially those that incorporate simulation elements and engaging narratives. These games offer safe and controlled environments where players can experience different perspectives and emotions, contributing to the development of empathetic skills that can be transferred to real life.

Cultural Impact

The world of video games plays a significant role in cultural dissemination, operating similarly to other entertainment markets by introducing and shaping new perceptions of reality. A clear example of this phenomenon is K-pop, which has expanded South Korean culture beyond its physical borders, attracting fans to the language, cuisine, and tourism of the country. Similarly, video games also export cultural elements, fostering the exchange of references between different societies (Borowiecki & Prieto-Rodriguez, 2015; Cerezo-Pizarro et al., 2023).

As with other forms of entertainment, the development of a game is heavily influenced by the cultural background of the company and its creators. Films and books often serve as inspiration for narratives and settings, reflecting in the storyline, character construction, and visual aesthetics. However, video games have the distinctive feature of real-time interaction between players from different parts of the world, promoting a more dynamic cultural exchange.

In this sense, the video game universe undergoes a process similar to what is observed in South America, where different cultures coexist. Similarly, how Brazilians were described by social scientist Darcy Ribeiro, in his book "O povo brasileiro," gamers do not necessarily share a common ethnic, cultural, or historical background; what unites them is language and in this virtual context, the shared experience within the virtual environment. Interestingly, the digital world has been the birthplace of a variety of new terms and dialects (Casado & Peña-Acuña, 2022). Thus, creating a unique symbolism derived from players' collective experience.

Despite this connectivity, there are still barriers that limit global interaction in video games. Many online servers are divided by geographic regions, restricting interaction possibilities to users within the same area or those using VPNs. Additionally, linguistic challenges persist, as not all players are proficient in foreign languages, which can hinder communication and, therefore, cultural exchange.

Game Industry Responsibility

The video game industry plays a central role in shaping the emotional and affective experiences of players, bearing responsibility for the impacts their products generate. In pursuit of prolonged engagement, many companies adopt game mechanics based on variable rewards and intermittent reinforcement, which can trigger intense emotional responses and even compulsive behaviors. Such mechanisms, present in loot boxes and microtransactions, activate reward circuits similar to those observed in pathological gambling, raising ethical concerns about the intentional design of potentially addictive games (King et al., 2019).

Furthermore, developers' responsibility extends to the representation of characters and narratives, as games can either reinforce or challenge social stereotypes. Inclusive representations have been associated with positive effects on players' self-esteem and sense of belonging, whereas the perpetuation of stigmas can cause negative emotional impacts, especially among minority groups (Shaw, 2014).

Another critical point concerns sensitive and violent content. Prolonged exposure to violence in games may desensitize players and alter their emotional responses, although the literature presents divergent findings regarding the extent of these effects (Anderson et al., 2010). In this context, the industry's responsibility involves ensuring appropriate content ratings and providing effective parental control tools—an area that is heavily regulated and whose importance is widely recognized worldwide.

The relationship between screen time and emotional well-being also highlights the need for corporate policies that promote healthy gaming habits. While some companies have implemented break reminders or screen time reports, such measures remain limited and often lack effectiveness (Przybylski et al., 2017). Encouraging a balance between player engagement and mental health is a crucial step in mitigating negative outcomes. Furthermore, the moderation of online communities and the protection of user data—including affective data collected through technologies such as facial recognition and behavioral analysis—represent emerging ethical frontiers. The collection and use of such data demand transparency and informed consent in order to safeguard players' privacy and emotional well-being (Lugrin et al., 2021).

Although some aspects of the video game industry's operations are already subject to more structured oversight, such as age ratings and the regulation of loot boxes in certain countries, there are still areas that require more effective monitoring. For example, organizations such as the Entertainment Software Rating Board (ESRB) and the Pan European Game Information (PEGI) provide ratings that guide consumers regarding the sensitive content of games, such as violence, offensive language, and adult themes (Nieborg & Poell, 2018). Furthermore, countries such as Belgium and the Netherlands have classified loot boxes as a form of gambling, subjecting them to gambling legislation (Xiao, 2021). In contrast, France, while acknowledging the potential emotional and psychological risks of this practice, chose not to legally classify them as gambling, adopting a self-regulation approach and requiring transparency, such as the disclosure of chances to obtain virtual items (Autorité Nationale des Jeux (ANJ), 2018). However, the oversight of mechanics that promote addiction or excessive use, such as continuous reward systems and daily challenges, remains in its early stages in many contexts. Another area requiring greater regulatory attention is the protection of sensitive data, especially in light of the emerging use of artificial intelligence to collect and interpret players' emotional and behavioral data, with few clear standards regarding consent and the ethical use of this information (Lugrin et al., 2021). The moderation of online communities continues to be a challenge, with frequent reports of toxic environments and harassment, with no universal guidelines that hold companies accountable for the emotional protection of their users (Blackburn & Kwak, 2014).

Conclusion

In conclusion, video games are far more than mere sources of entertainment—they are powerful emotional stimuli that shape players' experiences, interactions, and well-being. Whether through immersive storytelling, engaging mechanics, or social connectivity, games have the potential to evoke a vast array of emotions, from joy and excitement to frustration and empathy. Understanding these emotional dynamics is essential for both researchers and game developers, as it allows for the creation of experiences that maximize positive psychological outcomes while mitigating potential risks. As technology continues to advance and gaming culture expands, the study of emotional and affective responses to gaming will remain a critical field of research, offering deeper insights into how digital experiences influence human emotion and behavior.

References

Anderson, C. A., Shibuya, A., Ihori, N., Swing, E. L., Bushman, B. J., Sakamoto, A., Rothstein, H. R., & Saleem, M. (2010). Violent video game effects on aggression, empathy, and prosocial behavior in eastern and western countries: A meta-analytic review. *Psychological Bulletin, 136*(2), 151–173. https://doi.org/10.1037/a0018251. PMID: 20192553.

Autorité Nationale des Jeux (ANJ). (2018). *Loot boxes: Report on risks and regulation*. https://anj.fr/sites/default/files/2020-10/rapport-coffres-jeu-en.pdf

Bavelier, D., Green, C. S., Pouget, A., & Schrater, P. (2012). Brain plasticity through the life span: Learning to learn and action video games. *Annual Review of Neuroscience, 35*, 391–416. https://doi.org/10.1146/annurev-neuro-060909-152832

Beaumont, R., Walker, H., Weiss, J., et al. (2021). Randomized controlled trial of a video gaming-based social skills program for children on the autism spectrum. *Journal of Autism and Developmental Disorders, 51*, 3637–3650.

Blackburn, J., & Kwak, H. (2014). Stfu noob!: Predicting crowdsourced decisions on toxic behavior in online games. In *Proceedings of the 23rd international conference on World wide web* (pp. 877–888). ACM.

Borowiecki, K., & Prieto-Rodriguez, J. (2015). Video games playing: A substitute for cultural consumptions? *Journal of Cultural Economics, 39*, 239–258. https://doi.org/10.1007/S10824-014-9229-Y

Brilliant, T. D., Nouchi, R., & Kawashima, R. (2019). Does video gaming have impacts on the brain: Evidence from a systematic review. *Brain Sciences, 9*(10), 251. https://doi.org/10.3390/brainsci9100251. PMID: 31557907; PMCID: PMC6826942.

Burkitt, I. (2002). Complex emotions: Relations, feelings and images in emotional experience. *The Sociological Review, 50*(2_suppl), 151–167.

Butler, O., Herr, K., Willmund, G., Gallinat, J., Kühn, S., & Zimmermann, P. (2020). Trauma, treatment and Tetris: Video gaming increases hippocampal volume in male patients with combat-related posttraumatic stress disorder. *Journal of Psychiatry & Neuroscience, 45*(4), 279–287. https://doi.org/10.1503/jpn.190027. PMID: 32293830; PMCID: PMC7828932.

Caro, K., & Popovac, M. (2020). Gaming when things get tough? Examining how emotion regulation and coping self-efficacy influence gaming during difficult life situations. *Computers in Human Behavior, 111*, 106428.

Casado, J., & Peña-Acuña, B. (2022). Video games lexicon included in Spanish language: A multiple case study. *Linguo Didáctica, 1*, 15–35. https://doi.org/10.33776/linguodidactica.v1.7428

Cerezo-Pizarro, M., Revuelta-Domínguez, F., Guerra-Antequera, J., & Melo-Sánchez, J. (2023). The cultural impact of video games: A systematic review of the literature. *Education Sciences, 13*, 1116. https://doi.org/10.3390/educsci13111116

Cotonhoto, L. A., & Rossetti, C. B. (2016). The practice of electronic games by young children: What do recent studies say? *Revista Psicopedagogia, 33*(102), 346–357.

Ferguson, C. J., Smith, S. M., Miller-Stratton, H., Fritz, S., & Heinrich, E. (2013). Violent video games and aggression: Causal relationship or byproduct of family violence and intrinsic aggression? *Aggression and Violent Behavior, 18*(3), 318–325.

Ferguson, C. J., Gryshyna, A., Kim, J. S., Knowles, E., Nadeem, Z., Cardozo, I., Esser, C., Trebbi, V., & Willis, E. (2022). Video games, frustration, violence, and virtual reality: Two studies. *The*

British Journal of Social Psychology, 61(1), 83–99. https://doi.org/10.1111/bjso.12471. Epub 2021 Jun 11. PMID: 34114247.
Fernandes, D. M., Carramillo-Going, L., Lemos, D. I. M., Avoglia, H. R. C., & Alves, H. (2017). Electronic games, violence, and moral development. *Boletim – Academia Paulista de Psicologia, 37*(92), 94–114.
Fontoura, G. M. (2022a). Representation of minorities in video games: An analysis of characters and narratives. *Cultural Studies Notebook, 5*(3), 112–130.
Fontoura, G. M. (2022b). *Games for all: The representation of minorities in video games* [Undergraduate thesis, Federal University of Pampa]. UNIPAMPA Institutional Repository.
Frome, J. (2007). Eight ways videogames generate emotion. In *DiGRA conference proceedings* (pp. 831–835). DiGRA.
Gentile, D. A., Lynch, P. J., Linder, J. R., & Walsh, D. A. (2004). Violent video games: The effects on youth, and implications for policy and practice. *American Psychologist, 69*(1), 66–78.
Gentile, D. A., Anderson, C. A., Yukawa, S., Ihori, N., Saleem, M., Ming, L. K., Shibuya, A., Liau, A. K., Khoo, A., Bushman, B. J., Rowell Huesmann, L., & Sakamoto, A. (2009). The effects of prosocial video games on prosocial behaviors: International evidence from correlational, longitudinal, and experimental studies. *Personality and Social Psychology Bulletin, 35*(6), 752–763. https://doi.org/10.1177/0146167209333045
Granic, I., Lobel, A., & Engels, R. C. M. E. (2014). The benefits of playing video games. *American Psychologist, 69*(1), 66–78. https://doi.org/10.1037/a0034857
Greitemeyer, T., & Osswald, S. (2011). Playing video games cooperatively increases empathic concern. *Social Psychology, 42*(2), 92–98.
Halldorsson, B., Hill, C., Waite, P., Partridge, K., Freeman, D., & Creswell, C. (2021). Annual research review: Immersive virtual reality and digital applied gaming interventions for the treatment of mental health problems in children and young people: The need for rigorous treatment development and clinical evaluation. *Journal of Child Psychology and Psychiatry, 62*, 584–605. https://doi.org/10.1111/jcpp.13400
Herrera, F., Bailenson, J. N., Weisz, E., Ogle, E., & Zaki, J. (2018). Building long-term empathy: A large-scale comparison of traditional and virtual reality perspective-taking. *PLoS One, 13*(10), e0204494.
Jones, C., Scholes, L., Johnson, D., Katsikitis, M., & Carras, M. C. (2014). Gaming well: Links between videogames and flourishing mental health. *Frontiers in Psychology, 5*, 260. https://doi.org/10.3389/fpsyg.2014.00260
King, D. L., Delfabbro, P. H., & Griffiths, M. D. (2019). The convergence of gambling and gaming practices: A systematic review of the evidence. *Addictive Behaviors, 94*, 1–13.
Koepp, M., Gunn, R., Lawrence, A., et al. (1998). Evidence for striatal dopamine release during a video game. *Nature, 393*, 266–268. https://doi.org/10.1038/30498
Kowert, R., Vogelgesang, J., Festl, R., & Quandt, T. (2015). Psychosocial causes and consequences of online video game play. *Computers in Human Behavior, 45*(2015), 51–58. https://doi.org/10.1016/j.chb.2014.11.074
Kuss, D. J., & Griffiths, M. D. (2012). Internet gaming addiction: A systematic review of empirical research. *International Journal of Mental Health and Addiction, 10*(2), 278–296. https://doi.org/10.1007/s11469-011-9318-5
Lugrin, J. L., Latoschik, M. E., Habel, M., Roth, D., Seufert, C., & Grafe, S. (2021). Ethics and privacy in emotional data collection for games and virtual reality. *Computers in Entertainment, 19*(2), 1–15.
Mennecke, B. E., Triplett, J. L., Hassall, L. M., & Jordan, Z. (2010). Embodied social presence theory. In *Proceedings of the 43rd Hawaii international conference on system sciences*. IEEE.
Morgan, H., O'Donovan, A., Almeida, R., Lin, A., & Perry, Y. (2020). The role of the avatar in gaming for trans and gender diverse young people. *International Journal of Environmental Research and Public Health, 17*(22), 8617. https://doi.org/10.3390/ijerph17228617
Nieborg, D. B., & Poell, T. (2018). The platformization of cultural production: Theorizing the contingent cultural commodity. *New Media & Society, 20*(11), 4275–4292.

Oei, A. C., & Patterson, M. D. (2013). Enhancing cognition with video games: A multiple game training study. *PLoS One, 8*(3), e58546. https://doi.org/10.1371/journal.pone.0058546. Epub 2013 Mar 13. PMID: 23516504; PMCID: PMC3596277.

Panksepp, J. (2004). Affective consciousness: Core emotional feelings in animals and humans. *Consciousness and Cognition, 14*(1), 30–80.

Pichon, S., Bediou, B., Antico, L., Jack, R., Garrod, O., Sims, C., Green, C. S., Schyns, P., & Bavelier, D. (2021). Emotion perception in habitual players of action video games. *Emotion, 21*(6), 1324–1339. https://doi.org/10.1037/emo0000740

Przybylski, A. K., Weinstein, N., & Murayama, K. (2017). Internet gaming disorder: Investigating the clinical relevance of a new phenomenon. *American Journal of Psychiatry, 174*(3), 230–236.

Rivero, T. S., Querino, E. H. G., & Starling-Alves, I. (2012). Video games: Their impact on attention, perception, and executive functions. *Revista Neuropsicología Latinoamericana, 4*(3), 1–8.

Shaw, A. (2014). *Gaming at the edge: Sexuality and gender at the margins of gamer culture*. University of Minnesota Press.

Vuorre, M., Ballou, N., Hakman, T., Magnusson, K., & Przybylski, A. K. (2024). Affective uplift during video game play: A naturalistic case study. *ACM Games, 2*(3), Article 23, 14 pages. https://doi.org/10.1145/3659464

Walsh, O., Linehan, C., & Ryan, C. (2025). Is there evidence that playing games promotes social skills training for autistic children and youth? *Autism, 29*(2), 329–343. https://doi.org/10.1177/13623613241277309

Wang, G. Y., Simkute, D., & Griskova-Bulanova, I. (2023). Neurobiological link between stress and gaming: A scoping review. *Journal of Clinical Medicine, 12*(9), 3113. https://doi.org/10.3390/jcm12093113. PMID: 37176554; PMCID: PMC10179187.

Xiao, L. Y. (2021). Loot boxes: Regulating a digital gambling phenomenon. *International Journal of Mental Health and Addiction, 19*(1), 110–122.

Yee, N., & Bailenson, J. N. (2007). The proteus effect: The effect of transformed self-representation on behavior. *Human Communication Research, 33*(3), 271–290.

Chapter 10
Gaming and Emotional Regulation: Coping Mechanisms and Strategies

Ana Abutara, Douglas Amaral, Jordan Vieira, Natália Rocha, Melina Houlis, Pedro Makoto Uchida, and Lucas Murrins Marques

Abstract Video games have emerged as complex emotional ecosystems where players navigate a broad spectrum of experiences, offering both opportunities for emotional regulation and risks for emotional dysregulation. This chapter explores how gaming can foster adaptive coping mechanisms, such as resilience, problem-solving, and emotional flexibility, while also highlighting the dangers of maladaptive patterns like escapism and Internet Gaming Disorder (IGD). The chapter examines the influence of different game genres, gender differences in coping strategies, and the therapeutic potential of gaming for emotional health. Design features that support emotional resilience, such as adaptive difficulty, narrative engagement, and positive reinforcement, are discussed alongside risks related to excessive gaming and online toxicity. Recommendations are provided for players and developers to maximize gaming's benefits for emotional regulation while minimizing its potential harm. Ultimately, gaming, when approached mindfully, can serve as a powerful tool for emotional growth, resilience, and psychological well-being.

Keywords Emotional regulation · Coping strategies · Internet gaming disorder · Gaming and mental health · Therapeutic gaming

A. Abutara · J. Vieira · N. Rocha · M. Houlis
Mental Health Department, Santa Casa de São Paulo School of Medical Sciences, São Paulo, Brazil

D. Amaral
Mackenzie Presbyterian University, São Paulo, Brazil

P. M. Uchida
Pontifícia Universidade Católica de Campinas, Campinas, São Paulo, Brazil

L. M. Marques (✉)
Mental Health Department, Santa Casa de São Paulo School of Medical Sciences, São Paulo, Brazil

Artificial Intelligence Research Division, Infinity Doctor's Inc., Dover, DE, USA

L. Murrins Marques et al. (eds.), *Social and Affective Neuroscience of Gaming*, https://doi.org/10.1007/978-3-032-00325-6_10

Introduction

Emotional regulation refers to the ability to influence and manage one's emotional responses to various stimuli, a critical process for maintaining mental health and social well-being (Gross, 2015). This process encompasses coping strategies that allow individuals to modulate emotions at different stages—by directing attention, reinterpreting events, or modifying the emotional response itself. Effective emotional regulation is essential for navigating life's challenges, enabling individuals to remain resilient in the face of stress, frustration, or adversity. But with the rise of the digital world, particularly gaming, we must ask: how do games impact our ability to regulate emotions?

Video games, especially online ones, are designed to immerse players in complex, interactive environments where they must navigate challenges, solve problems, and interact with others. These dynamics naturally evoke a spectrum of emotions, from excitement and joy to frustration and disappointment. Given the emotionally engaging nature of gaming, it raises an important question: Does playing games influence our capacity to regulate emotions in everyday life? Can skills developed in managing in-game challenges—like controlling impulses or handling failures—translate into improved emotional regulation beyond the gaming context?

Additionally, certain gaming behaviors, such as rage-quitting or excessive frustration, may reinforce maladaptive emotional responses, making it harder for players to regulate their emotions outside of the gaming environment. Could frequent exposure to these negative emotional cycles in gaming impair emotional resilience and contribute to issues like anxiety or emotional dysregulation?

This chapter will explore these critical questions, examining the potential risks and benefits that gaming might pose to emotional regulation. We will consider both the therapeutic potential of gaming in fostering emotional resilience and coping mechanisms, as well as the dangers of maladaptive gaming behaviors that can lead to emotional dysregulation.

The Role of Gaming in Emotional Regulation

Emotional regulation is essential for psychological stability and social well-being, as it allows individuals to manage their responses to a variety of life challenges. In the context of gaming, the immersive and interactive qualities inherent to video games create an environment where emotions are continuously engaged and tested. Games can serve as a medium through which players not only encounter diverse emotional experiences but also practice strategies to manage those emotions. This raises important questions about how gaming might influence emotional regulation. By examining both the ways games encourage emotional engagement and their mixed emotional outcomes, we gain insight into the dual potential of gaming as a tool for positive growth and as a source of risk.

Emotional Engagement in Gaming

Emotional engagement is a powerful motivator in gaming, driving players to continue playing and invest more deeply in the game experience. Studies, such as the systematic review by Johnson et al. (2016), highlight that gamification—the application of game design elements in non-gaming contexts—can enhance health and well-being by tapping into intrinsic motivation. This type of motivation, fueled by the enjoyment and engagement present in well-designed games, is considered more sustainable for long-term behavior change than external rewards (Johnson et al., 2016; Ryan & Deci, 2000). In fact, gamification thrives on creating intrinsically motivating interventions, as these are more likely to yield enduring improvements in behavior and well-being.

The appeal of gamification, especially within health interventions, lies in its ability to integrate elements of game design that inherently support motivation (Deterding et al., 2011). Features like immediate feedback, progress indicators, goal-setting, challenges, social interaction, and customization contribute to an engaging and enjoyable gameplay experience. Johnson et al. (2016) identify numerous studies that show the positive effects of gamified interventions on health-related behaviors, particularly in areas such as physical activity, diet, medication adherence, and mental well-being. Nonetheless, their review also reveals that outcomes can be mixed or neutral, underscoring the importance of tailoring game elements to specific user needs and contexts for optimal effectiveness.

One critical factor in effective gamification is balancing challenge and reward. While challenges are necessary to sustain engagement, they should not be overwhelming, as excessive difficulty can lead to frustration and disengagement (Deterding et al., 2011). The appropriate level of challenge varies depending on the player's skills, prior experience, and personal goals, which suggests that adaptive and personalized gamified interventions are most effective (Johnson et al., 2016). Successful studies often incorporate adaptive difficulty levels or provide multiple levels of challenge to accommodate a range of skill levels, leading to more consistently positive results.

Moreover, social interaction and community engagement are significant components of emotional engagement in gaming. Many games foster a sense of belonging, competition, and cooperation through collaborative gameplay and competitive leaderboards (Johnson et al., 2016). Such features are especially beneficial in health-related interventions, where users can support one another in achieving shared goals. However, research also indicates that some gamified elements, like rewards, may be unsuitable in contexts requiring mindfulness and self-reflection (Ahtinen et al., 2013). Ultimately, the effectiveness of gamified interventions depends on carefully designing these elements to meet the motivational and psychological needs of users, with future studies ideally employing consistent methodologies and comprehensive measures to refine our understanding of gamification's impact across diverse contexts.

Positive and Negative Emotional Outcomes

While gaming can provide an avenue for emotional regulation, its effects on emotional well-being are not uniformly positive. Recent research indicates that games can have both beneficial and detrimental impacts on emotional health, depending on factors such as an individual's pre-existing emotional state, the type of game, and the depth of engagement (Lin et al., 2023).

One of the potential emotional benefits of gaming is its ability to cultivate a sense of control and mastery. Many games require players to develop strategies, overcome challenges, and make decisions, which can foster a sense of agency and boost self-esteem, potentially reducing feelings of helplessness. For individuals with anxiety or depression, achieving goals within a game can provide relief and a sense of accomplishment, making gaming a valuable outlet for positive reinforcement.

However, the very characteristics that make games engaging can also exacerbate emotional challenges. Some individuals with emotional dysregulation may turn to gaming as an escape from negative emotions or as a method of self-soothing (Guerrini-Usubini et al., 2023). This can lead to excessive gaming, where players neglect essential aspects of their lives, ultimately intensifying feelings of isolation, loneliness, and depression. Such patterns of behavior can contribute to Internet Gaming Disorder (IGD), characterized by excessive and uncontrolled gaming that impairs daily functioning. Lin et al. (2023) found that emotional dysregulation positively predicted IGD through the mediating effects of negative affect and metacognitions about online gaming, accounting for a significant proportion of IGD cases.

Additionally, certain games may expose players to violent or distressing content, which can negatively affect emotional health, particularly for those with trauma histories or mental health challenges. Exposure to intense, potentially re-traumatizing content may worsen symptoms and disrupt emotional stability. As a result, while gaming holds promise for enhancing emotional regulation, it is crucial to be aware of these risks. Assessing a player's emotional vulnerabilities, understanding game mechanics, and recognizing potential adverse effects are essential for maximizing the emotional benefits of gaming while minimizing its drawbacks.

Coping Mechanisms in Gaming

Gaming environments provide players with a range of scenarios that elicit both positive and negative emotions. As players navigate these experiences, they naturally develop coping mechanisms to manage the associated challenges, stressors, and frustrations. The strategies players adopt can help regulate their emotions within the game and, at times, may even translate to improved coping outside the gaming context. However, while some strategies encourage adaptive emotional regulation, others can reinforce maladaptive behaviors, leading to potential psychological and emotional challenges.

Types of Coping Strategies in Gaming

Gaming offers a unique environment where players encounter a wide range of emotional and social scenarios, from intense competition to collaborative problem-solving. Through these experiences, players naturally develop various coping strategies that enable them to handle in-game challenges and, in some cases, translate these skills into real-life emotional regulation (Granic et al., 2014). Games present situations requiring resilience, strategic thinking, and adaptability, which are essential skills for coping with stress and adversity (Kühn et al., 2019). The interactive nature of gaming, especially in online settings, also provides opportunities for social engagement, which can serve as a valuable outlet for emotional expression and stress relief (Cole & Griffiths, 2007).

In many games, players are immersed in socially dynamic settings where they engage with others through teamwork, communication, and competition. These interactions encourage the development of social coping strategies, such as building alliances, resolving conflicts, and navigating interpersonal dynamics (De Freitas & Griffiths, 2008). For example, multiplayer games often require cooperative strategies and clear communication, which foster skills in negotiation and collaboration that can enhance emotional intelligence and resilience (Cole & Griffiths, 2007).

On a cognitive and behavioral level, games that involve problem-solving and decision-making encourage players to manage frustration, maintain focus, and exercise patience (Boot et al., 2008). This process can lead to a sense of accomplishment, as players learn to overcome obstacles and adapt to new challenges, which in turn can improve self-esteem, elevate mood, and promote emotional balance (Ryan et al., 2006). The positive reinforcement loop that well-designed games offer contributes to intrinsic motivation, making gaming a powerful medium for building positive coping mechanisms that players may also apply beyond the virtual world.

Moreover, gaming can foster creativity and flexible thinking, as players often need to think outside the box to succeed. Whether navigating complex puzzles or strategizing in fast-paced scenarios, players develop cognitive skills that help them manage stress by reframing challenges as opportunities for growth (Adachi & Willoughby, 2013). These adaptive strategies enhance players' capacity to handle stress, frustration, and setbacks both in and out of the game, offering valuable tools for emotional regulation.

However, while gaming can nurture adaptive coping strategies, it also has the potential to reinforce maladaptive patterns. For some players, the boundary between using games as a form of stress relief and as a means of escapism can become blurred. Games that allow players to evade real-world responsibilities or avoid challenges may inadvertently foster dependence on gaming as an avoidance mechanism (Kardefelt-Winther, 2014). This dual potential of gaming highlights the importance of a balanced approach, where players can maximize the positive aspects of gaming while remaining vigilant against the risks of maladaptive behaviors.

Adaptive Versus Maladaptive Coping

Adaptive coping strategies in gaming include mechanisms that promote self-regulation, helping players manage stressors effectively. By practicing self-control and resilience, players may develop skills that carry over to real-life situations, using games as tools for cognitive and behavioral growth. Yet, the escapism that gaming can offer poses a significant risk. For some, the immersive nature of games becomes a means to avoid real-life challenges and suppress negative emotions, leading to an over-reliance on gaming as a coping mechanism. This form of avoidance can evolve into maladaptive coping, where gaming becomes less about skill-building and more about escape, which may contribute to problematic patterns, such as Internet Gaming Disorder (IGD), as outlined in the DSM-V. IGD, often associated with mood disorders, is characterized by excessive gaming that disrupts personal and social functioning.

Maladaptive coping in gaming typically involves unhealthy responses to adversity, which may exacerbate issues like depression and anxiety. Examples include prolonged gaming sessions that interfere with daily routines, self-blame when performance expectations are unmet, and an inability to collaborate effectively with others. Additionally, some players may experience emotional numbing and escape from stressful realities by fully immersing themselves in the virtual world. Such behaviors can create a vicious cycle of stress and dependency, where gaming serves as both the refuge and the cause of emotional turmoil.

There is a fine line between engaging with games and developing a dependency on them. A study conducted in Australia in 2022 examined 403 young gamers (Slack et al., 2022), identifying key differences between those with healthy gaming habits and those who exhibited signs of gaming disorder. The findings revealed that players who used gaming as an escape from real-life adversities often displayed symptoms of depression and anxiety, with escapism and emotional numbing emerging as the primary maladaptive strategies associated with psychological distress. This study highlights the importance of recognizing when gaming shifts from a source of enjoyment to a mechanism for avoidance.

Coping Strategies Across Different Genres

Coping strategies in gaming are also influenced by gender differences, with distinct patterns observed between male and female players in their emotional responses and regulation strategies. Males, e.g., often exhibit higher impulsivity, aggressive behaviors, and a stronger inclination toward reward-driven mechanisms, which can increase dependency risks. Females, on the other hand, may experience feelings of loneliness and negative mood states that predispose them to affective disorders, potentially making recovery from IGD more challenging (Dong & Potenza, 2022).

A recent neuroimaging study from China (2024) investigated the neural distinctions between males and females with IGD, using MRI scans to compare groups of healthy and IGD-affected individuals. The study found significant differences in brain regions associated with circadian rhythms and emotional processing, particularly in areas responsible for impulsivity and aggression in males. Men with IGD exhibited higher overall brain activity in these areas, suggesting a neurobiological basis for dysregulated behavior. Conversely, females with IGD showed reduced neural connectivity, particularly in areas linked to attention regulation, with a noted overlap in patterns associated with ADHD. These findings underscore how gender differences shape coping mechanisms and highlight the need for tailored interventions to address the distinct ways male and female players handle stress, attention, and decision-making in gaming contexts.

In sum, the diversity of coping mechanisms in gaming illustrates both the potential benefits and risks of using games as a medium for emotional regulation. While some strategies encourage adaptive behaviors that foster resilience and self-regulation, others may lead to dependency and emotional dysregulation, particularly when gaming is used as a means of escape. Understanding these mechanisms, including the influence of gender, can provide a foundation for promoting healthier gaming habits and supporting emotional well-being.

Gaming as a Tool for Emotional Self-Regulation

Emotional regulation is the ability to manage and respond effectively to emotional experiences, allowing individuals to navigate complex feelings in constructive ways (Toh & Kirschner, 2023). This skill involves recognizing emotions, reflecting on them, and employing strategies to modulate their intensity and impact. In daily life, individuals often adapt their coping strategies based on the specific challenges they face, and research has shown that video games can serve as a valuable medium for practicing these skills. Games provide immersive environments where players can engage in a variety of scenarios, which may help them manage emotions such as anxiety and frustration, ultimately contributing to enhanced emotional resilience and well-being (Villani et al., 2018).

Therapeutic Benefits of Gaming

According to Self-Determination Theory (SDT), developed by Deci and Ryan, individuals have three core psychological needs—competence, autonomy, and relatedness—that, when satisfied, intrinsically motivate them to engage in activities for their own enjoyment rather than external rewards (Ryan & Deci, 2000). Video games can effectively meet these needs within their diverse environments. The first need, competence, is fulfilled as players learn and improve their skills progressively within the game. Autonomy is fostered by allowing players to control their in-game actions and choices, creating a sense of ownership and control over their experiences. Finally, relatedness is satisfied as players interact with others, whether through in-game characters or fellow players, helping them feel connected and fostering a sense of belonging. Integrating SDT principles into game design may thus contribute to an environment that supports emotional well-being, making games a viable tool for mental health interventions.

Through immersion in virtual worlds, players also practice coping with negative emotions such as stress and anxiety. Achieving goals within a game provides a positive sense of accomplishment, reinforcing confidence and motivation (Sarsenbayeva et al., 2020). This experience can be extended into real life, as players may begin to seek similar feelings of reward, approaching daily challenges with increased resilience and optimism. Conversely, experiencing losses or setbacks within a game can create temporary frustration, but games provide a safe environment for individuals to process these emotions. Players can learn from their mistakes, express frustrations, and try again without real-world consequences or judgment, which builds emotional resilience that may translate to everyday life.

A 2024 meta-analysis by researchers at the School of Psychology in Hubei, China, demonstrated the effectiveness of video games in improving mental health across various disorders (Zhan et al., 2024). Focusing on "serious games"—games explicitly designed for cognitive and emotional benefits—the study found that these games provide interactive experiences that help players develop new skills and manage emotions more effectively. The study also highlights the accessibility and reduced stigma associated with gaming as a mental health tool, compared to traditional therapies.

Further research from Finland and the United Kingdom supports these findings, showing that video games can aid mental health by improving adherence to treatments and making interventions more appealing (Lukka & Palva, 2023). Games have been found to alleviate symptoms of depression and anxiety, as well as to offer cognitive training, physical activity, and relaxation, all of which can enhance mood. Additionally, social interaction within games may reduce feelings of isolation, while interactive narratives promote emotional regulation. As such, video games offer potential as a therapeutic intervention, fostering personal growth and well-being.

When examining specific age groups, studies show that games effectively train cognitive skills in adults and the elderly. For instance, Anguera et al. (2017) found that video games improve cognitive control in older adults, which has implications for mental health in older populations. Likewise, research on adults with schizophrenia indicated that digital games could enhance cognitive function, making them promising tools in the treatment of severe mental health disorders (Kimhy et al., 2015). In children and teenagers, video games have shown particular benefits for those with Autism Spectrum Disorder (ASD), offering greater engagement and improved social interaction compared to traditional therapies (Beaumont et al., 2021). Another study conducted in Pittsburgh highlighted that video-game-based interventions were more widely accepted among children and teenagers with anxiety than non-game approaches (Pramana et al., 2018).

Game Design Elements Supporting Emotional Regulation

Video games incorporate specific design elements that can support and strengthen players' ability to regulate their emotions. These elements—such as game control, reward systems, and story involvement—provide structured environments that encourage players to confront, manage, and learn from emotional experiences within the safety of a virtual world. By repeatedly encountering and responding to challenges and setbacks, players can develop resilience and effective coping skills that extend into their daily lives.

A key component of this emotional regulation lies in game control features. Video games often allow players to save progress, retry levels, and experiment with different approaches to overcome obstacles. This sense of control gives players the freedom to correct mistakes without fear of real-world consequences, which can be a powerful tool for learning to manage frustration and disappointment. By facing challenges within the structured parameters of a game, players are encouraged to reevaluate their strategies and emotions, fostering a mindset that turns setbacks into learning experiences (Villani et al., 2018). This aspect of game control also helps players build resilience, as they learn to cope with frustration by confronting obstacles repeatedly until they succeed. This practice not only strengthens emotional endurance but also fosters a problem-solving mindset that is transferable to real-world situations.

Another important element is the reward system, which recognizes and reinforces player achievements, regardless of their size. By delivering positive feedback through rewards—such as points, achievements, or visual cues—games promote a cycle of motivation and engagement. Studies indicate that when players are rewarded for persistence and progress, even in the face of failure, they are more likely to continue engaging with the game and improving their performance (Granic et al., 2014). This cycle of positive reinforcement encourages players to focus on progress rather than perfection, cultivating an attitude of resilience and perseverance. The reward system in games thus helps individuals develop a constructive approach to setbacks, as they learn to appreciate incremental progress. This mindset is valuable in real-life contexts, where sustained effort and incremental growth are often more effective than immediate success (Granic et al., 2014).

Additionally, story involvement or narrative engagement in video games can play a substantial role in emotional self-regulation. Many games feature intricate storylines where players form connections with characters or become deeply invested in the plot, allowing them to explore and express a range of emotions. Role-playing games (RPGs), for example, let players create unique avatars and make choices that affect the storyline, enabling them to experience scenarios from different perspectives (Villani et al., 2018). This level of engagement can lead to emotional growth, as players navigate complex situations and reflect on their emotional responses. The immersion in a game's narrative allows players to experiment with different identities and explore emotional reactions in a controlled setting, which can foster empathy, self-awareness, and emotional regulation. Moreover, collaborative and social aspects of such games encourage interpersonal skills, as players work together or communicate to achieve shared goals, enhancing their social resilience and adaptability.

Research suggests that these design elements can make video games powerful tools for emotional self-regulation by allowing players to face, process, and adapt to emotions in a safe and structured environment. A well-designed game balances these elements to encourage emotional growth and resilience, turning gameplay into an experience that promotes well-being beyond entertainment.

Challenges and Risks: When Gaming Becomes a Maladaptive Coping Mechanism

While gaming can serve as an effective tool for emotional regulation, it also carries potential risks, particularly when used as an escape from real-life challenges. When gaming becomes a primary coping strategy for managing negative emotions, it may foster maladaptive behaviors, leading to issues such as addiction, emotional dysregulation, and social isolation. These risks are particularly pronounced among individuals who struggle to regulate their emotions outside of the gaming environment, as they may rely on gaming to avoid confronting real-world stressors. This section will explore the potential dangers associated with excessive gaming and how maladaptive coping mechanisms can contribute to mental health challenges.

Gaming Addiction and Emotional Dysregulation

Internet gaming disorder (IGD), or problematic gaming in the virtual realm, is characterized by excessive involvement in online games, causing significant distress and dysfunction in various areas of life, much like other addictive disorders (American Psychiatric Association, 2022; Ko & Yen, 2024). IGD has been linked to mood disorders, such as depression and anxiety, with research demonstrating a clear association between problematic gaming and emotional distress across different age groups and populations (González-Bueso et al., 2018; Ostinelli et al., 2021).

Studies have shown that adolescents and young adults who engage in problematic gaming behaviors often experience elevated levels of anxiety, depression, and overall emotional instability. Excessive gaming has been associated with lower scores in physical health, psychological well-being, social relationships, and environmental quality—four key areas evaluated in quality-of-life assessments

(González-Bueso et al., 2018). These findings indicate that problematic gaming can significantly impact both mental and physical health, particularly among vulnerable populations who may lack alternative coping strategies.

According to Loton et al. (2016), the relationship between gaming addiction and impaired mental health is, in part, explained by maladaptive coping strategies, such as withdrawal and resignation. Marques et al. (2023) also found that escapism as a coping mechanism to the real world is an important predictor of IGD. The motivation to play as a characteristic to identify IGD seems to be essential to distinguish between high engagement in gaming and actual gaming addiction. While high engagement may involve extended play, it does not inherently lead to mental health impairment, whereas gaming addiction tends to correlate more directly with emotional and social dysfunction. Consequently, monitoring gaming habits and implementing preventive strategies are crucial, especially for adolescents and individuals with pre-existing health conditions who may be at greater risk.

Moreover, excessive gaming as an escape mechanism often exacerbates difficulties in emotional regulation. Research suggests that individuals with IGD tend to suppress emotions rather than employ healthier cognitive reappraisal strategies, which could help them process negative emotions more constructively (Yen et al., 2017). This reliance on escapism as a coping mechanism can lead to increased emotional dysregulation, as gaming becomes an ineffective tool for managing complex emotional states. Maladaptive strategies like rumination, self-blame, and catastrophizing have also been positively associated with problematic gaming, with studies indicating that rumination, in particular, plays a significant role among male gamers (Estupiñá et al., 2024). Positive reappraisal, on the other hand, has been shown to act as a protective factor, highlighting the need to promote healthier coping strategies in the context of gaming (Kökönyei et al., 2019). Emotional dysregulation, driven by emotional suppression and a lack of skills to manage emotions, remains a key factor in problematic gaming behaviors (Blasi et al., 2019).

Online Interactions and Emotional Outcomes

The social dynamics of online gaming environments can have both positive and negative effects on players' emotional well-being. While online games provide platforms for social connection, they can also foster toxic behavior that impacts mental health and contributes to emotional stress. Massively Multiplayer Online Role-Playing Games (MMORPGs) are known to facilitate strong friendships and emotional bonds, allowing players to express themselves and find support in ways they may not experience in offline life. These virtual interactions can offer emotional support, particularly for individuals who lack social connections or face depression, and may even improve social well-being for those who experience social isolation in real life (Cole & Griffiths, 2007).

However, the same online interactions can also contribute to feelings of dependency, making it difficult for some players to regulate their gaming time effectively.

Self-regulation is crucial in online gaming environments to avoid negative outcomes associated with over-reliance on these social interactions (Seay & Kraut, 2007). Recognizing the potential for harmful interactions, researchers have examined in-game support tools designed to reduce toxicity. Such tools not only penalize negative behavior but also offer emotional support for players affected by toxicity, which can help decrease stress and foster a more positive gaming environment (Reid et al., 2022).

Additionally, there are screening tools available to identify individuals with IGD, such as The Gaming Disorder Identification Test (GADIT), which can assist in detecting early signs of problematic gaming. Identifying at-risk individuals is essential for enabling targeted interventions that support healthier gaming habits and prevent the escalation of maladaptive behaviors (Chan et al., 2024). However, few studies have reported the consistent use of these screening tools, underlining the need for further research and implementation to better support mental health in online gaming communities.

Strategies for Promoting Healthy Emotional Regulation Through Gaming

Promoting healthy emotional regulation through gaming involves both player awareness and intentional design by game developers. By choosing and engaging with games that foster positive emotional experiences, players can cultivate healthier habits and strategies for managing emotions. Simultaneously, game developers have a significant role in designing games that encourage emotional well-being, social interaction, and personal growth. This section outlines strategies aimed at supporting players and game developers in maximizing the emotional benefits of gaming while minimizing potential risks.

Recommendations for Players

For players seeking to improve emotional regulation through gaming, selecting games that promote relaxation and positive emotions can be beneficial. Games that encourage problem-solving and create a sense of achievement—such as Portal 2 or Angry Birds—are particularly effective for boosting mood and alleviating stress (Granic et al., 2014). These types of games help players manage negative emotions like anxiety and frustration, providing an emotionally supportive experience. Additionally, many games require players to adapt to new challenges, which can enhance cognitive flexibility and resilience. Strategy games and role-playing games (RPGs), for instance, encourage players to develop cognitive reappraisal skills, a valuable strategy for emotional regulation that can help players reinterpret difficult situations constructively in real life (Granic et al., 2014).

Choosing games with positive social interactions is another effective approach to fostering healthy emotional regulation. Games that facilitate cooperative or supportive interactions, such as collaborative multiplayer games, can help players develop emotional bonds and access social support, both of which are essential for emotional well-being (Johannes et al., 2021). Participating in gaming communities that encourage communication and teamwork can reduce feelings of loneliness or isolation, promoting a greater sense of connection and social belonging. For players, these interactions can also provide opportunities to practice empathy and improve interpersonal skills within a virtual environment, potentially translating to enhanced social resilience outside of the game.

Balancing gameplay with other life activities is essential for maintaining both physical and mental health. To prevent excessive screen time and potential Internet addiction, players are encouraged to set boundaries on their gaming hours and engage in other fulfilling activities, such as physical exercise, social gatherings, or hobbies (Fineberg et al., 2022). By monitoring gameplay time and establishing limits, players can enjoy the emotional benefits of gaming without compromising other aspects of their well-being.

Lastly, players can benefit from choosing games that promote personal growth and self-discovery. Games that emphasize overcoming challenges and building skills can create enriching emotional experiences, such as the feeling of "fiero"—intense pride after successfully overcoming a difficult challenge—or achieving a state of "flow," where the player becomes fully immersed in an activity (Granic et al., 2014). These emotionally rewarding experiences contribute to a sense of accomplishment, enhancing players' self-efficacy and resilience as they apply these skills to real-world challenges.

Recommendations for Game Developers

Game developers play an essential role in promoting healthy emotional regulation through thoughtfully designed games. By focusing on elements that support physical and emotional well-being, developers can create experiences that encourage positive emotional engagement and resilience. One key approach is to design games that incorporate physical movement and social interaction, as these features have been shown to support both physical and mental health (Granic et al., 2014). Games that promote physical activities, like fitness-based games, not only improve physical health but also help players manage stress, contributing to overall emotional well-being.

The use of gamification techniques—such as rewards, challenges, and competitive elements—is another powerful tool for promoting healthy behaviors. By incorporating these elements, developers can encourage players to pursue positive activities and maintain motivation (Johannes et al., 2021). Reward systems and achievements create a sense of accomplishment and emotional satisfaction, reinforcing players' intrinsic motivation to engage in healthy gaming practices. These elements foster resilience by creating a structure in which players receive positive feedback for effort and progress.

Developing engaging narratives is also crucial for supporting players' mental health. Serious games aimed at emotional and psychological well-being should feature meaningful storylines and relatable characters, enabling players to explore and understand their emotions in a safe environment (Granic et al., 2014). When players identify with in-game characters, they may become more reflective about their emotional experiences, helping them to build empathy, cope with stress, and apply these insights to real-life situations.

Customization and adaptability are further strategies for enhancing emotional engagement. Allowing players to personalize their gameplay experience based on individual preferences and needs can create positive emotional responses and sustain interest in the game. This adaptability can be particularly beneficial in serious games or therapeutic settings, where tailoring the experience to meet individual needs can promote both engagement and emotional regulation.

Finally, fostering self-reflection within the gaming experience can contribute positively to emotional health. Developers are encouraged to design game mechanics that promote introspection, prompting players to consider their emotional responses and reactions throughout gameplay. Games that facilitate self-reflection help players develop greater self-awareness and emotional resilience, empowering them to understand and regulate their emotions more effectively.

Conclusion

This chapter has examined the dual role of gaming in emotional regulation, highlighting its potential to foster resilience, adaptive coping skills, and well-being. When approached thoughtfully, gaming can serve as a valuable tool for emotional growth, meeting players' psychological needs for autonomy, competence, and relatedness. Studies show that well-designed games encourage problem-solving, social engagement, and emotional reflection, which can enhance emotional intelligence across various age groups and backgrounds. However, gaming's benefits come with risks. For some, excessive gaming can become a maladaptive escape, contributing to emotional dysregulation, anxiety, and depression, particularly in cases of Internet Gaming Disorder (IGD).

Balancing gaming's advantages with its potential pitfalls requires active involvement from both players and developers. Players can cultivate a healthier gaming experience by selecting games that promote relaxation and positive emotions, setting time limits, and balancing gameplay with other activities. For developers, incorporating meaningful narratives, adaptive mechanics, and tools for self-reflection can enhance the emotional benefits of gaming while minimizing risks.

Ultimately, when gaming is used mindfully, it has the potential to support emotional resilience and contribute to personal growth in ways that extend beyond the screen.

Acknowledgements AA is support by São Paulo Research Foundation (FAPESP, #24/09587-9). JV is support by São Paulo Research Foundation (FAPESP, #24/09578-0). MH is support by São Paulo Research Foundation (FAPESP, #24/22303-0).

References

Adachi, P. J. C., & Willoughby, T. (2013). More than just fun and games: The longitudinal relationships between strategic video games, self-reported problem solving skills, and academic grades. *Journal of Youth and Adolescence, 42*(7), 1041–1052. https://doi.org/10.1007/S10964-013-9913-9

Ahtinen, A., Mattila, E., Välkkynen, P., Kaipainen, K., Vanhala, T., Ermes, M., Sairanen, E., Myllymäki, T., & Lappalainen, R. (2013). Mobile mental wellness training for stress management: Feasibility and design implications based on a one-month field study. *JMIR mHealth and uHealth, 1*(2), e11. https://doi.org/10.2196/MHEALTH.2596

American Psychiatric Association. (2022). *Diagnostic and statistical manual of mental disorders*. APA. https://doi.org/10.1176/APPI.BOOKS.9780890425787

Anguera, J. A., Gunning, F. M., & Areán, P. A. (2017). Improving late life depression and cognitive control through the use of therapeutic video game technology: A proof-of-concept randomized trial. *Depression and Anxiety, 34*(6), 508–517. https://doi.org/10.1002/DA.22588

Beaumont, R., Walker, H., Weiss, J., & Sofronoff, K. (2021). Randomized controlled trial of a video gaming-based social skills program for children on the autism spectrum. *Journal of Autism and Developmental Disorders, 51*(10), 3637–3650. https://doi.org/10.1007/s10803-020-04801-z

Blasi, M. D., Giardina, A., Giordano, C., Coco, G. L., Tosto, C., Billieux, J., & Schimmenti, A. (2019). Problematic video game use as an emotional coping strategy: Evidence from a sample of MMORPG gamers. *Journal of Behavioral Addictions, 8*(1), 25–34. https://doi.org/10.1556/2006.8.2019.02

Boot, W. R., Kramer, A. F., Simons, D. J., Fabiani, M., & Gratton, G. (2008). The effects of video game playing on attention, memory, and executive control. *Acta Psychologica, 129*(3), 387–398. https://doi.org/10.1016/J.ACTPSY.2008.09.005

Chan, G. C. K., Saunders, J. B., Stjepanović, D., Mcclure-Thomas, C., Connor, J., Hides, L., Wood, A., King, D., Siste, K., Long, J., & Leung, J. K. (2024). The gaming disorder identification test (GADIT)—A screening tool for gaming disorder based on ICD-11. *Journal of Behavioral Addictions, 13*(3), 729. https://doi.org/10.1556/2006.2024.00038

Cole, H., & Griffiths, M. D. (2007). Social interactions in massively multiplayer online role-playing gamers. *CyberPsychology and Behavior, 10*(4), 575–583. https://doi.org/10.1089/CPB.2007.9988

De Freitas, S., & Griffiths, M. (2008). The convergence of gaming practices with other media forms: What potential for learning? A review of the literature. *Learning, Media and Technology, 33*(1), 11–20. https://doi.org/10.1080/17439880701868796

Deterding, S., Khaled, R., Nacke, L., & Dixon, D. (2011). Gamification: Toward a definition. In *Proceedings of the 2011 annual conference extended abstracts on human factors in computing systems* (pp. 1–5). ACM.

Dong, G. H., & Potenza, M. N. (2022). Considering gender differences in the study and treatment of internet gaming disorder. *Journal of Psychiatric Research, 153*, 25–29. https://doi.org/10.1016/J.JPSYCHIRES.2022.06.057

Estupiñá, F. J., Bernaldo-de-Quirós, M., Vallejo-Achón, M., Fernández-Arias, I., & Labrador, F. (2024). Emotional regulation in gaming disorder: A systematic review. *The American Journal on Addictions, 33*(6), 605–620. https://doi.org/10.1111/AJAD.13621

Fineberg, N. A., Menchón, J. M., Hall, N., Dell'Osso, B., Brand, M., Potenza, M. N., Chamberlain, S. R., Cirnigliaro, G., Lochner, C., Billieux, J., Demetrovics, Z., Rumpf, H. J., Müller, A., Castro-Calvo, J., Hollander, E., Burkauskas, J., Grünblatt, E., Walitza, S., Corazza, O., et al. (2022). Advances in problematic usage of the internet research – A narrative review by experts from the European network for problematic usage of the internet. *Comprehensive Psychiatry, 118*, 152346. https://doi.org/10.1016/J.COMPPSYCH.2022.152346

González-Bueso, V., Santamaría, J. J., Fernández, D., Merino, L., Montero, E., & Ribas, J. (2018). Association between internet gaming disorder or pathological video-game use and comorbid psychopathology: A comprehensive review. *International Journal of Environmental Research and Public Health, 15*(4), 668. https://doi.org/10.3390/IJERPH15040668

Granic, I., Lobel, A., & Engels, R. C. M. E. (2014). The benefits of playing video games. *The American Psychologist, 69*(1), 66–78. https://doi.org/10.1037/a0034857

Gross, J. J. (2015). Emotion regulation: Current status and future prospects. *Psychological Inquiry, 26*(1), 1–26. https://doi.org/10.1080/1047840X.2014.940781

Guerrini-Usubini, A., Cattivelli, R., Scarpa, A., Musetti, A., Varallo, G., Franceschini, C., & Castelnuovo, G. (2023). The interplay between emotion dysregulation, psychological distress, emotional eating, and weight status: A path model. *International Journal of Clinical and Health Psychology, 23*(1), 100338. https://doi.org/10.1016/J.IJCHP.2022.100338

Johannes, N., Vuorre, M., & Przybylski, A. K. (2021). Video game play is positively correlated with well-being. *Royal Society Open Science, 8*(2), 202049. https://doi.org/10.1098/RSOS.202049

Johnson, D., Deterding, S., Kuhn, K.-A., Staneva, A., Stoyanov, S., & Hides, L. (2016). Gamification for health and wellbeing: A systematic review of the literature. *Internet Interventions, 6*, 89–106. https://doi.org/10.1016/j.invent.2016.10.002

Kardefelt-Winther, D. (2014). A conceptual and methodological critique of internet addiction research: Towards a model of compensatory internet use. *Computers in Human Behavior, 31*(1), 351–354. https://doi.org/10.1016/J.CHB.2013.10.059

Kimhy, D., Vakhrusheva, J., Bartels, M. N., Armstrong, H. F., Ballon, J. S., Khan, S., Chang, R. W., Hansen, M. C., Ayanruoh, L., Lister, A., Castrén, E., Smith, E. E., & Sloan, R. P. (2015). The impact of aerobic exercise on brain-derived neurotrophic factor and neurocognition in individuals with schizophrenia: A single-blind, randomized clinical trial. *Schizophrenia Bulletin, 41*(4), 859–868. https://doi.org/10.1093/SCHBUL/SBV022

Ko, C.-H., & Yen, J.-Y. (2024). Internet gaming disorder. In *Tasman's psychiatry* (pp. 2949–2969). Springer. https://doi.org/10.1007/978-3-030-51366-5_164

Kökönyei, G., Kocsel, N., Király, O., Griffiths, M. D., Galambos, A., Magi, A., Paksi, B., & Demetrovics, Z. (2019). The role of cognitive emotion regulation strategies in problem gaming among adolescents: A nationally representative survey study. *Frontiers in Psychiatry, 10*, 430621. https://doi.org/10.3389/FPSYT.2019.00273/BIBTEX

Kühn, S., Gallinat, J., & Mascherek, A. (2019). Effects of computer gaming on cognition, brain structure, and function: A critical reflection on existing literature. *Dialogues in Clinical Neuroscience, 21*(3), 319. https://doi.org/10.31887/DCNS.2019.21.3/SKUEHN

Lin, S., Tan, L., Chen, X., Liao, Z., Li, Y., Tang, Y., Shi, Y., Hao, J., Wang, X., Huang, Q., & Shen, H. (2023). Emotion dysregulation and internet gaming disorder in young people: Mediating effects of negative affect and metacognitions. *Journal of Affective Disorders, 341*, 104–111. https://doi.org/10.1016/J.JAD.2023.08.077

Loton, D., Borkoles, E., Lubman, D., & Polman, R. (2016). Video game addiction, engagement and symptoms of stress, depression and anxiety: The mediating role of coping. *International Journal of Mental Health and Addiction, 14*(4), 565–578. https://doi.org/10.1007/S11469-015-9578-6

Lukka, L., & Palva, J. M. (2023). The development of game-based digital mental health interventions: Bridging the paradigms of health care and entertainment. *JMIR Serious Games, 11*(1), e42173. https://doi.org/10.2196/42173

Marques, L. M., Uchida, P. M., Aguiar, F. O., Kadri, G., Santos, R. I. M., & Barbosa, S. P. (2023). Escaping through virtual gaming—What is the association with emotional, social, and mental health? A systematic review. *Frontiers in Psychiatry, 14*, 1257685. https://doi.org/10.3389/FPSYT.2023.1257685/BIBTEX

Ostinelli, E. G., Zangani, C., Giordano, B., Maestri, D., Gambini, O., D'Agostino, A., Furukawa, T. A., & Purgato, M. (2021). Depressive symptoms and depression in individuals with internet gaming disorder: A systematic review and meta-analysis. *Journal of Affective Disorders, 284*, 136–142. https://doi.org/10.1016/J.JAD.2021.02.014

Pramana, G., Parmanto, B., Lomas, J., Lindhiem, O., Kendall, P. C., & Silk, J. (2018). Using mobile health gamification to facilitate cognitive behavioral therapy skills practice in child anxiety treatment: Open clinical trial. *JMIR Serious Games, 6*(2), e9. https://doi.org/10.2196/games.8902

Reid, E., Mandryk, R. L., Beres, N. A., Klarkowski, M., & Frommel, J. (2022). Feeling good and in control: In-game tools to support targets of toxicity. *Proceedings of the ACM on Human-Computer Interaction, 6*, 235. https://doi.org/10.1145/3549498

Ryan, R. M., & Deci, E. L. (2000). Self-determination theory and the facilitation of intrinsic motivation, social development, and well-being. *American Psychologist, 55*(1), 68–78. https://doi.org/10.1037/0003-066X.55.1.68

Ryan, R. M., Rigby, C. S., & Przybylski, A. (2006). The motivational pull of video games: A self-determination theory approach. *Motivation and Emotion, 30*(4), 347–363. https://doi.org/10.1007/S11031-006-9051-8

Sarsenbayeva, Z., Tag, B., Yan, S., Kostakos, V., & Goncalves, J. (2020). Using video games to regulate emotions. In *ACM international conference proceeding series* (pp. 755–759). ACM. https://doi.org/10.1145/3441000.3441035

Seay, A. F., & Kraut, R. E. (2007). Project massive: Self-regulation and problematic use of online gaming. In *Conference on human factors in computing systems – Proceedings* (pp. 829–838). ACM. https://doi.org/10.1145/1240624.1240749

Slack, J. D., Delfabbro, P., & King, D. L. (2022). Toward a delineation of the differences between high engagement and problem gaming. *Addictive Behaviors Reports, 16*, 100462. https://doi.org/10.1016/J.ABREP.2022.100462

Toh, W., & Kirschner, D. (2023). Developing social-emotional concepts for learning with video games. *Computers and Education, 194*, 104708. https://doi.org/10.1016/J.COMPEDU.2022.104708

Villani, D., Carissoli, C., Triberti, S., Marchetti, A., Gilli, G., & Riva, G. (2018). Videogames for emotion regulation: A systematic review. *Games for Health Journal, 7*(2), 85–99. https://doi.org/10.1089/G4H.2017.0108/ASSET/IMAGES/LARGE/FIGURE1.JPEG

Yen, J. Y., Yeh, Y. C., Wang, P. W., Liu, T. L., Chen, Y. Y., & Ko, C. H. (2017). Emotional regulation in young adults with internet gaming disorder. *International Journal of Environmental Research and Public Health, 15*(1), 30. https://doi.org/10.3390/IJERPH15010030

Zhan, J., Liu, C., Wang, Z., Cai, Z., & He, J. (2024). Effects of game-based digital interventions for mental disorders: A meta-analysis. *Journal of Affective Disorders, 362*, 731–741. https://doi.org/10.1016/J.JAD.2024.07.095

Chapter 11
Cognitive Benefits and Challenges of Gaming

Siddhesh Zadey, Brendan Kelleher, and Jeel Vasa

Abstract Video games have amassed an immense socio-cultural impact, with direct benefits and adverse impacts on human cognition. In this chapter, we provide a high-level picture of the cognitive benefits and challenges of gaming. Literature on cognitive benefits and challenges of gaming is heterogeneous and complex due to several reasons. Studies have shown improvements across various cognitive domains, such as attention, memory, executive functions, and spatial abilities. These benefits have made video games a popular candidate for alternative and adjunctive therapies for a wide range of psychiatric, developmental, and neurocognitive disorders.

While the positive impacts of gaming are noteworthy, there is also a growing body of research on pathological gaming behaviors, particularly Internet gaming disorder (IGD). Preliminary findings on IGD show that real-time strategy and role-playing video games, such as massively multiplayer online role-playing games, are most prevalent among pathological play. The combination of game factors, along with risk factors, influences the positive or negative cognitive impact of gaming across populations. Future research should focus on studying diverse populations, standardizing the game features and cognitive constructs under study, and using sophisticated methods for robust causal insights.

S. Zadey (✉)
Association for Socially Applicable Research (ASAR), Pune, Maharashtra, India

Department of Epidemiology, Columbia University Mailman School of Public Health, New York, NY, USA

GEMINI Research Center, Duke University School of Medicine, Durham, NC, USA
e-mail: sidzadey@asarforindia.org

B. Kelleher
GEMINI Research Center, Duke University School of Medicine, Durham, NC, USA

Massachusetts General Hospital, Boston, MA, USA

Vanderbilt University School of Medicine, Nashville, TN, USA

J. Vasa
Association for Socially Applicable Research (ASAR), Pune, Maharashtra, India

Department of Psychiatry, All India Institute of Medical Sciences, Nagpur, Maharashtra, India

L. Murrins Marques et al. (eds.), *Social and Affective Neuroscience of Gaming*,
https://doi.org/10.1007/978-3-032-00325-6_11

Keywords Cognitive benefits · Internet gaming disorder · Video games · Mental health · Neuropsychology

Introduction

Gaming, including video games, has significantly impacted our cultures globally over the last five decades (Cerezo-Pizarro et al., 2023). Consequently, the benefits and adversities of gaming at social and psychological levels have been a matter of wide public interest and controversy. These instances highlight why studying the impacts of video games on cognitive health, especially in adolescents and young adults, is especially important.

This chapter aims to provide an overview of the cognitive benefits of gaming, as well as challenges due to unhealthy gaming behaviors and disorders. Our goal is not to provide a systematic and exhaustive summary of the literature on the topic. Rather, we focus on the "big" picture, largely utilizing existing systematic and narrative reviews and meta-analysis studies, followed by randomized controlled trials (RCTs) and laboratory experiments with human participants that use brain imaging data. Our goal is to provide a starting point to the readers based on the best possible information on the topic.

The chapter is organized as follows. Section "Cognitive Benefits of Gaming" describes the cognitive benefits of video games across different cognitive constructs. This section also provides a subsection on how games have been leveraged to design therapeutic interventions for psychiatric and neurocognitive disorders. Section "Cognitive Challenges of Gaming" describes the adverse effects of unhealthy or problematic gaming on different cognitive constructs, drawing on studies in people with internet gaming disorder (IGD). Section "Conclusions" concludes with limitations in the literature and future research directions.

Cognitive Benefits of Gaming

Overall Benefits of Gaming

Studies on the cognitive effects of video gaming show significant improvements across various domains, such as attention, memory, executive functions, and spatial abilities.

Two meta-analyses investigated the impact of video games on overall cognitive functions, including auditory processing, executive functions, motor skills, spatial imagery, and visual processing (Powers et al., 2013). In these analyses, quasi-experimental studies showed a moderate improvement in information processing. Experiments with randomization in comparison groups showed improvements, but the effects were smaller. These findings emphasize that gaming can improve cognitive functions, but the magnitude of the effects depends on the domain assessed and the methodological rigor of the studies.

Attention

Attention is a fundamental cognitive process that encompasses various subdomains, including visual-spatial, auditory, focused, transient, and divided attention. A total of 73 outcome measures assessed these forms of attention. Of these, 42 showed no significant changes, 30 demonstrated significant improvements, including sustained attention and top-down attention, and one showed deterioration in general attention when measured by a maze chase task measuring correct moves per second (Boivin et al., 2016; Sampalo et al., 2023). Improvements were observed across multiple measures of attention, cognitive control, and processing speed. Additionally, certain video games and immersive "exergames" (see Table 11.1 for glossary) can provide temporary boosts to attention in healthy adults, comparable to benefits from meditation or exercise. Five of 13 experiments reported positive effects on attention, specifically in executive attention, sustained attention, and top-down attention (Barton et al., 2020).

Working Memory (WM)

Gaming, particularly action-oriented genres, engages WM through challenges involving rapid information processing and multitasking. While comparing experienced video game players (VGPs) and non-gamers (NVGPs), VGPs were faster and more accurate in updating working memory (Colzato et al., 2013). This suggests that playing first-person shooter games enhances the flexible updating of task-relevant information without increasing impulsivity. Analyses of those with gaming disorder/problem gaming vs non-gamers consistently showed gamers outperforming non-gamers across verbal, visuospatial, and N-back tasks (Waris

Table 11.1 Glossary of terms

Term	Definition
Exergames	Video games that integrate physical activity with gameplay, requiring movements like running or jumping. Used in fitness, rehabilitation, and cognitive training (Chan et al., 2024; Connolly et al., 2012).
Serious games	Games designed for purposes beyond entertainment, such as education, training, or therapy, leveraging game mechanics to enhance learning and decision-making (Granic et al., 2014).
Biofeedback games	Games that use physiological data (e.g., heart rate, brain activity) to provide real-time feedback, helping players regulate physiological or psychological states. Often used for mental health and stress management (Pallavicini et al., 2009).
Brain games	A non-immersive, user-friendly form of CCT that enhances cognitive function through engaging, adaptable, and competitive gameplay. It prioritizes user motivation and entertainment while incorporating cognitive training elements (Kletzel et al., 2021).
Computerized cognitive training (CCT)	A structured, science-based digital program designed to target specific cognitive functions (e.g., memory, attention). It may lack the engagement and adaptability found in brain gaming (Kletzel et al., 2021).

et al., 2019; Wilms et al., 2013). The consistent link between video gaming and working memory in both an extreme group analysis and a full-sample analysis suggests that gaming can boost cognitive performance.

Visual Attention and Short-Term Memory

Video games enhance visual attention and the encoding of information into short-term memory. However, studies directly addressing this hypothesis are limited. One study with 42 young male participants, grouped into non-players (<2 h/month), casual players (4–8 h/month), and experienced players (>15 h/month), examined the effects of gaming on attention and memory (Wilms et al., 2013). While gaming did not improve visual short-term memory capacity, it increased the speed of visual information encoding. Greater improvements correlated with more gaming, indicating that gaming enhances basic attentional functioning with potential applications in cognitive rehabilitation.

Implicit Memory Associations and Problem Gaming

Implicit memory processes, which operate without conscious awareness, play a crucial role in learning and habit formation. Expert gamers demonstrate superior implicit memory associations compared to non-gamers (Boot et al., 2008). Action video game players exhibit faster response times, likely due to the reinforcement of associative learning mechanisms. Neuroimaging studies reveal enhanced connectivity in brain regions associated with procedural learning (Corsi, 1972). While these mechanisms support cognitive benefits, they can also contribute to maladaptive behaviors.

Applications of Benefits

Considering the above-mentioned cognitive benefits, video games are increasingly being used as therapeutic tools to overcome specific cognitive, emotional, and behavioral challenges (Dewhirst et al., 2022).

Attention Deficit Hyperactivity Disorder (ADHD)

ADHD, characterized by deficits in executive functioning and self-regulation, has been a significant focus of game-based interventions (Benzing & Schmidt, 2019). Exergames or active video games, serious games, and biofeedback games (see Table

11.1 for definitions) enhance executive functioning, response inhibition, and arousal regulation. For instance, EndeavorRx, an FDA-approved intervention, is known to improve multitasking in children (Oh et al., 2024). Plan-it Commander, a role-playing game, engages children with ADHD in structured tasks and challenges that reinforce planning and working memory through in-game rewards. Such game-based digital therapeutics offer promising adjunctive interventions for ADHD, complementing pharmacological and behavioral treatments.

Intellectual Disability (ID)

Gaming interventions for individuals with IDs primarily target executive functions and adaptive behaviors to enhance autonomy, social skills, and independent living (Stančin et al., 2020). Serious games target executive functions, including planning, decision-making, and social cognition (Arffa, 2007; Shapoval et al., 2022). Unlike ADHD treatments, interventions for Intellectual Disability (ID) have yet to meet regulatory standards for efficacy and safety, highlighting the challenge of developing effective treatments due to the wide variation in cognitive abilities among individuals with ID (Nazarova et al., 2022).

Autism Spectrum Disorders (ASD)

Gaming interventions have emerged as promising tools for improving social skills, emotional recognition, motor coordination, and adaptive functioning in individuals with ASD (Vajawat et al., 2021). The Picture Exchange Communication System (PECS), a widely used intervention for nonverbal and minimally verbal individuals with ASD, facilitates communication through symbol exchange (The Picture Exchange Communication System (PECS), n.d.). The digital adaptation of PECS within serious games offers several advantages, such as structured turn-taking, request initiation, and sentence formation (Almurashi et al., 2022). However, current digital PECS applications often focus on early-phase communication, such as requesting, but lack full integration of later phases that involve commenting, responding to questions, and sentence structuring. Some studies have explored the use of smart glasses for PECS-based communication, offering real-time visual prompts and social cues, but challenges related to usability and long-term adoption persist (Battaglia & Mcdonald, 2015). In addition to communication benefits, gaming interventions have been found to enhance motor coordination, cognitive flexibility, and attention regulation. Serious games that incorporate PECS also contribute to better visual scanning, fine motor skills, and executive functioning (Chaidi et al., 2024; Laamarti et al., 2014). Games that use multi-sensory engagement techniques show promise in improving sustained attention in ASD (Lima et al., 2020).

Schizophrenia

Although exergames remain experimental, evidence suggests improvements in aerobic fitness among individuals with schizophrenia, with significant increases observed in exergamers compared to those receiving treatment as usual—particularly in walking speed and aerobic capacity (Bang-Kittilsen et al., 2020, 2021; Kimhy et al., 2015; Leutwyler et al., 2015; Roberts et al., 2021; Shimizu et al., 2017). Video game-based interventions may further enhance patient motivation and adherence to treatment, supporting physical activity in a population prone to sedentary behavior (Shimizu et al., 2017). Studies have noted improvements following exergame use with enhancements in global cognitive functioning (Bang-Kittilsen et al., 2020, 2021; Kimhy et al., 2015). Increased cerebral blood flow was detected in the dorsolateral prefrontal cortex (dlPFC) after 3 months of exergame use, suggesting potential enhancement of executive function (Bang-Kittilsen et al., 2021; Shimizu et al., 2017). Additionally, aerobic exercise, incorporated in exergames, has been associated with increased neurotrophin levels, further contributing to cognitive benefits (Kimhy et al., 2015). Video games have also been shown to improve facial recognition, a function associated with temporal lobe activity, highlighting their potential for targeting multiple cognitive domains relevant to schizophrenia (Gülkesen et al., 2017; Isleyen et al., 2014).

Beyond cognition, video gaming appears to enhance social functioning and quality of life. While previous literature provided limited data on this aspect, recent findings indicate that exergame use leads to improvements in motivation and daily activity levels in people with schizophrenia (Kimhy et al., 2015; Shimizu et al., 2017). Both cognitive remediation and video game interventions have been associated with increased self-efficacy in daily life and social tasks, possibly through reinforcement of success, mastery, and vicarious learning (Bandura, 1977; Bryce et al., 2018; Medalia & Choi, 2010).

Depression

Video game-based interventions have shown promise in alleviating depressive symptoms, though research in this area remains relatively new. Reviews have highlighted the positive effects of video games on depression, with findings showing that the majority of studies on exergames reported significant symptom reduction (Primack et al., 2012). Serious game-based interventions also have encouraging outcomes, though results are inconclusive and primarily focused on younger populations (Pine et al., 2020). One reason why video games may be effective in treating depression is their ability to engage users in a way that feels more rewarding than clinical interventions, thus increasing adherence, especially among adolescents and young adults. Some video games integrate therapeutic strategies, such as cognitive behavioral therapy (CBT), which has well-established efficacy in depression treatment (Tark et al., 2019). Digital adaptations of CBT have been successfully implemented, further supporting the potential of video games as a viable therapeutic medium (Zhang et al., 2019).

Exergames present potential benefits for depression, as increased physical activity correlates with lower depression scores due to the proven antidepressant effects of aerobic exercise (Christ et al., 2020; Galper et al., 2006; Pinto Pereira et al., 2014; Porras-Segovia et al., 2019). Given that sedentary lifestyles can contribute to both depression and metabolic disorders, exergames provide an accessible alternative for promoting physical activity, particularly for individuals prone to social isolation (Carson et al., 2016; Morres et al., 2019). Studies suggest that exercise improves depressive symptoms by enhancing neural regeneration and reducing inflammation and oxidative stress (Rodriguez-Ayllon et al., 2019; Russo-Neustadt et al., 2004; Winter et al., 2007).

Anxiety Disorders

The impact of exergames and serious gaming interventions on anxiety-related disorders has been extensively studied, with findings highlighting both their potential benefits and limitations. While both exergames and usual care reduced anxiety, no added benefit was found over non-exercise controls (Alves et al., 2018). Sensitivity analyses confirmed consistent results for exergames but varied effects across populations and protocols (Collado-Mateo et al., 2017; Ruivo et al., 2017; Vieira et al., 2018; Voon et al., 2016; Xu et al., 2016). No enhancement of usual care effects was observed with exergames (Rossi & Tirapegui, 2004). The anxiolytic effects of exergames may be linked to increased tryptophan and serotonin produced by linking exercising to gaming (Collado-Mateo et al., 2017; Joca et al., 2003; Ruivo et al., 2017; Takács & Stauder, 2016; Vieira et al., 2018). Non-pharmacological treatments also reduce medication costs and side effects. Social interaction in exergames may further support adherence (Rosenberg et al., 2010; Wagener et al., 2012; Yuen et al., 2011).

Healthy Aging and Dementia

For older adults and individuals with cognitive decline, gaming can provide opportunities to maintain or improve cognitive health. These interventions focus on learning, memory, and social engagement to support healthy aging and manage dementia-related challenges. Games like exergames and cognitive training games target learning, memory, language, perceptual-motor ability, attention, executive function, and social cognition (Chan et al., 2024; Sokolov et al., 2020). A meta-analysis of RCTs found a significant positive effect of video games on general cognition in people with MCI/dementia, with an estimated improvement of 1.64 points over 1 year on the Mini-Mental State Examination (MMSE), surpassing the minimal clinically important difference. While this improvement is meaningful, it does not imply a reversal of cognitive decline but rather a potential slowing of deterioration or temporary enhancement of cognitive function (Ferreira-Brito et al., 2021).

However, small sample sizes, risk of bias, and the use of cognitive screening tools for measuring outcomes reduce confidence in these findings. Functional capacity, an essential factor for independent living, was assessed in only two studies, and no progression/conversion to dementia measures were included, limiting the understanding of long-term impacts. The confidence level in using technology was evaluated in only a few studies, despite its crucial role in intervention adherence (Ferreira-Brito et al., 2021).

Gaming interventions show promise in cognitive rehabilitation and dementia care. A review of 77 studies concluded that video games and virtual reality (VR) exergames have positive effects on cognitive function and mental health in older adults (Eckert et al., 2024). Small sample sizes and reliance on custom-built applications over commercial software limit scalability, highlighting the need for larger trials and adaptable gaming solutions (Huckans et al., 2013; Kletzel et al., 2021). While some meta-analyses have shown computerized cognitive training benefits for global cognition and working memory, others suggest limited effects, especially for individuals with more severe cognitive impairments (Li et al., 2022).

Although video games cannot replace standard treatments such as psychotherapy and pharmacotherapy, they present strategic advantages as adjunctive interventions (Table 11.2), particularly in making mental health support more accessible and cost-effective (Gelenberg, 2010; Higgins et al., 2011; Kayambu et al., 2015). Video games and similar complementary therapies may be more beneficial for mild disease stages and could support stable phases of more severe conditions.

Despite their potential, video game-based interventions require further research, particularly in assessing long-term efficacy, as users frequently change games rapidly (Ma, 2015). Future research should incorporate validated functional capacity assessments, monitor disease progression, and refine interventions to enhance accessibility and scalability in clinical and home settings. Expanding access to video game therapies for marginalized communities is also important.

Cognitive Challenges of Gaming

While various experimental and observational approaches recognize the positive impacts of gaming on attention, working memory, and visual processing, there is also a growing body of research on pathological gaming behaviors, particularly Internet gaming disorder (IGD). IGD is a recognized mental health disorder in both the Diagnostic and Statistical Manual of Mental Disorders, 5th Edition and the 11th revision of the International Classification of Diseases (APA, 2013; World Health Organization, 2019). Theoretical models offer valuable insights into the development and subtypes of IGD, explaining how IGD arises and highlighting cognitive and behavioral vulnerabilities that predispose individuals to problematic gaming. Models such as the Cognitive-Behavioral, Compensatory Use, and Interaction of Person-Affect-Cognition-Execution (I-PACE) frameworks (Table 11.3) address key aspects of IGD pathogenesis, including stress, impulsivity, and coping motives (Ropovik et al., 2023). The risk factors and IGD subtypes are presented in Box.

Table 11.2 Overview of the tasks/assessment questionnaires used for attention, working memory, and impulsivity

Task/assessment	Measures	How it measures	Findings
Electroencephalography (EEG)	Brain electrical activity	Detects neural activity patterns related to attention.	Significant improvements observed.
ADHD-T questionnaire	Attention deficit and hyperactivity	Self-reported or observed measures of attention and behavior.	Significant improvements noted.
Stroop task	Cognitive flexibility and attention	Inhibition of automatic responses while focusing on task-relevant features.	Improvements in both executive and attentional components.
Symbol digits modalities test	Processing speed and attention	Measures ability to pair symbols with numbers under time constraints.	Improvements noted in task performance.
Focused/distributed spatial attention	Visual attention distribution	Tasks focusing on local vs. global attentional allocation.	Enhanced distribution and focus noted in visual attention tasks.
Test of everyday attention	Functional attention in daily tasks	Simulates real-world attentional demands, e.g., map reading or auditory monitoring.	Positive effects on functional attention from serious games.
Toulouse–Pieron test	Sustained visual attention	Measures concentration through symbol identification tasks.	Positive effects on sustained attention observed.
Tests of variables of attention	Sustained and selective attention	Tracks errors and reaction times to visual/auditory stimuli.	Improvements in attentional performance.
Test of attentional performance	Focused and divided attention	Evaluates responses to simultaneous tasks requiring high attention.	Notable improvement in multitasking and divided attention.
Neglect test	Visual-spatial neglect	Assesses deficits in attending to one side of space.	Improvement noted in attentional tasks.
Go/No-Go task	Response inhibition and attention	Tests ability to suppress actions in response to specific cues.	Improvement in inhibition control.
Speed subtask	Processing speed	Measures reaction time to stimuli under timed conditions.	Enhanced speed in attentional processing noted.
Flanker task	Executive control and selective attention	Suppression of irrelevant information while focusing on target stimuli.	Positive effects on executive attention observed.
Sustained attention to response task (SART)	Sustained attention	Tracks ability to maintain focus and avoid impulsive errors.	Improved ability to sustain focus.

(continued)

Table 11.2 (continued)

Task/assessment	Measures	How it measures	Findings
Paced auditory serial addition test (PASAT)	Auditory attention and working memory	Requires adding sequentially presented auditory numbers.	Enhanced auditory attention performance.
Psychomotor vigilance task (PVT)	Sustained attention and reaction time	Tracks response times to visual stimuli over a sustained period.	Significant improvement in vigilance and response times.
Attentional blink task (ABT)	Temporal attention and processing	Tests ability to detect a second target amidst a rapid sequence of stimuli.	Improvements in temporal attention.
Useful field of view (UFOV)	Visual attention capacity	Evaluates processing speed and spatial attention across a visual field.	Increased capacity in visual field attention.
Digit-span backward task	Working memory	Tests the ability to recall a sequence of numbers in reverse order.	Significant improvement in working memory performance.
Theory of visual attention (TVA)	Visual short-term memory and encoding	Measures capacity and speed of visual information processing.	Improved encoding speed but no change in short-term memory capacity.
Enumeration task	Attention span and processing speed	Evaluates how many items can be simultaneously processed visually.	Speed of processing improved with more gaming hours.
Attention network test (ANT)	Alerting, orienting, and executive control	Assesses components of attention using cues and target responses.	Positive improvements in all attentional components.
N-back task	Working memory monitoring/updating	Tests the ability to track and update information across varying levels of difficulty.	Video game players showed faster and more accurate working memory updating.
Stop-signal paradigm	Response inhibition and impulsivity	Measures the ability to halt a pre-planned response upon encountering a stop signal.	No difference between gamers and non-gamers.

In this section, we examine how pathological gamers compare to healthy regular gamers and non-gamers in regard to cognition. Using cognitive task results, alongside brain activation findings from simultaneous neuroimaging (such as magnetic resonance imaging or MRI), we hope to provide an overview of how IGD impacts cognition in contrast to the beneficial results observed outside of IGD. Two main trends exist in the vast majority of IGD findings, which are highlighted throughout the section. First, many of the central neural pathways and cognitive functions that

Table 11.3 Integrated models of IGD pathogenesis

Model	What it attributes IGD to	Implications for treatment
Cognitive-behavioral model	Stress, impulsivity, and escape/coping motivations (Dong & Potenza, 2014).	Treatment should focus on stress management, impulse control, and restructuring maladaptive thoughts and behaviors.
Compensatory use model	Escape, coping, fantasy, achievement, competition, and social motives (Kardefelt-Winther, 2014).	Treatments could involve addressing underlying emotional issues, improving real-life coping mechanisms, and social skills.
I-PACE model	Depression, anxiety, ADHD, impulsivity, loneliness, low self-esteem (Brand et al., 2016).	Treatment may involve psychological interventions for depression, anxiety, and self-esteem, alongside impulse control strategies.

are positively impacted by regular gaming often show a significantly different effect in IGD. Second, the circuitry and cognitive constructs affected in IGD are very similar to other substance and non-substance (i.e., gambling) dependence and addictions (Gilman et al., 2015; Madden et al., 2009).

Reward Valuation and Risk Assessment

People with IGD exhibit altered reward circuitry, affecting their cravings for gaming and the ability to learn realistic reward values. Additionally, they often exhibit excessive risk-taking behavior, together contributing to impulsive and often harmful decision-making, especially in terms of gaming behavior. These alterations can be measured across various cognitive paradigms, including reward learning, probability discounting, and risk-taking.

Reward Learning: Cue Reactivity

Cue reactivity paradigms examine how people respond to certain stimuli of interest (such as games) compared to neutral stimuli, assessing learned reward prediction. In cue reactivity paradigms, people with IGD show not only higher self-reported craving for games but also higher cue-induced activations in the striatum compared to regular gamers and non-gaming controls (Dong et al., 2017b; Liu et al., 2017). The striatum—in particular, the ventral striatum (VS), containing the nucleus accumbens (NAc)—is a key facilitator of cravings and reward learning. Thus, activations of the striatum in response to gaming cues implicate alterations of the brain's reward system in IGD, reflecting similar findings in substance use and gambling populations. Additionally, the duration since IGD diagnosis is positively associated with dorsal striatum (DS) activation, suggesting a shift from ventral to dorsal processing of reward cues in the striatum (Liu et al., 2017). Beyond the striatum, people with

IGD also show irregular hyperactivity in the prefrontal cortex (PFC) and precuneus, regions that are further implicated in reward learning, such as action monitoring and planning goal-oriented behavior (Dong et al., 2017b).

Reward Valuation: Probability Discounting

The probability of discounting tasks measures reward valuation by allowing individuals to select between options of varying risks and rewards. Across multiple studies evaluating probability discounting, people with IGD consistently prefer the higher-risk options to fixed ones and show significantly quicker reaction times in making choices compared to controls (Dong & Potenza, 2016; Lin et al., 2015; Wang et al., 2019). These risk-taking behaviors are also similar to those in substance use disorders and pathological gambling (Lin et al., 2015). In terms of brain activation, the inferior frontal gyrus (IFG), a region involved in subjective risk evaluation, is significantly hypoactive compared to controls during probability discounting. This hypoactivity is directly correlated with riskier decision-making in pathological gamers (Dong & Potenza, 2016; Lin et al., 2015). The precuneus also exhibits decreased activity during probability discounting. This may be due to its role in risk aversion, as it is more likely to be activated when people are facing potential losses and positively correlates with reaction time, indicating the deliberation of potential losses (Lin et al., 2015; Mohr et al., 2010).

Other regions affected in IGD reward valuation include the anterior and posterior regions of the cingulate cortex (ACC and PCC) and the medial and sagittal gyri of the temporal lobe (MTG and STG) (Dong & Potenza, 2016). On a larger scale, individuals with IGD display altered functional connectivity in the brain's reward circuitry while showing weaker connections in the anterior salience network, including the ACC (Wang et al., 2019). Altogether, people with IGD demonstrate abnormal risk and reward assessment, leading them to make more impulsive decisions that can negatively impact their lives.

Risk and Impulsivity Scales

Beyond paradigms, self-report scales such as the Barratt Impulsivity Scale-11 (BIS) further indicate that individuals with IGD show increased impulsive decision-making, illustrating that cognitively, IGD is characterized not only by an inability to properly gauge reward but also potential losses (Chen et al., 2015; Ko et al., 2015). This result correlates with reduced functional connectivity between brain regions involved in the corticolimbic circuit, particularly the amygdala, dlPFC, and orbital frontal cortex (OFC). Changes within this circuit also reflect a reduced capacity to evaluate potential threats and losses, further disinhibiting impulsive gaming behaviors.

Cognitive Control

As a central moderator of goal-directed behavior, the cognitive control system modulates the operation of other cognitive and emotional systems, an essential component of healthy everyday functioning (Cognitive Control—National Institute of Mental Health (NIMH), n.d.). In IGD, various sources of evidence suggest that pathological gaming is associated with deficits in cognitive control, leading to dysfunctional goal-seeking and limited self-control.

Response Inhibition: Go/No-Go Task

The Go/No-Go task evaluates one's cognitive ability to quickly inhibit certain unfavorable responses, a skill that is essential to day-to-day life. Individuals with IGD show markedly poorer response inhibition compared to controls, which can also be attributed to functional alterations in cognitive control circuitry (Chen et al., 2015). In particular, the supplementary motor area (SMA) and pre-SMA, regions that are critical for selecting appropriate behaviors and inhibiting others, are shown to be hypoactive in individuals with IGD (Chen et al., 2015). Reduced activation in the SMA may indicate how IGD persists through detriments in self-control, especially regarding gaming. The SMA also communicates with other nodes in the frontal and parietal cortices to enact cognitive control, and there is evidence that these regions also function differently in IGD during No-Go trials (Ding et al., 2014).

Response Selection and Performance Monitoring: Stroop Task

The Stroop Task evaluates cognitive control capabilities, also measuring one's ability to monitor their actions, detect, and correct errors in the process. IGD subjects exhibit higher response times to incongruent trials (also known as the "Stroop Effect") compared to both regular gamers and non-gamers while simultaneously exhibiting lower activation in the medial frontal gyrus (MFG) (Dong et al., 2017a). The MFG most notably contains the dorsolateral prefrontal cortex (dlPFC), a central hub for cognitive control and general executive function. Collectively, individuals with IGD not only show impaired risk and reward circuitry but also lack sufficient cognitive control capabilities to regulate maladaptive behavior.

Perception and Understanding of Self

Beyond the domains of reward-seeking, risk evaluation, and cognitive control, IGD also creates tangible alterations in social understanding, particularly in recognizing one's self-concept, which encompasses a person's abilities, passions, goals,

Table 11.4 Overview of the cognitive constructs, task results, and brain regions that are negatively impacted in Internet gaming disorder

Cognitive construct	Paradigm	Results	Brain regions
Reward learning	Cue-reactivity	Higher craving for games	Striatum, prefrontal cortex (PFC), precuneus
Reward valuation	Probability discounting	Prefer riskier options to fixed ones and quicker choice-making	Inferior frontal gyrus (IFG), precuneus, cingulate cortex, medial temporal gyrus (MTG), superior temporal gyrus (STG)
Impulsivity	BIS	Increased impulsivity	Amygdala, dorsolateral prefrontal cortex (dlPFC), orbitofrontal cortex (OFC)
Response inhibition	Go/No-Go task	Poorer response inhibition	Supplementary motor area (SMA), pre-SMA, fronto-parietal areas
Response selection and performance monitoring	Stroop task	Higher response times to incongruent trials	dlPFC, inferior parietal lobule (IPL), cingulate cortex, MTG
Perception and understanding of self	Self-reflection tasks	Impaired understanding of self and negative self-evaluation	Angular gyrus, IPL

feelings, principles, and behaviors (Table 11.4). When performing self-reflection tasks, for example, individuals with IGD exhibit impairments in recognizing and understanding their own emotions while also evaluating themselves more negatively in terms of body image and social standing compared to controls (Kim et al., 2018; Leménager et al., 2016). These results are also correlated with significant alterations in the activation of the angular gyrus (AG), a region located in the inferior parietal lobule (IPL) that is closely tied to identification processing and empathy. These results suggest a self-concept shift in IGD towards negative self-evaluation and a lack of emotional understanding and expressiveness. Similar behavioral results have been observed in substance abuse populations (Yan et al., 2020).

Conclusions

Literature on cognitive benefits and challenges of gaming is heterogeneous and complex due to differences in aims, methods, populations of interest, and other reasons. As noted, gaming, when moderate, can have several benefits for enhancing attention, working memory, and executive functions. These benefits have been leveraged for designing therapeutic interventions for a wide range of psychiatric and neurocognitive disorders, as well as for ensuring cognitive health in aging populations. However, literature, mostly among adolescents and young adults, relating to IGD presents substantial neuropsychological evidence on adverse cognitive effects due to problematic gaming behaviors.

Just as the structure of a game can predispose its cognitive benefits, it may also be influential in driving pathological gaming behaviors. Video game features can be divided in various ways, each of which can potentially impact the game's potential benefit versus risk (D. King et al., 2010b). Social features between players, e.g., can determine the extent to which a player can form a virtual social identity or assess their standing compared to other players, making them feel more personally invested in the game (Vorderer et al., 2003). Meanwhile, games that allow players to create a virtual identity, such as an avatar, can lead to further immersion in pursuit of enhancing one's virtual identity (Blinka, 2008). Reward and punishment schedules on an event- or meta-level also shape the addictiveness of a game, with a variable ratio reinforcement schedule leading to the most strongly-wired reward-seeking behaviors (Chumbley & Griffiths, 2006). Games using near misses or losing events that are very similar to winning ones also increase excitement and motivation, similar to gambling (Griffiths, 1991). The physical controls and presentation features can also increase a game's appeal to the player (Chappell et al., 2006; D. King et al., 2010a).

While the research on game types and IGD is preliminary, findings show that real-time strategy and role-playing video games, such as massively multiplayer online role-playing games, are most prevalent among pathological play, compared to action and other games such as phone games (Eichenbaum et al., 2015). Since these game types integrate many of the core principles outlined above, it is evident that game factors, such as well-designed reward schedules, social incentives, and exciting aesthetics, can drive pathological gaming behavior. The combination of game factors, along with interpersonal and intrapersonal risk factors (see Box), influences the positive or negative cognitive impact of gaming across populations.

Inferences from current literature are limited in wider application and rigor due to several reasons. First, most studies are set in high-income countries and focus on young populations. As video games become popular among people of different ages and those living around the world, studies in those populations would be important for ensuring representativeness in the literature. Second, better standardization of game structures is required for comparability across studies and for isolating how specific features of the games are associated with certain cognitive benefits or challenges. Finally, in future studies, greater emphasis should be placed on study designs and analytic techniques that can provide causal insights into how games influence cognition.

Box: IGD Risk Factors and Subtypes

Intrapersonal, interpersonal, and environmental risk factors:

Interpersonal and environmental risk factors play roles in IGD development. Family dynamics, parent-children relationships, peer association, school engagement and academic performance, the presence of supporting environments at school or home, and other social factors contribute to IGD risk because they can facilitate a person's development of affective, cognitive, and behavioral responses related to IGD (Gao et al., 2022). However, beyond these, individual styles of processing information and coping responses might be more proximal factors of a behavior or health status. In other words, these intrapersonal and environmental factors may play as mediators or underlying mechanisms in explaining the relationships from higher socio-ecological level factors to individual IGD.

(continued)

Stress is a key risk factor for IGD in adolescents and young adults. However, the extent to which stress leads to IGD depends significantly on its interaction with intrapersonal states and internal cognitive processes. For people predisposed to psychological stress or discomfort due to maladaptive personality traits, gaming can evolve into problematic behaviors, heightening the risk of developing IGD (Gao et al., 2022). For example, a teenager with low self-esteem and high levels of daily stress may turn to online gaming as a way to escape their problems, eventually spending excessive hours in virtual environments and neglecting real-life responsibilities. Recent findings underscore that intrapersonal factors hold greater significance than interpersonal and environmental influences in IGD pathogenesis (Zhuang et al., 2023). Consequently, many prominent models explaining IGD revolve around the interplay of stress and individual cognitive vulnerabilities, highlighting the importance of targeting these domains in prevention and intervention strategies.

IGD Subtypes

Subtypes of IGD identified through the integrated models of pathogenesis (see Table 2) shed light on the types of games preferred and the cognitive features involved.

Internalizing Subtype: The internalizing subtype is characterized by negative personal affect and escapism as primary motivators for gaming. This subtype is strongly associated with psychological vulnerabilities, including loneliness, anxiety, depression, and social vulnerability (Zhuang et al., 2023). Protective factors like self-esteem can mitigate these risks. A cohort study of adolescents has noted that poor psychological well-being often precedes pathological gaming behaviors (Lemmens et al., 2011). Similarly, longitudinal studies indicate that pre-existing symptoms of depression and anxiety positively predict future gaming behaviors and IGD (Gao et al., 2022; Teng et al., 2021).

High levels of comorbidity are observed in this subtype, where gaming is used as a maladaptive coping strategy for mood modification or escaping unsatisfactory realities (Lee et al., 2017). Internalizing symptoms often include withdrawal, tolerance, escape, and preoccupation. While gaming for escapism may initially serve as an adaptive coping mechanism, it can become problematic when adverse moods stem from unresolved internalizing disorders like depression or anxiety.

The self-medication hypothesis suggests that people use gaming to regulate painful emotional states and low self-esteem due to inadequate functional coping strategies (Bäcklund et al., 2024). Escapism and emotion regulation motives are prominent in questionnaires assessing gaming behaviors, including introjected regulation, escapism, and coping (Bäcklund et al., 2024). This underscores the importance of addressing underlying psychological vulnerabilities when designing interventions for the internalizing subtype of IGD.

Externalizing Subtype: It is distinguished by conflict-driven and impulsive behaviors. Gamers within this subtype often use gaming to release aggressive impulses or seek sensations to alleviate boredom (Lee et al., 2017). A "social-related" symptom cluster is associated with this subtype, encompassing

(continued)

conflict, persistence, deception, problem behaviors, and displacement (Liu et al., 2017b). The externalizing subtype exhibits outward changes, often linked to conditions such as ADHD and conduct disorder, characterized by defiance of authority and impulsivity.

Aggressiveness and impulsivity are central to this subtype, with aggression showing a moderate correlation with IGD (Zhuang et al., 2023). Protective factors such as agreeableness mitigate the risks associated with aggression. Social stressors, often stemming from turbulent or unstable environments, amplify the likelihood of aggressive behavior, further exacerbating IGD. For example, the Social Information Processing model suggests that aggressive individuals misinterpret social cues, harbor attributional biases, and lack problem-solving skills. These deficits drive them toward gaming as a virtual escape from unsatisfactory real-life interactions (Larkin et al., 2013).

Risky and impulsive behaviors also play a role in the externalizing subtype. Although not as strongly correlated with IGD as aggression, factors such as low self-control, low conscientiousness, and impulsivity contribute to IGD development (Gao et al., 2022). Excessive gaming can further impair executive control, creating a vicious cycle where impulsivity both drives and results from IGD (Bäcklund et al., 2024). Neuroimaging studies support this, showing decreased orbitofrontal function in individuals with IGD, which parallels ADHD-related impulsivity (Wartberg et al., 2020).

Moreover, ADHD symptoms, including attention hyperactivity and inattention, significantly predict IGD (Zhuang et al., 2023). Cross-lagged panel studies affirm that hyperactivity and inattention predispose adolescents to IGD, while excessive gaming exacerbates impulsivity—a core feature of ADHD—thus intensifying the disorder (Gao et al., 2022). Integrating conflict, aggression, and impulsive clusters into the externalizing subtype provides a comprehensive understanding of its pathogenesis, guiding more targeted intervention.

Social subtype: Social dynamics can contribute to IGD development. The social cluster emphasizes the role of gaming as a platform for social interaction, with some researchers even proposing a distinct social subtype of IGD (Lee et al., 2017). Social motives are among the primary reasons adolescents engage in gaming, often surpassing skills enhancement and coping motives (Lee et al., 2017). Preference for online social interaction reflects the belief that online interactions are safer, more effective, and more comfortable than face-to-face communication (King & Delfabbro, 2014). This inclination can partially predict IGD, as social features embedded within games amplify their addictive potential (Tullett-Prado et al., 2021). Integrating these social dimensions into interventions is crucial for a holistic approach to managing IGD.

References

Almurashi, H., Bouaziz, R., Alharthi, W., Al-Sarem, M., Hadwan, M., & Kammoun, S. (2022). Augmented reality, serious games and picture exchange communication system for people with ASD: Systematic literature review and future directions. *Sensors (Basel, Switzerland), 22*(3), 1250. https://doi.org/10.3390/s22031250

Alves, M. L. M., Mesquita, B. S., Morais, W. S., Leal, J. C., Satler, C. E., & Dos Santos Mendes, F. A. (2018). Nintendo WiiTM versus Xbox KinectTM for assisting people with Parkinson's disease. *Perceptual and Motor Skills, 125*(3), 546–565. https://doi.org/10.1177/0031512518769204

APA. (2013). *Diagnostic and statistical manual of mental disorders: DSM-5TM* (5th ed., pp. xliv, 947). American Psychiatric Publishing, Inc. https://doi.org/10.1176/appi.books.9780890425596

Arffa, S. (2007). The relationship of intelligence to executive function and non-executive function measures in a sample of average, above average, and gifted youth. *Archives of Clinical Neuropsychology: The Official Journal of the National Academy of Neuropsychologists, 22*(8), 969–978. https://doi.org/10.1016/j.acn.2007.08.001

Bäcklund, C., Sörman, D. E., Gavelin, H. M., Király, O., Demetrovics, Z., & Ljungberg, J. K. (2024). Comparing psychopathological symptoms, life satisfaction, and personality traits between the WHO and APA frameworks of gaming disorder symptoms: A psychometric investigation. *Scandinavian Journal of Psychology, 65*(4), 665–682. https://doi.org/10.1111/sjop.13010

Bandura, A. (1977). Self-efficacy: Toward a unifying theory of behavioral change. *Psychological Review, 84*(2), 191–215. https://doi.org/10.1037/0033-295X.84.2.191

Bang-Kittilsen, G., Egeland, J., Langerud Holmen, T., Andersen, E., Ulleberg, P., Torgersen, T. B., Mordal, J., & Engh, J. (2020). O4.3. Shared and differential neurocognitive effects comparing high-intensity interval training and active video-gaming in schizophrenia – Results from an RCT. *Schizophrenia Bulletin, 46*(Suppl 1), S9–S10. https://doi.org/10.1093/schbul/sbaa028.020

Bang-Kittilsen, G., Egeland, J., Holmen, T. L., Bigseth, T. T., Andersen, E., Mordal, J., Ulleberg, P., & Engh, J. A. (2021). High-intensity interval training and active video gaming improve neurocognition in schizophrenia: A randomized controlled trial. *European Archives of Psychiatry and Clinical Neuroscience, 271*(2), 339–353. https://doi.org/10.1007/s00406-020-01200-4

Barton, A. C., Sheen, J., & Byrne, L. K. (2020). Immediate attention enhancement and restoration from interactive and immersive technologies: A scoping review. *Frontiers in Psychology, 11*, 2050. https://doi.org/10.3389/fpsyg.2020.02050

Battaglia, D., & Mcdonald, M. (2015). Effects of the picture exchange communication system (PECS) on maladaptive behavior in children with autism spectrum disorders (ASD): A review of the literature. *Journal of the American Academy of Special Education Professionals*, 8–20. https://www.researchgate.net/publication/274635346_Effects_of_the_Picture_Exchange_Communication_System_PECS_on_Maladaptive_Behavior_in_Children_with_Autism_Spectrum_Disorders_ASD_A_Review_of_the_Literature

Benzing, V., & Schmidt, M. (2019). The effect of exergaming on executive functions in children with ADHD: A randomized clinical trial. *Scandinavian Journal of Medicine & Science in Sports, 29*(8), 1243–1253. https://doi.org/10.1111/sms.13446

Blinka, L. (2008). The relationship of players to their avatars in MMORPGs: Differences between adolescents, emerging adults and adults. *Cyberpsychology: Journal of Psychosocial Research on Cyberspace, 2*(1), Article 1. https://cyberpsychology.eu/article/view/4211

Boivin, M. J., Nakasujja, N., Sikorskii, A., Opoka, R. O., & Giordani, B. (2016). A randomized controlled trial to evaluate if computerized cognitive rehabilitation improves neurocognition in Ugandan children with HIV. *AIDS Research and Human Retroviruses, 32*(8), 743–755. https://doi.org/10.1089/aid.2016.0026

Boot, W. R., Kramer, A. F., Simons, D. J., Fabiani, M., & Gratton, G. (2008). The effects of video game playing on attention, memory, and executive control. *Acta Psychologica, 129*(3), 387–398. https://doi.org/10.1016/j.actpsy.2008.09.005

Brand, M., Young, K. S., Laier, C., Wölfling, K., & Potenza, M. N. (2016). Integrating psychological and neurobiological considerations regarding the development and maintenance of specific internet-use disorders: An Interaction of Person-Affect-Cognition-Execution (I-PACE) model. *Neuroscience and Biobehavioral Reviews, 71*, 252–266. https://doi.org/10.1016/j.neubiorev.2016.08.033

Bryce, S. D., Rossell, S. L., Lee, S. J., Lawrence, R. J., Tan, E. J., Carruthers, S. P., & Ponsford, J. L. (2018). Neurocognitive and self-efficacy benefits of cognitive remediation in schizophrenia: A randomized controlled trial. *Journal of the International Neuropsychological Society: JINS, 24*(6), 549–562. https://doi.org/10.1017/S1355617717001369

Carson, V., Hunter, S., Kuzik, N., Gray, C. E., Poitras, V. J., Chaput, J.-P., Saunders, T. J., Katzmarzyk, P. T., Okely, A. D., Connor Gorber, S., Kho, M. E., Sampson, M., Lee, H., &

Tremblay, M. S. (2016). Systematic review of sedentary behaviour and health indicators in school-aged children and youth: An update. *Applied Physiology, Nutrition, and Metabolism = Physiologie Appliquee, Nutrition Et Metabolisme, 41*(6 Suppl 3), S240–S265. https://doi.org/10.1139/apnm-2015-0630

Cerezo-Pizarro, M., Revuelta-Domínguez, F.-I., Guerra-Antequera, J., & Melo-Sánchez, J. (2023). The cultural impact of video games: A systematic review of the literature. *Education Sciences, 13*(11), Article 11. https://doi.org/10.3390/educsci13111116

Chaidi, I., Pergantis, P., Drigas, A., & Karagiannidis, C. (2024). Gaming platforms for people with ASD. *Journal of Intelligence, 12*(12), 122. https://doi.org/10.3390/jintelligence12120122

Chan, J. Y. C., Liu, J., Chan, A. T. C., & Tsoi, K. K. F. (2024). Exergaming and cognitive functions in people with mild cognitive impairment and dementia: A meta-analysis. *npj Digital Medicine, 7*(1), 1–9. https://doi.org/10.1038/s41746-024-01142-4

Chappell, D., Eatough, V., Davies, M. N. O., & Griffiths, M. (2006). EverQuest—It's just a computer game right? An interpretative phenomenological analysis of online gaming addiction. *International Journal of Mental Health and Addiction, 4*(3), 205–216. https://doi.org/10.1007/s11469-006-9028-6

Chen, C.-Y., Huang, M.-F., Yen, J.-Y., Chen, C.-S., Liu, G.-C., Yen, C.-F., & Ko, C.-H. (2015). Brain correlates of response inhibition in internet gaming disorder. *Psychiatry and Clinical Neurosciences, 69*(4), 201–209. https://doi.org/10.1111/pcn.12224

Christ, C., Schouten, M. J., Blankers, M., van Schaik, D. J., Beekman, A. T., Wisman, M. A., Stikkelbroek, Y. A., & Dekker, J. J. (2020). Internet and computer-based cognitive behavioral therapy for anxiety and depression in adolescents and young adults: Systematic review and meta-analysis. *Journal of Medical Internet Research, 22*(9), e17831. https://doi.org/10.2196/17831

Chumbley, J., & Griffiths, M. (2006). Affect and the computer game player: The effect of gender, personality, and game reinforcement structure on affective responses to computer game-play. *Cyberpsychology & Behavior: The Impact of the Internet, Multimedia and Virtual Reality on Behavior and Society, 9*(3), 308–316. https://doi.org/10.1089/cpb.2006.9.308

Cognitive Control—National Institute of Mental Health (NIMH). (n.d.). Retrieved May 1, 2025, from https://www.nimh.nih.gov/research/research-funded-by-nimh/rdoc/constructs/cognitive-control

Collado-Mateo, D., Dominguez-Muñoz, F. J., Adsuar, J. C., Garcia-Gordillo, M. A., & Gusi, N. (2017). Effects of exergames on quality of life, pain, and disease effect in women with fibromyalgia: A randomized controlled trial. *Archives of Physical Medicine and Rehabilitation, 98*(9), 1725–1731. https://doi.org/10.1016/j.apmr.2017.02.011

Colzato, L. S., van den Wildenberg, W. P. M., Zmigrod, S., & Hommel, B. (2013). Action video gaming and cognitive control: Playing first person shooter games is associated with improvement in working memory but not action inhibition. *Psychological Research, 77*(2), 234–239. https://doi.org/10.1007/s00426-012-0415-2

Connolly, T. M., Boyle, E. A., MacArthur, E., Hainey, T., & Boyle, J. M. (2012). A systematic literature review of empirical evidence on computer games and serious games. *Computers & Education, 59*(2), 661–686. https://doi.org/10.1016/j.compedu.2012.03.004

Corsi, P. M. (1972). *Human memory and the medial temporal region of the brain.* McGill University. https://escholarship.mcgill.ca/concern/theses/05741s554

Dewhirst, A., Laugharne, R., & Shankar, R. (2022). Therapeutic use of serious games in mental health: Scoping review. *British Journal of Psychiatry Open, 8*(2), e37. https://doi.org/10.1192/bjo.2022.4

Ding, W., Sun, J., Sun, Y.-W., Chen, X., Zhou, Y., Zhuang, Z., Li, L., Zhang, Y., Xu, J., & Du, Y. (2014). Trait impulsivity and impaired prefrontal impulse inhibition function in adolescents with internet gaming addiction revealed by a Go/No-Go fMRI study. *Behavioral and Brain Functions, 10*, 20. https://doi.org/10.1186/1744-9081-10-20

Dong, G., & Potenza, M. N. (2014). A cognitive-behavioral model of internet gaming disorder: Theoretical underpinnings and clinical implications. *Journal of Psychiatric Research, 58*, 7–11. https://doi.org/10.1016/j.jpsychires.2014.07.005

Dong, G., & Potenza, M. N. (2016). Risk-taking and risky decision-making in internet gaming disorder: Implications regarding online gaming in the setting of negative consequences. *Journal of Psychiatric Research, 73*, 1–8. https://doi.org/10.1016/j.jpsychires.2015.11.011

Dong, G., Li, H., Wang, L., & Potenza, M. N. (2017a). Cognitive control and reward/loss processing in internet gaming disorder: Results from a comparison with recreational internet game-users. *European Psychiatry: The Journal of the Association of European Psychiatrists, 44*, 30–38. https://doi.org/10.1016/j.eurpsy.2017.03.004

Dong, G., Wang, L., Du, X., & Potenza, M. N. (2017b). Gaming increases craving to gaming-related stimuli in individuals with internet gaming disorder. *Biological Psychiatry: Cognitive Neuroscience and Neuroimaging, 2*(5), 404–412. https://doi.org/10.1016/j.bpsc.2017.01.002

Eckert, M., Ostermann, T., Ehlers, J. P., & Hohenberg, G. (2024). Dementia and video games: Systematic and bibliographic review. *Healthcare (Basel, Switzerland), 12*(22), 2310. https://doi.org/10.3390/healthcare12222310

Eichenbaum, A., Kattner, F., Bradford, D., Gentile, D. A., & Green, C. S. (2015). Role-playing and real-time strategy games associated with greater probability of internet gaming disorder. *Cyberpsychology, Behavior, and Social Networking, 18*(8), 480–485. https://doi.org/10.1089/cyber.2015.0092

Ferreira-Brito, F., Ribeiro, F., Aguiar de Sousa, D., Costa, J., Caneiras, C., Carriço, L., & Verdelho, A. (2021). Are video games effective to promote cognition and everyday functional capacity in mild cognitive impairment/dementia patients? A meta-analysis of randomized controlled trials. *Journal of Alzheimer's Disease: JAD, 84*(1), 329–341. https://doi.org/10.3233/JAD-210545

Galper, D. I., Trivedi, M. H., Barlow, C. E., Dunn, A. L., & Kampert, J. B. (2006). Inverse association between physical inactivity and mental health in men and women. *Medicine and Science in Sports and Exercise, 38*(1), 173–178. https://doi.org/10.1249/01.mss.0000180883.32116.28

Gao, Y.-X., Wang, J.-Y., & Dong, G.-H. (2022). The prevalence and possible risk factors of internet gaming disorder among adolescents and young adults: Systematic reviews and meta-analyses. *Journal of Psychiatric Research, 154*, 35–43. https://doi.org/10.1016/j.jpsychires.2022.06.049

Gelenberg, A. J. (2010). A review of the current guidelines for depression treatment. *The Journal of Clinical Psychiatry, 71*(7), e15. https://doi.org/10.4088/JCP.9078tx1c

Gilman, J. M., Calderon, V., Curran, M. T., & Evins, A. E. (2015). Young adult cannabis users report greater propensity for risk-taking only in non-monetary domains. *Drug and Alcohol Dependence, 147*, 26–31. https://doi.org/10.1016/j.drugalcdep.2014.12.020

Granic, I., Lobel, A., & Engels, R. C. M. E. (2014). The benefits of playing video games. *American Psychologist, 69*(1), 66–78. https://doi.org/10.1037/a0034857

Griffiths, M. (1991). Psychobiology of the near-miss in fruit machine gambling. *The Journal of Psychology, 125*(3), 347–357. https://doi.org/10.1080/00223980.1991.10543298

Gülkesen, K. H., Işleyen, F., Cinemre, B., Samur, M. K., Kaya, S. Ş., & Zayim, N. (2017). A web-based game for teaching facial expressions to schizophrenic patients. *Applied Clinical Informatics, 8*(3), 719–730. https://doi.org/10.4338/ACI-2016-10-RA-0172

Higgins, J. P. T., Altman, D. G., Gøtzsche, P. C., Jüni, P., Moher, D., Oxman, A. D., Savovic, J., Schulz, K. F., Weeks, L., Sterne, J. A. C., & Cochrane Bias Methods Group, & Cochrane Statistical Methods Group. (2011). The Cochrane Collaboration's tool for assessing risk of bias in randomised trials. *BMJ (Clinical Research Ed.), 343*, d5928. https://doi.org/10.1136/bmj.d5928

Huckans, M., Hutson, L., Twamley, E., Jak, A., Kaye, J., & Storzbach, D. (2013). Efficacy of cognitive rehabilitation therapies for mild cognitive impairment (MCI) in older adults: Working toward a theoretical model and evidence-based interventions. *Neuropsychology Review, 23*(1), 63–80. https://doi.org/10.1007/s11065-013-9230-9

Isleyen, F., Gulkesen, K. H., Cinemre, B., Samur, M. K., Zayim, N., & Sen Kaya, S. (2014). Evaluation of the usability of a serious game aiming to teach facial expressions to schizophrenic patients. *Studies in Health Technology and Informatics, 205*, 662–666.

Joca, S. R. L., Padovan, C. M., & Guimarães, F. S. (2003). [Stress, depression and the hippocampus]. *Revista Brasileira De Psiquiatria (Sao Paulo, Brazil: 1999), 25*(Suppl 2), 46–51. https://doi.org/10.1590/s1516-44462003000600011

Kardefelt-Winther, D. (2014). A conceptual and methodological critique of internet addiction research: Towards a model of compensatory internet use. *Computers in Human Behavior, 31*, 351–354. https://doi.org/10.1016/j.chb.2013.10.059

Kayambu, G., Boots, R., & Paratz, J. (2015). Early physical rehabilitation in intensive care patients with sepsis syndromes: A pilot randomised controlled trial. *Intensive Care Medicine, 41*(5), 865–874. https://doi.org/10.1007/s00134-015-3763-8

Kim, M.-K., Jung, Y. H., Kyeong, S., Shin, Y.-B., Kim, E., & Kim, J.-J. (2018). Neural correlates of distorted self-concept in individuals with internet gaming disorder: A functional MRI study. *Frontiers in Psychiatry, 9*, 330. https://doi.org/10.3389/fpsyt.2018.00330

Kimhy, D., Vakhrusheva, J., Bartels, M. N., Armstrong, H. F., Ballon, J. S., Khan, S., Chang, R. W., Hansen, M. C., Ayanruoh, L., Lister, A., Castrén, E., Smith, E. E., & Sloan, R. P. (2015). The impact of aerobic exercise on brain-derived neurotrophic factor and neurocognition in individuals with schizophrenia: A single-blind, randomized clinical trial. *Schizophrenia Bulletin, 41*(4), 859–868. https://doi.org/10.1093/schbul/sbv022

King, D., Delfabbro, P., & Griffiths, M. (2010a). The role of structural characteristics in problem video game playing: A review. *Cyberpsychology: Journal of Psychosocial Research on Cyberspace, 4*(1), Article 1. https://cyberpsychology.eu/article/view/4229

King, D., Delfabbro, P., & Griffiths, M. (2010b). Video game structural characteristics: A new psychological taxonomy. *International Journal of Mental Health and Addiction, 8*(1), 90–106. https://doi.org/10.1007/s11469-009-9206-4

King, D. L., & Delfabbro, P. H. (2014). The cognitive psychology of Internet gaming disorder. *Clinical Psychology Review, 34*(4), 298–308. https://doi.org/10.1016/j.cpr.2014.03.006

Kletzel, S. L., Sood, P., Negm, A., Heyn, P. C., Krishnan, S., Machtinger, J., Hu, X., & Devos, H. (2021). Effectiveness of brain gaming in older adults with cognitive impairments: A systematic review and meta-analysis. *Journal of the American Medical Directors Association, 22*(11), 2281–2288.e5. https://doi.org/10.1016/j.jamda.2021.05.022

Ko, C.-H., Hsieh, T.-J., Wang, P.-W., Lin, W.-C., Yen, C.-F., Chen, C.-S., & Yen, J.-Y. (2015). Altered gray matter density and disrupted functional connectivity of the amygdala in adults with internet gaming disorder. *Progress in Neuro-Psychopharmacology & Biological Psychiatry, 57*, 185–192. https://doi.org/10.1016/j.pnpbp.2014.11.003

Laamarti, F., Eid, M., & El Saddik, A. (2014). An overview of serious games. *International Journal of Computer Games Technology, 2014*(1), 358152. https://doi.org/10.1155/2014/358152

Larkin, P., Jahoda, A., & MacMahon, K. (2013). The Social Information Processing model as a framework for explaining frequent aggression in adults with mild to moderate intellectual disabilities: A systematic review of the evidence. *Journal of Applied Research in Intellectual Disabilities: JARID, 26*(5), 447–465. https://doi.org/10.1111/jar.12031

Lee, S.-Y., Lee, H. K., & Choo, H. (2017). Typology of Internet gaming disorder and its clinical implications. *Psychiatry and Clinical Neurosciences, 71*(7), 479–491. https://doi.org/10.1111/pcn.12457

Leménager, T., Dieter, J., Hill, H., Hoffmann, S., Reinhard, I., Beutel, M., Vollstädt-Klein, S., Kiefer, F., & Mann, K. (2016). Exploring the neural basis of avatar identification in pathological internet gamers and of self-reflection in pathological social network users. *Journal of Behavioral Addictions, 5*(3), 485–499. https://doi.org/10.1556/2006.5.2016.048

Lemmens, J. S., Valkenburg, P. M., & Peter, J. (2011). Psychosocial causes and consequences of pathological gaming. *Computers in Human Behavior, 27*(1), 144–152. https://doi.org/10.1016/j.chb.2010.07.015

Leutwyler, H., Hubbard, E., Cooper, B., & Dowling, G. (2015). The impact of a videogame-based pilot physical activity program in older adults with schizophrenia on subjectively and objectively measured physical activity. *Frontiers in Psychiatry, 6*, 180. https://doi.org/10.3389/fpsyt.2015.00180

Li, F., Parsons, J., Peri, K., Yu, A., & Cheung, G. (2022). Effects of cognitive interventions on quality of life among adults with mild cognitive impairment: A systematic review and meta-analysis of randomised controlled trials. *Geriatric Nursing (New York, N.Y.), 47*, 23–34. https://doi.org/10.1016/j.gerinurse.2022.06.009

Lima, J. L., Axt, G., Teixeira, D. S., Monteiro, D., Cid, L., Yamamoto, T., Murillo-Rodriguez, E., & Machado, S. (2020). Exergames for children and adolescents with autism spectrum disorder: An overview. *Clinical Practice and Epidemiology in Mental Health, 16*, 1–6. https://doi.org/10.2174/1745017902016010001

Lin, X., Zhou, H., Dong, G., & Du, X. (2015). Impaired risk evaluation in people with internet gaming disorder: fMRI evidence from a probability discounting task. *Progress in Neuro-Psychopharmacology & Biological Psychiatry, 56*, 142–148. https://doi.org/10.1016/j.pnpbp.2014.08.016

Liu, L., Yip, S. W., Zhang, J.-T., Wang, L.-J., Shen, Z.-J., Liu, B., Ma, S.-S., Yao, Y.-W., & Fang, X.-Y. (2017). Activation of the ventral and dorsal striatum during cue reactivity in internet gaming disorder. *Addiction Biology, 22*(3), 791–801. https://doi.org/10.1111/adb.12338

Liu, L., Yip, S. W., Zhang, J.-T., Wang, L.-J., Shen, Z.-J., Liu, B., Ma, S.-S., Yao, Y.-W., & Fang, X.-Y. (2017b). Activation of the ventral and dorsal striatum during cue reactivity in Internet gaming disorder. *Addiction Biology, 22*(3), 791–801. https://doi.org/10.1111/adb.12338

Ma, Y. (2015). The discrepancy between video game purchase and consumption in a content distribution network. In *Proceedings of the 2015 joint international mechanical, electronic and information technology conference* (pp. 249–252). Atlantis Press. https://doi.org/10.2991/jimet-15.2015.46

Madden, G. J., Petry, N. M., & Johnson, P. S. (2009). Pathological gamblers discount probabilistic rewards less steeply than matched controls. *Experimental and Clinical Psychopharmacology, 17*(5), 283–290. https://doi.org/10.1037/a0016806

Medalia, A., & Choi, J. (2010). Motivational enhancements in schizophrenia. In *Neurocognition and social cognition in schizophrenia patients* (Vol. 177, pp. 158–172). Karger Publishers. https://karger.com/books/book/2682/chapter/5759736/Motivational-Enhancements-in-Schizophrenia

Mohr, P. N. C., Biele, G., & Heekeren, H. R. (2010). Neural processing of risk. *The Journal of Neuroscience: The Official Journal of the Society for Neuroscience, 30*(19), 6613–6619. https://doi.org/10.1523/JNEUROSCI.0003-10.2010

Morres, I. D., Hatzigeorgiadis, A., Stathi, A., Comoutos, N., Arpin-Cribbie, C., Krommidas, C., & Theodorakis, Y. (2019). Aerobic exercise for adult patients with major depressive disorder in mental health services: A systematic review and meta-analysis. *Depression and Anxiety, 36*(1), 39–53. https://doi.org/10.1002/da.22842

Nazarova, V. A., Sokolov, A. V., Chubarev, V. N., Tarasov, V. V., & Schiöth, H. B. (2022). Treatment of ADHD: Drugs, psychological therapies, devices, complementary and alternative methods as well as the trends in clinical trials. *Frontiers in Pharmacology, 13*, 1066988. https://doi.org/10.3389/fphar.2022.1066988

Oh, S., Choi, J., Han, D. H., & Kim, E. (2024). Effects of game-based digital therapeutics on attention deficit hyperactivity disorder in children and adolescents as assessed by parents or teachers: A systematic review and meta-analysis. *European Child & Adolescent Psychiatry, 33*(2), 481–493. https://doi.org/10.1007/s00787-023-02174-z

Pallavicini, F., Algeri, D., Repetto, C., Gorini, A., & Riva, G. (2009). Biofeedback, virtual reality and mobile phones in the treatment of generalized anxiety disorder (GAD): A phase-2 controlled clinical trial. *Journal of Cybertherapy and Rehabilitation, 2*(4), 315–327.

Pine, R., Fleming, T., McCallum, S., & Sutcliffe, K. (2020). The effects of casual videogames on anxiety, depression, stress, and low mood: A systematic review. *Games for Health Journal, 9*(4), 255–264. https://doi.org/10.1089/g4h.2019.0132

Pinto Pereira, S. M., Geoffroy, M.-C., & Power, C. (2014). Depressive symptoms and physical activity during 3 decades in adult life: Bidirectional associations in a prospective cohort study. *JAMA Psychiatry, 71*(12), 1373–1380. https://doi.org/10.1001/jamapsychiatry.2014.1240

Porras-Segovia, A., Rivera, M., Molina, E., López-Chaves, D., Gutiérrez, B., & Cervilla, J. (2019). Physical exercise and body mass index as correlates of major depressive disorder in community-dwelling adults: Results from the PISMA-ep study. *Journal of Affective Disorders, 251*, 263–269. https://doi.org/10.1016/j.jad.2019.01.050

Powers, K. L., Brooks, P. J., Aldrich, N. J., Palladino, M. A., & Alfieri, L. (2013). Effects of video-game play on information processing: A meta-analytic investigation. *Psychonomic Bulletin & Review, 20*(6), 1055–1079. https://doi.org/10.3758/s13423-013-0418-z

Primack, B. A., Carroll, M. V., McNamara, M., Klem, M. L., King, B., Rich, M., Chan, C. W., & Nayak, S. (2012). Role of video games in improving health-related outcomes: A systematic review. *American Journal of Preventive Medicine, 42*(6), 630–638. https://doi.org/10.1016/j.amepre.2012.02.023

Roberts, M. T., Lloyd, J., Välimäki, M., Ho, G. W., Freemantle, M., & Békefi, A. Z. (2021). Video games for people with schizophrenia. *The Cochrane Database of Systematic Reviews, 2*(2), CD012844. https://doi.org/10.1002/14651858.CD012844.pub2

Rodriguez-Ayllon, M., Cadenas-Sánchez, C., Estévez-López, F., Muñoz, N. E., Mora-Gonzalez, J., Migueles, J. H., Molina-García, P., Henriksson, H., Mena-Molina, A., Martínez-Vizcaíno, V., Catena, A., Löf, M., Erickson, K. I., Lubans, D. R., Ortega, F. B., & Esteban-Cornejo, I. (2019).

Role of physical activity and sedentary behavior in the mental health of preschoolers, children and adolescents: A systematic review and meta-analysis. *Sports Medicine (Auckland, N.Z.), 49*(9), 1383–1410. https://doi.org/10.1007/s40279-019-01099-5

Ropovik, I., Martončik, M., Babinčák, P., Baník, G., Vargová, L., & Adamkovič, M. (2023). Risk and protective factors for (internet) gaming disorder: A meta-analysis of pre-COVID studies. *Addictive Behaviors, 139*, 107590. https://doi.org/10.1016/j.addbeh.2022.107590

Rosenberg, D., Depp, C. A., Vahia, I. V., Reichstadt, J., Palmer, B. W., Kerr, J., Norman, G., & Jeste, D. V. (2010). Exergames for subsyndromal depression in older adults: A pilot study of a novel intervention. *The American Journal of Geriatric Psychiatry: Official Journal of the American Association for Geriatric Psychiatry, 18*(3), 221–226. https://doi.org/10.1097/JGP.0b013e3181c534b5

Rossi, L., & Tirapegui, J. (2004). [Serotoninergic system and its implications on physical exercise]. *Arquivos Brasileiros de Endocrinologia e Metabologia, 48*(2), 227–233. https://doi.org/10.1590/s0004-27302004000200004

Ruivo, J. M. A. D. S., Karim, K., O'Shea, R., Oliveira, R. C. S., Keary, L., O'Brien, C., & Gormley, J. P. (2017). In-class active video game supplementation and adherence to cardiac rehabilitation. *Journal of Cardiopulmonary Rehabilitation and Prevention, 37*(4), 274–278. https://doi.org/10.1097/HCR.0000000000000224

Russo-Neustadt, A. A., Alejandre, H., Garcia, C., Ivy, A. S., & Chen, M. J. (2004). Hippocampal brain-derived neurotrophic factor expression following treatment with reboxetine, citalopram, and physical exercise. *Neuropsychopharmacology: Official Publication of the American College of Neuropsychopharmacology, 29*(12), 2189–2199. https://doi.org/10.1038/sj.npp.1300514

Sampalo, M., Lázaro, E., & Luna, P.-M. (2023). Action video gaming and attention in young adults: A systematic review. *Journal of Attention Disorders, 27*(5), 530–538. https://doi.org/10.1177/10870547231153878

Shapoval, S., Gimeno-Santos, M., Mendez Zorrilla, A., Garcia-Zapirain, B., Guerra-Balic, M., Signo-Miguel, S., & Bruna-Rabassa, O. (2022). Serious games for executive functions training for adults with intellectual disability: Overview. *International Journal of Environmental Research and Public Health, 19*(18), 11369. https://doi.org/10.3390/ijerph191811369

Shimizu, N., Umemura, T., Matsunaga, M., & Hirai, T. (2017). An interactive sports video game as an intervention for rehabilitation of community-living patients with schizophrenia: A controlled, single-blind, crossover study. *PLoS One, 12*(11), e0187480. https://doi.org/10.1371/journal.pone.0187480

Sokolov, A. A., Collignon, A., & Bieler-Aeschlimann, M. (2020). Serious video games and virtual reality for prevention and neurorehabilitation of cognitive decline because of aging and neurodegeneration. *Current Opinion in Neurology, 33*(2), 239–248. https://doi.org/10.1097/WCO.0000000000000791

Stančin, K., Hoić-Božić, N., & Skočić Mihić, S. (2020). Using digital game-based learning for students with intellectual disabilities—A systematic literature review. *Informatics in Education, 19*, 323–341. https://doi.org/10.15388/infedu.2020.15

Takács, J., & Stauder, A. (2016). [The role of regular physical activity in the prevention and intervention of symptoms of anxiety and anxiety disorders]. *Psychiatria Hungarica: A Magyar Pszichiatriai Tarsasag Tudomanyos Folyoirata, 31*(4), 327–337.

Tark, R., Metelitsa, M., Akkermann, K., Saks, K., Mikkel, S., & Haljas, K. (2019). Usability, acceptability, feasibility, and effectiveness of a gamified mobile health intervention (Triumf) for pediatric patients: Qualitative study. *JMIR Serious Games, 7*(3), e13776. https://doi.org/10.2196/13776

Teng, Z., Pontes, H. M., Nie, Q., Griffiths, M. D., & Guo, C. (2021). Depression and anxiety symptoms associated with internet gaming disorder before and during the COVID-19 pandemic: A longitudinal study. *Journal of Behavioral Addictions, 10*(1), 169–180. https://doi.org/10.1556/2006.2021.00016

The Picture Exchange Communication System (PECS). (n.d.). Retrieved April 30, 2025, from https://nationalautismresources.com/the-picture-exchange-communication-system-pecs/?srsltid=AfmBOoqN57SBvAKvJnJL4lEz5X5O1tT3RU8INjk9WSfMTK9cDPiDfhVG

Tullett-Prado, D., Stavropoulos, V., Mueller, K., Sharples, J., & Footitt, T. A. (2021). Internet Gaming Disorder profiles and their associations with social engagement behaviours. *Journal of Psychiatric Research, 138*, 393–403. https://doi.org/10.1016/j.jpsychires.2021.04.037

Vajawat, B., Varshney, P., & Banerjee, D. (2021). Digital gaming interventions in psychiatry: Evidence, applications and challenges. *Psychiatry Research, 295*, 113585. https://doi.org/10.1016/j.psychres.2020.113585

Vieira, Á., Melo, C., Machado, J., & Gabriel, J. (2018). Virtual reality exercise on a home-based phase III cardiac rehabilitation program, effect on executive function, quality of life and depression, anxiety and stress: A randomized controlled trial. *Disability and Rehabilitation. Assistive Technology, 13*(2), 112–123. https://doi.org/10.1080/17483107.2017.1297858

Voon, K., Silberstein, I., Eranki, A., Phillips, M., Wood, F. M., & Edgar, D. W. (2016). Xbox KinectTM based rehabilitation as a feasible adjunct for minor upper limb burns rehabilitation: A pilot RCT. *Burns: Journal of the International Society for Burn Injuries, 42*(8), 1797–1804. https://doi.org/10.1016/j.burns.2016.06.007

Vorderer, P., Hartmann, T., & Klimmt, C. (2003). Explaining the enjoyment of playing video games: The role of competition. In *Proceedings of the second international conference on entertainment computing* (pp. 1–9). ACM.

Wagener, T. L., Fedele, D. A., Mignogna, M. R., Hester, C. N., & Gillaspy, S. R. (2012). Psychological effects of dance-based group exergaming in obese adolescents. *Pediatric Obesity, 7*(5), e68–e74. https://doi.org/10.1111/j.2047-6310.2012.00065.x

Wang, Z., Liu, X., Hu, Y., Zheng, H., Du, X., & Dong, G. (2019). Altered brain functional networks in internet gaming disorder: Independent component and graph theoretical analysis under a probability discounting task. *CNS Spectrums, 24*(5), 544–556. https://doi.org/10.1017/S1092852918001505

Waris, O., Jaeggi, S. M., Seitz, A. R., Lehtonen, M., Soveri, A., Lukasik, K., Söderström, U., Hoffing, R. A. C., & Laine, M. (2019). Video gaming and working memory: A large-scale cross-sectional correlative study. *Computers in Human Behavior, 97*, 94–103. https://doi.org/10.1016/j.chb.2019.03.005

Wartberg, L., Kriston, L., & Thomasius, R. (2020). Internet gaming disorder and problematic social media use in a representative sample of German adolescents: Prevalence estimates, comorbid depressive symptoms and related psychosocial aspects. *Computers in Human Behavior, 103*, 31–36. https://doi.org/10.1016/j.chb.2019.09.014

Wilms, I. L., Petersen, A., & Vangkilde, S. (2013). Intensive video gaming improves encoding speed to visual short-term memory in young male adults. *Acta Psychologica, 142*(1), 108–118. https://doi.org/10.1016/j.actpsy.2012.11.003

Winter, B., Breitenstein, C., Mooren, F. C., Voelker, K., Fobker, M., Lechtermann, A., Krueger, K., Fromme, A., Korsukewitz, C., Floel, A., & Knecht, S. (2007). High impact running improves learning. *Neurobiology of Learning and Memory, 87*(4), 597–609. https://doi.org/10.1016/j.nlm.2006.11.003

World Health Organization. (2019). *International classification of diseases 11th revision (ICD-11)*. WHO.

Xu, X., Li, J., Pham, T. P., Salmon, C. T., & Theng, Y.-L. (2016). Improving psychosocial well-being of older adults through exergaming: The moderation effects of intergenerational communication and age cohorts. *Games for Health Journal, 5*(6), 389–397. https://doi.org/10.1089/g4h.2016.0060

Yan, F., Costello, M., & Allen, J. (2020). Self-perception and relative increases in substance use problems in early adulthood. *Journal of Drug Issues, 50*(4), 538–549. https://doi.org/10.1177/0022042620941812

Yuen, H. K., Holthaus, K., Kamen, D. L., Sword, D. O., & Breland, H. L. (2011). Using Wii Fit to reduce fatigue among African American women with systemic lupus erythematosus: A pilot study. *Lupus, 20*(12), 1293–1299. https://doi.org/10.1177/0961203311412098

Zhang, A., Borhneimer, L. A., Weaver, A., Franklin, C., Hai, A. H., Guz, S., & Shen, L. (2019). Cognitive behavioral therapy for primary care depression and anxiety: A secondary meta-analytic review using robust variance estimation in meta-regression. *Journal of Behavioral Medicine, 42*(6), 1117–1141. https://doi.org/10.1007/s10865-019-00046-z

Zhuang, X., Zhang, Y., Tang, X., Ng, T. K., Lin, J., & Yang, X. (2023). Longitudinal modifiable risk and protective factors of internet gaming disorder: A systematic review and meta-analysis. *Journal of Behavioral Addictions, 12*(2), 375–392. https://doi.org/10.1556/2006.2023.00017

Part IV
Social Dynamics and Community in Gaming

Chapter 12
Social Interaction and Connectivity in Gaming Communities

Pedro Makoto Uchida, Lara Poor, Gabriel Kadri, Raphael Santos, Gabriel Rêgo, and Lucas Murrins Marques

Abstract The rise of online gaming has transformed the gaming experience into a complex social phenomenon, where communities emerge around shared interests, collaboration, and competitive play. This chapter explores the social dimensions of gaming communities, analyzing how online games facilitate social interaction, community formation, and identity construction. Differences between multiplayer and single-player communities are examined, highlighting their distinct social structures and cultural dynamics. Challenges such as toxicity, addiction, cyberbullying, and the limitations of online socialization are discussed, along with the potential benefits of supportive environments, particularly for marginalized groups like the LGBTQ+ community. By comparing online and real-world communities, we offer a nuanced perspective on how digital gaming environments impact social connectivity, emotional well-being, and the broader cultural landscape. The findings underscore the importance of fostering inclusive, healthy online spaces that balance virtual and offline interactions to promote individual and collective flourishing.

The original version of the chapter has been revised. A correction to this chapter can be found at https://doi.org/10.1007/978-3-032-00325-6_19

P. M. Uchida (✉) · L. Poor
Pontifícia Universidade Católica de Campinas, Campinas, São Paulo, Brazil

G. Kadri · R. Santos
Mental Health Department, Santa Casa de São Paulo School of Medical Sciences, São Paulo, Brazil

G. Rêgo
Human Developmental Sciences Graduate Program, Mackenzie Presbyterian University, São Paulo, Brazil

National Institute of Science and Technology in Social and Affective Neuroscience, INCT-SANI, São Paulo, Brazil

L. M. Marques
Mental Health Department, Santa Casa de São Paulo School of Medical Sciences, São Paulo, Brazil

Artificial Intelligence Research Division, Infinity Doctor's Inc., Dover, DE, USA

L. Murrins Marques et al. (eds.), *Social and Affective Neuroscience of Gaming*,
https://doi.org/10.1007/978-3-032-00325-6_12

Keywords Gaming communities · Social interaction · Multiplayer and single-player games · Online socialization · LGBTQ+ in gaming

Introduction

The gaming landscape has evolved beyond isolated, individual experiences, transforming into complex social environments where players interact, collaborate, and form meaningful connections. As digital gaming continues to grow, so does its impact on social dynamics, with communities forming around shared interests, competitive play, and narrative engagement. Within these virtual spaces, social connections are formed that not only mirror real-world relationships but also often transcend geographical boundaries, creating inclusive or exclusive communities based on shared gaming experiences.

The interplay between social dynamics and gaming communities is multifaceted. Online multiplayer games, in particular, serve as a powerful platform for socialization, fostering relationships that range from friendships to support networks, and even rivalries. The digital architecture of these platforms shapes how players connect, communicate, and collaborate, often blending traditional social norms with new, emergent forms of interaction unique to the gaming world (Cole & Griffiths, 2007).

Understanding these dynamics requires a careful examination of how gaming influences social connectivity and vice versa. The previous Part III, explored the intricate relationship between gaming and well-being/mental health aspects, shedding light on both the positive and negative effects of gaming on emotional health and cognitive processes. Building upon these themes, the current section shifts focus toward the social domain, examining how gaming not only impacts the individual player but also the communities that arise and thrive within these digital worlds.

This chapter sets the stage for this exploration by delving into the formation, structure, and dynamics of gaming communities. From the contrast between multiplayer and single-player communities to the unique social experiences they offer, we explore how gaming facilitates social interaction, community building, and even serves as a form of escape or refuge. Whether focusing on supportive communities or toxic subcultures, the aim is to unravel the complex web of social interactions that shape the gaming experience. Ultimately, this examination sheds light on the potential of gaming as a social phenomenon that extends beyond mere entertainment to influence real-world social connectivity, empathy, and belonging.

What Are Gaming Communities?

Video game communities are intricate networks of social interaction, uniting individuals who share an interest in gaming. These communities offer an environment for collaboration, competition, and exchange of information. Beyond mere entertainment, they represent dynamic social systems where participants build connections, share cultural values, and develop skills through their engagement. Unlike traditional online groups, gaming communities are inherently interactive, driven by the immersive and participatory nature of video games, which fosters deeper relationships and shared experiences (Taylor, 2006; Pearce, 2011).

The origins of gaming communities trace back to the 1970s, coinciding with the rise of arcade culture and early home consoles such as the Atari 2600. At the time, interactions between players occurred mainly in physical spaces, emphasizing local connections (Kent, 2001). As computer games and the internet advanced, these communities transitioned to virtual spaces, enabling connections that transcended geographic boundaries. Online multiplayer games of the 1990s, such as Doom and Quake, catalyzed this shift by offering unprecedented opportunities for players to collaborate and compete in real time across vast distances (Kowert & Oldmeadow, 2013).

By the early 2000s, MMORPGs like World of Warcraft and EverQuest solidified the concept of large-scale virtual gaming communities, integrating structured gameplay with social frameworks. Players not only participated in complex missions but also built long-lasting relationships within these virtual worlds (Pearce, 2011). The emergence of streaming platforms like Twitch and YouTube, alongside communication tools like Discord and Reddit, further expanded these communities, fostering subcultures, professional gaming networks, and fan-driven content creation (Johnson, 2017).

The social implications of gaming communities are profound and multifaceted. On one hand, these environments foster socialization by creating spaces where individuals with shared interests can meet, interact, and form lasting bonds. Participation in cooperative gameplay has been shown to enhance communication, problem-solving skills, and teamwork, often transcending cultural and linguistic barriers (Shaw, 2011). These interactions can foster a sense of belonging and contribute to personal development, particularly in collaborative and creative gaming scenarios.

The diversity of interactions within gaming communities offers unique opportunities for cultural exchange, creative expression, and identity formation. Players engage in activities that not only enhance their gaming experiences but also contribute to broader digital literacy and technological fluency (Taylor, 2006). Moreover, these communities play a pivotal role in shaping new forms of media consumption, influencing industries such as streaming, e-sports, and game development.

How Online Games Foster Social Interaction and Community Building and Differences Between Multiplayer and Single Player Gaming Communities

Online games play a significant role in fostering community formation by providing players with mechanisms for social interaction, collaboration, and relationship building. Single-player games are a game mode characterized by the player taking on the role of the protagonist, setting out on an individual journey without real-time interaction with other human players. In this context, players face challenges, solve puzzles, explore environments, and advance through the narrative presented by the game. This mode emphasizes an immersive and personalized experience, allowing players to engage deeply with the storyline and the game world. With technological advancements, game developers strive to innovate within single-player games, creating more immersive and dynamic worlds, featuring branching narratives that depend on player choices. Additionally, the use of artificial intelligence (where non-human characters make different decisions based on the main player's actions) and realistic graphics enhances the uniqueness of the experience (Tang et al., 2020).

In contrast, multiplayer games are a game mode defined by simultaneous interactions among multiple players, whether in a cooperative, competitive, or asymmetrical manner. Here, the essence of the gaming experience lies in interacting with other human players, which can occur through voice chat, text, or other forms of digital communication. It is important to note that many games aim to reach the broadest audience possible by offering both single-player and multiplayer modes, allowing players to enjoy both experiences within a single title. The creation of these hybrid modes seeks to meet the diverse objectives of various audiences, whether it be the narrative immersion of single-player or the competitiveness and cooperation of multiplayer. Massively multiplayer online games (MMOGs), particularly MMORPGs, are effective in this regard, as they require players to cooperate to complete missions and overcome challenges, creating a sense of teamwork and shared purpose (Hsiao & Chiou, 2012a; Warmelink & Siitonen, 2013). These collaborative environments often lead to the formation of stable social units, such as clans or guilds, where players interact regularly, build relationships, and develop a sense of belonging (Hsiao & Chiou, 2012b; Ang, 2011).

Although each mode of gameplay has its distinct objectives and characteristics, fans and enthusiasts from different modes often come together, either virtually or in person, forming large gaming communities. Each gaming community has its unique traits, but they can generally be summarized based on how they are organized. Much of what exists today in gaming communities derives from the original objectives and origins of each game mode. Singleplayer communities, while lacking direct interaction within the game, are incredibly active. Key objectives of these communities include collaborating by sharing tips and strategies to help members progress through the game's story. Additionally, these communities serve as fertile ground for discussions about narratives and plots, as well as philosophical debates and personal experience sharing (Albrechtslund, 2010; de Wildt & Aupers, 2019). Many players also dedicate hours to uncovering the minutiae and hidden secrets crafted by developers, known as easter eggs, fostering detailed discussions about every aspect of the game. Lastly, a competitive spirit often emerges in single-player settings, with players competing over high scores or who can "complete" the game in the shortest time, leading to indirect competitions, known as Speed Runs.

On the other hand, multiplayer communities organize themselves according to the nature and origin of the games, focusing more on cooperation and competition. In these communities, interactions center around sharing tactics, strategies, and forming teams, further enhancing competitiveness. As in any gaming environment, some individuals attempt to cheat, making discussions about toxic behavior and cheating increasingly common as they urge developers to dedicate more resources to anti-cheat systems. Guilds exemplify how games support community structures, enabling players to organize, assign roles, establish social norms, and work together toward common objectives (Hsiao & Chiou, 2012a; Koivisto, 2003; Park & Chung, 2011).

Game mechanics and environments further encourage both cooperation and competition, fostering sub-communities like guilds and supporting player-driven content, which strengthens the overall community (Koivisto, 2003). In-game communication tools such as chat and messaging systems also facilitate social interaction, allowing players to connect, coordinate activities, and forge social bonds (Hsu & Lu, 2007; Crenshaw & Nardi, 2016). Another essential element of community building in games is the sharing of game knowledge and experiences. Experienced players often mentor newcomers, fostering social ties and a collective identity (Hsiao & Chiou, 2012b; Gandolfi et al., 2021). Players with greater game knowledge frequently assume leadership roles, such as guild leaders, shaping community norms, and influencing development (Hsiao & Chiou, 2012b).

The design of games is instrumental in community formation, with reward systems that incentivize collaboration. Group-based rewards promote distributive justice and motivate players to cooperate, thereby strengthening community ties (Yuan et al., 2007). Furthermore, the architecture of games, defined by their mechanics and interfaces, influences social behaviors and the formation of network graphs that depict player interactions. This often results in the emergence of strongly bonded groups, essential for the community's stability and growth (Kirman, 2010). Outside the game environment, related platforms such as wikis, forums, and streaming services extend the sense of community. These platforms allow players to further connect, discuss strategies, share experiences, and build communities that transcend the boundaries of the game itself (Gandolfi et al., 2021; Loh & Kretschmer, 2022).

With the rise of eSports, the social, cultural, and economic impacts of the competitive landscape have become evident. Multiplayer communities, once scattered, have begun to organize themselves into dedicated fan bases, rallying to support teams and players much like traditional sports. The professionalization of players and the growth of these fan bases have also driven the structuring and organization of gaming communities. A notable example is the "Tribo do Gaules," a community formed around the Brazilian streamer Gaules, which stands out in the Counter-Strike scene. With a strong fanbase that follows streams and supports Brazilian teams in major tournaments, "A Tribo" operates like an organized fan group, establishing itself on platforms like Twitch, where thousands gather daily to interact, learn, and share moments around a common passion. This organization reflects the maturation of gaming communities within the eSports arena, transforming fans and players—who are now public figures—into integral parts of a professional and dedicated structure that showcases the commitment and seriousness of the competitive gaming industry.

Comparing Online and Real-Life Communities

Communities formed in online games differ from real-world communities in their interactions, formation, and the emotional and social dynamics involved (Etzioni & Etzioni, 1997; Warmelink & Siitonen, 2013). While both types of communities share similarities, such as group formation and social bonds around common interests (Barnett & Coulson, 2010), there are key distinctions in how these bonds are created and maintained.

In online games, communities often emerge based on shared objectives within the game rather than personal relationships, which are a core aspect of real-life communities (Barnett & Coulson, 2010). These virtual communities transcend geographical boundaries, allowing people from diverse backgrounds to connect and collaborate asynchronously, unlike real-life communities that depend on physical proximity and synchronous interactions. This broad reach of online communities offers inclusivity, particularly for individuals who are homebound due to illness or disability, and the ability to explore new identities without the limitations of physical appearance (Etzioni & Etzioni, 1997). However, the social dynamics in virtual settings tend to be more impersonal, focused on shared goals rather than deep personal connections (Memmi, 2008).

Real communities, on the other hand, are grounded in face-to-face interactions that facilitate emotional exchanges and social accountability. These interactions foster a sense of immediate connection and responsibility among members, which is often lacking in the more abstract social dynamics of online communities. Although virtual communities can accommodate large numbers of participants and retain strong collective memories, they generally lack the emotional depth and long-term involvement seen in real-world communities (Etzioni & Etzioni, 1997). Nevertheless, online and real-life communities do share fundamental similarities. Both are formed based on common interests and goals, and each can influence the other. In fact, some communities, referred to as "convergent communities," blend virtual and real interactions, creating hybrid social spaces where members engage both online and offline (Snodgrass et al., 2011).

Virtual worlds, especially in MMORPGs, showcase how accepted behaviors and valued traits strengthen online communities, mirroring the socialization patterns of real-world networks (Rea, 2010). Furthermore, the resilience of online communities, as demonstrated by the Uru diaspora, reveals that even when virtual spaces dissolve, members find new platforms to maintain their social bonds, much like diasporic communities in the real world (Pearce, 2011). A key advantage of online communities is their ability to provide platforms for self-exploration and identity building. Through avatars and game interactions, players can explore different facets of their identities, a process also observed in real-life communities where individuals express their identities through affiliations and social roles (Park & Chung, 2011; Costa Pinto et al., 2015). Moreover, online gaming environments serve as important spaces for social support, allowing players to receive both practical assistance and emotional encouragement, much like real communities (O'Connor et al.,

2015; Crenshaw & Nardi, 2016). However, the transient nature of virtual communities is evident, as seen in games like World of Warcraft, where approximately 25% of guilds disband monthly, highlighting their fragility compared to the more stable bonds of real-world settings (Hsiao & Chiou, 2012b).

Another significant difference lies in communication and interaction styles. In online games, communication is mediated through technology, which limits the availability of social cues and makes it difficult to interpret tone and intention (Yuan et al., 2007). This contrasts with the richness of face-to-face interaction in real-life communities, where nonverbal signals and social nuances play a crucial role in communication (Jia et al., 2015).

Additionally, online communities can vary greatly in size, ranging from small guilds to massive networks that span multiple servers. While real-world communities also vary in scale, they tend to be smaller and geographically concentrated, which strengthens the social ties among members (Ang, 2011; Badrinarayanan et al., 2014). The impact of these communities on real life also differs. Although friendships formed in online games can extend into the offline world, the broader influence of virtual communities on everyday life is often limited compared to real-life communities, which directly shape individuals' relationships, opportunities, and sense of belonging (Ang, 2011; Costa Pinto et al., 2015). However, the potential of online communities to create modified virtual environments that foster emotional involvement cannot be overlooked, as they offer a unique blend of social memory and inclusivity that complements the more affective and accountable nature of real communities (Sharkov & Kirillina, 2022).

In conclusion, while online and real-life communities differ in their structure and social dynamics, they are not mutually exclusive. Instead, they complement each other by offering a combination of strengths—online communities provide inclusivity and flexibility, while real-life communities excel in emotional depth and social accountability. Understanding the nuances of both types of communities is essential in appreciating their roles in shaping social interactions in the modern world (Etzioni & Etzioni, 1997; Carter, 2004).

Challenges of Socialization in Online Games

While online gaming offers valuable opportunities for interaction and community building, it also presents significant challenges compared to real-life socialization. One primary concern is the risk of addiction. Many players, especially children and adolescents, may become excessively immersed in gaming, often at the expense of their real-life responsibilities and relationships. This can result in irritability, a decline in social manners, and other negative behavioral shifts (Kaptsis et al., 2016; Yamada et al., 2021).

Though online games facilitate social connections, they can paradoxically contribute to social isolation when players prioritize virtual interactions over face-to-face relationships. This over-reliance on virtual socialization may decrease perceived social support and heighten feelings of loneliness, creating a disconnect between the virtual and real worlds (Hickerson & Mowen, 2013; Seay & Kraut, 2007).

Moreover, online gaming environments often expose players, particularly adolescents, to risks such as cyberbullying and harassment. The anonymity in these spaces exacerbates the potential for victimization, raising safety concerns that require careful monitoring, especially for younger players (Boccio & Leal, 2022). Competitive environments in online games also present challenges. Toxic behavior, including harassment, insults, and sabotage, is common in these settings, making the gaming experience unpleasant and socially unsafe for many players (Kou & Gui, 2021).

It is also important to recognize that the social skills acquired in online gaming environments do not always transfer smoothly to real-world settings. The absence of visual and auditory cues in online communication can make interpreting tone and intention difficult, creating challenges for players when engaging in face-to-face interactions (Warmelink & Siitonen, 2013). Despite these challenges, online games can provide an important outlet for social support, particularly for individuals who feel isolated in real life (O'Connor et al., 2015).

Ultimately, while the socialization opportunities presented by online gaming are valuable, they must be balanced with real-world interactions to mitigate negative impacts. Understanding the individual context, the type of game, and the dynamics of the online community is essential for promoting healthy social engagement in gaming.

Toxic Groups in the Online Cmmunity

As discussed earlier, seeking refuge in online gaming communities can provide temporary relief but may hinder individuals from addressing the underlying challenges of their daily lives, potentially resulting in a cycle of dependency. Many maladaptive social interactions online appear to stem from these individuals struggling to adapt to the challenges of the real world. People with low life satisfaction, frustration with the world, sedentary lifestyles, and diminished self-clarity are often associated with digital escapism (Marques et al., 2023). Studies also show a significant correlation between escapism, the lack of online support for real-world problems, and Internet gaming disorder (IGD) (Kaczmarek & Drążkowski, 2014; Lin et al., 2021; Marques et al., 2023).

The knowledge about the predictors and symptoms of IGD is extensive, with the disorder being linked to various mental health issues (Humphreys et al., 2019). Research reveals that IGD is significantly associated with toxic behavior, such as reduced empathy, and toxic attitudes within gaming communities (Gandolfi et al., 2023). Toxic behavior is more prevalent in certain types of games, such as Multiplayer Online Battle Arena (MOBA), first-person shooters (FPS), and MMORPGs. Which are some of the industry's most popular genres, attracting millions of players. According to (Kristiansen et al., 2023) there is a higher incidence of types of conflicts in specific online gaming styles. Competitive online gaming environments, for example, often exhibit higher levels of toxicity compared to more casual games. These toxic behaviors directly impact social connections within the game, thus hindering healthy socialization of a large number of gaming communities.

With the increasing accessibility of virtual games, especially for children, new concerns have arisen regarding developmental impacts and the rise in cyberbullying cases. Research on young people suggests that prolonged and pathological use of online games from an early age may exacerbate changes in brain functioning, emphasizing the need for prevention and early intervention (Sugaya, 2019). Studies, such as Cole and Griffiths (2007) show that one of the most significant predictors of IGD is the time spent gaming, which becomes more relevant given the negative correlation between age and average time spent on online games.

Cyberbullying has a profoundly negative impact on mental health. Common consequences of this form of harassment include depression, suicidal ideation, anxiety, hostility or aggressive behaviors, substance use and abuse, low self-esteem, peer relationship difficulties, emotional distress, loneliness, and a significant reduction in life satisfaction (Kwan et al., 2020). According to Trepte et al. (2011), the negative effects of online gaming on children at risk of excessive gaming can be mitigated through activities that foster offline friendships, promote physical and social proximity, and strengthen familiarity during gameplay, supporting the formation of social capital.

Given the above, it is equally important to highlight the opportunities and benefits that online gaming communities can offer. For example, MMORPGs can sometimes be toxic environments that attract individuals struggling to adapt to the real world; they also provide spaces for socialization and the promotion of mental health (Kristiansen et al., 2023; Marques et al., 2023; O'Connor et al., 2015). Numerous studies highlight the benefits of social interaction in these games, especially in MMOs. Cole and Griffiths (2007) found that 76.2% of women and 74.7% of men in the study reported having made "good friendships within the game" referring to MMORPGs. Additionally, Reer and Krämer (2018) suggest that social outcomes in MMOs and FPS games vary depending on the player's motivation and gaming habits. Thus, how a player engages with the virtual world, his environment, and motivation greatly affect both positive and negative aspects of the experience.

In conclusion, to balance online socialization with real-world interactions and mitigate negative impacts, there should be early detection of individuals exhibiting characteristics related to toxic traits, enabling timely intervention, effective treatment, and preventive measures, ultimately fostering a healthier and more inclusive online community.

LGBTQ+ Community in the Gaming World

Although video games are increasingly common among young people, sexual and gender diverse youth (SGDY) are sometimes a neglected population in studies on the subject (Di Cesare et al., 2023). The emergence of this theme reveals both positive and negative aspects in the relationship between this population and video games: on one hand, online gaming communities can promote a sense of belonging and important social support for marginalized groups; on the other hand, they may expose them to discrimination through digital microaggressions and hate speech, with mental health implications (Di Cesare et al., 2023; Gillin & Signorella, 2024).

Studies show that social support is essential for the formation of gender and sexual orientation identities. Not only that but also they suggest that SGDY's involvement with virtual LGBTQ+ communities has important impacts on their lives. Often, young people virtually encounter, for the first time, others with whom they identify and share similar experiences, creating an affirming space (Li et al., 2023). The promotion of safe spaces makes it possible to find an escape from stigma, violence, and even rejection and marginalization originating from family and offline relationships (Austin et al., 2020).

As evidenced by Marques et al. (2023), the motivation for escapism, that is, the desire to escape reality through video games, among transgender people, is sometimes related to the need to live their gender identities and represent themselves through avatars. For McKenna et al. (2024), avatar customization is an attractive feature and can be fundamental for young people to explore and experiment with new identities online. Thus, by entering these communities, the possibility of being authentic and accepted by peers fosters an alternative and powerful socio-emotional environment for experimentation (Austin et al., 2020).

The LGBTQ+ Community has been growing in the gaming world over the years. Despite still having low representation, new narratives, characters, elements, and locations have emerged, including greater diversity in productions (Shaw & Friesem, 2016). Additionally, new immersive technologies intensify the formation of online LGBTQ+ communities through interactions and embodied experiences and can also be used for voice and communication training, helping align voice with gender identity for people suffering from gender dysphoria (Li et al., 2023; Smith et al., 2023).

Communities foster social interaction and a sense of hope associated with the desire to live, as well as the willingness to give back, that is, to help other individuals in the community who have gone through similar challenges (Austin et al., 2020). A variety of studies have identified numerous benefits, most notably the establishment of support networks and the provision of emotional support and guidance applicable to both online and offline contexts (Li et al., 2023). This phenomenon is particularly noteworthy, as such dynamics are seldom observed in studies not centered on LGBTQ+ populations.

Interestingly, the different modes of verbal and non-verbal communication available in VR also enhance the feeling of co-presence between individuals, making virtual interactions resemble face-to-face interactions and intensifying user engagement in the same space, which strengthens relationships (Li et al., 2023). Undoubtedly, video games spark creativity, benefiting personal development and the exercise of coping strategies. However, they also intensely expose these young people to violence, potentially contributing to maladaptive coping behaviors (Di Cesare et al., 2023). In a study that evaluated Chinese students described as young men who have sex with men, a prevalence of 13% for Internet Gaming Disorder was observed. A strong association with depressive behaviors and addictive behaviors was also observed in these cases, along with a positive correlation with loneliness and hopelessness and a negative correlation with self-control.

Other studies show that lesbian, gay, and bisexual populations are more vulnerable to mental disorders compared to heterosexual individuals (King et al., 2008). According to Broman and Hakansson (2018), problematic gaming and internet use may be more common in non-heterosexual populations, with the exception of gambling, which is not associated with sexual orientation. Researchers suggest that lesbian, gay, and bisexual individuals may experience significantly greater psychological distress related to this. Furthermore, although marginalized groups are more likely to be exposed to hostile environments in games, both heterosexual players and LGBTQ+ individuals are targeted by negative comments related to sexual orientation and gender identity, which reinforces the violence cultivated in virtual environments (Gillin & Signorella, 2024). Regarding this population specifically, both the development of problematic gaming behavior and the low representation of diversity in games, as well as the limited adaptiveness of coping skills developed through video gaming, present challenges (Di Cesare et al., 2023).

In light of this, it is notable that research emphasizes the potential of video games for sexual and gender diverse youth, while also mentioning the existence of behaviors and environments harmful to their mental health. Not many articles were found on negative outcomes, indicating that this is still an underexplored topic and one with limited access to information. Therefore, more researchers need to investigate how video games can impact them, bringing to light negative aspects in greater depth. This is a vulnerable population both in offline and online life, making it crucial to consider sexual orientation and gender identity to understand their specificities, which often differ from the cisheteronormative population.

Conclusion

In conclusion, gaming communities represent a unique intersection of social connectivity and digital innovation, blending opportunities for meaningful interaction with challenges that demand critical attention. These communities not only foster collaboration, creativity, and identity exploration but also face issues such as toxicity, exclusion, and the risks of over-reliance on virtual interactions. The dual nature of these spaces highlights the importance of intentional game design and supportive community structures that promote inclusivity and emotional well-being. By leveraging their strengths while addressing their vulnerabilities, gaming communities have the potential to move beyond entertainment, becoming influential platforms for personal growth, social engagement, and cultural exchange in the digital era.

Acknowledgements GGR is supported by CNPq - INCT (National Institute of Science and Technology on Social and Affective Neuroscience, grant n. 406463/2022-0).

References

Albrechtslund, A.-M. (2010). Gamers telling stories understanding narrative practices in an online community. *Convergence: The International Journal of Research Into New Media Technologies, 16*, 112–124. https://doi.org/10.1177/1354856509348773

Ang, C. S. (2011). Interaction networks and patterns of guild community in massively multiplayer online games. *Social Network Analysis and Mining, 1*, 341–353. https://api.semanticscholar.org/CorpusID:23760488

Austin, A., Craig, S., Navega, N., & McInroy, L. (2020). It's my safe space: The life-saving role of the internet in the lives of transgender and gender diverse youth. *International Journal of Transgender Health, 21*, 1–12. https://doi.org/10.1080/15532739.2019.1700202

Badrinarayanan, V., Sierra, J., & Taute, H. (2014). Determinants and outcomes of online brand tribalism: Exploring communities of massively multiplayer online role playing games (MMORPGs). *Psychology & Marketing, 31*, 853–870. https://doi.org/10.1002/mar.20739

Barnett, J., & Coulson, M. (2010). Virtually real: A psychological perspective on massively multiplayer online games. *Review of General Psychology, 14*, 167–179. https://doi.org/10.1037/a0019442

Boccio, C., & Leal, W. (2022). Does socializing in the virtual world impact victimization in the real world? *Journal of Interpersonal Violence, 38*, 088626052211099. https://doi.org/10.1177/08862605221109922

Broman, N., & Hakansson, A. (2018). Problematic gaming and internet use but not gambling may be overrepresented in sexual minorities—A pilot population web survey study. *Frontiers in Psychology, 9*, 2184. https://doi.org/10.3389/fpsyg.2018.02184

Carter, D. (2004). Living in virtual communities: Making friends online. *Journal of Urban Technology, 11*, 109–125. https://doi.org/10.1080/10630730500064448

Cole, H., & Griffiths, M. D. (2007). Social interactions in massively multiplayer online role-playing gamers. *Cyberpsychology & Behavior, 10*(4), 575–583. https://doi.org/10.1089/cpb.2007.9988

Costa Pinto, D., Reale, G., Segabinazzi, R., & Rossi, C. (2015). Online identity construction: How gamers redefine their identity in experiential communities. *Journal of Consumer Behaviour, 14*, 399–409. https://doi.org/10.1002/cb.1556

Crenshaw, N., & Nardi, B. A. (2016). "It was more than just the game, it was the community": Social affordances in online games. In *2016 49th Hawaii international conference on system sciences (HICSS)* (pp. 3781–3790). IEEE. https://api.semanticscholar.org/CorpusID:5756493

de Wildt, L., & Aupers, S. (2019). Pop theology: Forum discussions on religion in videogames. *Information, Communication & Society, 23*, 1–19. https://doi.org/10.1080/1369118X.2019.1577476

Di Cesare, D., Craig, S., Brooks, A., & Doll, K. (2023). Setting the game agenda: Reviewing the emerging literature on video gaming and psychological well-being of sexual and gender diverse youth. *Games and Culture, 19*, 155541202311788. https://doi.org/10.1177/15554120231178883

Etzioni, A., & Etzioni, O. (1997). Communities: Virtual vs. real. *Science, 277*(5324), 295. https://doi.org/10.1126/science.277.5324.295

Gandolfi, E., Ferdig, R., & Soyturk, I. (2021). Exploring the learning potential of online gaming communities: An application of the Game Communities of Inquiry Scale. *New Media & Society, 25*, 146144482110271. https://doi.org/10.1177/14614448211027171

Gillin LE, Signorella ML. (2024). Attitudes Toward Sexual Orientation and Gender Identity in Online Multiplayer Gaming Spaces. *Psychol*, 127(6):3066–3088. https://doi.org/10.1177/00332941231153798. Epub 2023 Jan 23. PMID: 36688329; PMCID: PMC11529123

Gandolfi, E., Ferdig, R., Krause, K., Copus, A., Ostrowski-Delahanty, S., & Alemagno, S. (2023). Problematic gaming at a crossroad: Exploring the interplay between internet gaming disorder, toxic attitudes, and empathy in digital entertainment. *Games and Culture, 20*, 633–656. https://doi.org/10.1177/15554120231211991

Hickerson, B., & Mowen, A. (2013). Behavioral and psychological involvement of online video gamers: Building blocks or building walls to socialization? *Loisir et Société/Society and Leisure, 35*, 79–103. https://doi.org/10.1080/07053436.2012.10707836

Hsiao, C.-C., & Chiou, J.-S. (2012a). The effects of a player's network centrality on resource accessibility, game enjoyment, and continuance intention: A study on online gaming communities. *Electronic Commerce Research and Applications, 11*, 75–84. https://doi.org/10.1016/j.elerap.2011.10.001

Hsiao, C.-C., & Chiou, J.-S. (2012b). The impact of online community position on online game continuance intention: Do game knowledge and community size matter? *Information & Management, 49*, 292–300. https://doi.org/10.1016/j.im.2012.09.002

Hsu, C.-L., & Lu, H.-P. (2007). Consumer behavior in online game communities: A motivational factor perspective. *Computers in Human Behavior, 23*, 1642–1659. https://doi.org/10.1016/j.chb.2005.09.001

Humphreys, G., Higuchi, S., Achab, S., Poznyak, V., & O'Brien, C. (2019). Sharpening the focus on gaming disorder. *Bulletin of the World Health Organisation, 97*, 382–383. https://doi.org/10.2471/BLT.19.0o20619

Jia, L. (Adele), Shen, S., Bovenkamp, R., Iosup, A., Kuipers, F., & Epema, D. (2015). Socializing by gaming. *ACM Transactions on Knowledge Discovery from Data, 10*, 1–29. https://doi.org/10.1145/2736698

Johnson, M. R., & Woodcock, J. (2017). 'It's like the gold rush': *the lives and careers of professional video game streamers on Twitch.tv. Information, Communication & Society*, 22(3), 336–351. https://doi.org/10.1080/1369118X.2017.1386229

Kaczmarek, L., & Drążkowski, D. (2014). MMORPG escapism predicts decreased well-being: Examination of gaming time, game realism beliefs, and online social support for offline problems. *Cyberpsychology, Behavior, and Social Networking, 17*, 298–302. https://doi.org/10.1089/cyber.2013.0595

Kaptsis, D., King, D., Delfabbro, P., & Gradisar, M. (2016). Withdrawal symptoms in internet gaming disorder: A systematic review. *Clinical Psychology Review, 43*, 58–66. https://doi.org/10.1016/j.cpr.2015.11.006

Kent, S. L. (2001). *The ultimate history of video games: From Pong to Pokemon—The story behind the craze touched our lives and changed the world* (1st ed.). Three Rivers Press.

King, M., Semlyen, J., Tai, S., Killaspy, H., Osborn, D., Popelyuk, D., & Nazareth, I. (2008). A systematic review of mental disorder, suicide, and deliberate self harm in lesbian, gay and bisexual people. *BMC Psychiatry, 8*, 70. https://doi.org/10.1186/1471-244X-8-70

Kirman, B. (2010). Emergence and playfulness in social games. In *Proceedings of the 14th international academic MindTrek conference: Envisioning future media environments* (pp. 71–77). ACM. https://doi.org/10.1145/1930488.1930504

Koivisto, E. M. I. (2003, September). Supporting communities in massively multiplayer online role-playing games by game design. In *DiGRA conference.* https://api.semanticscholar.org/CorpusID:1847404

Kou, Y., & Gui, X. (2021). Flag and flaggability in automated moderation the case of reporting toxic behavior in an online game community. In *CHI conference on human factors in computing systems (CHI '21)* (pp. 1–12). ACM. https://doi.org/10.1145/3411764.3445279

Kowert, R., & Oldmeadow, J. A. (2013). (A)Social reputation: Exploring the relationship between online video game involvement and social competence. *Computers in Human Behavior, 29*(4), 1872–1878. https://doi.org/10.1016/j.chb.2013.03.003

Kristiansen, A., Lindberg, F., & Tempelhaug, A. (2023). Trouble in virtual heaven: Origin and consequences of social conflict in online consumption communities. *Journal of Consumer Culture, 23*(3), 575–596. https://doi.org/10.1177/14695405221127347

Li, L., Freeman, G., Schulenberg, K., & Acena, D. (2023). "We cried on each other's shoulders": How LGBTQ+ individuals experience social support in social virtual reality. In *Proceedings of the 2023 CHI conference on human factors in computing systems* (pp. 1–16). ACM. https://doi.org/10.1145/3544548.3581530

Lin PC, Yen JY, Lin HC, Chou WP, Liu TL, Ko CH. (2021). Coping, Resilience, and Perceived Stress in Individuals with Internet Gaming Disorder in Taiwan. *Int J Environ Res Public Health* Feb 11;18(4):1771. https://doi.org/10.3390/ijerph18041771. PMID: 33670343; PMCID: PMC7917620

Loh, J., & Kretschmer, T. (2022). Online communities on competing platforms: Evidence from game wikis. *Strategic Management Journal, 44*, 441–476. https://doi.org/10.1002/smj.3442

Kwan I, Dickson K, Richardson M, MacDowall W, Burchett H, Stansfield C, Brunton G, Sutcliffe K, Thomas J. 2020. Cyberbullying and Children and Young People's Mental Health: *A Systematic Map of Systematic Reviews. Cyberpsychol Behav Soc Netw*, Feb; 23(2):72–82. https://doi.org/10.1089/cyber.2019.0370. Epub 2020 Jan 23. PMID: 31977251; PMCID: PMC7044782

Marques, L. M., Uchida, P. M., Aguiar, F. O., Kadri, G., Santos, R. I. M., & Barbosa, S. P. (2023). Escaping through virtual gaming—What is the association with emotional, social, and mental health? A systematic review. *Frontiers in Psychiatry, 14*, 1257685. https://doi.org/10.3389/fpsyt.2023.1257685

McKenna, J. L., Wang, Y. C., Williams, C. R., McGregor, K., & Boskey, E. R. (2022). "You can't be deadnamed in a video game": Transgender and gender diverse adolescents' use of video game avatar creation for genderaffirmation and exploration. *Journal of LGBT Youth* , 21(1), 29–49. https://doi.org/10.1080/19361653.2022.2144583

Memmi, D. (2008). The nature of virtual communities. In S. Gill (Ed.), *Cognition, communication and interaction: Transdisciplinary perspectives on interactive technology* (pp. 70–82). Springer. https://doi.org/10.1007/978-1-84628-927-9_5

O'Connor, E., Longman, H., White, K., & Obst, P. (2015). Sense of community, social identity and social support among players of massively multiplayer online games (MMOGs): A qualitative analysis. *Journal of Community & Applied Social Psychology, 25*, 459–473. https://doi.org/10.1002/casp.2224

Pai, L., Yen, J.-Y., Lin, H.-C., Chou, W.-P., Liu, Z., & Ko, C.-H. (2021). Coping, resilience and perceived stress in individuals with internet gaming disorder in Taiwan. *International Journal of Environmental Research and Public Health, 18*, 1771. https://doi.org/10.3390/ijerph18041771

Park, S., & Chung, N. (2011). Mediating roles of self-presentation desire in online game community commitment and trust behavior of massive multiplayer online role-playing games. *Computers in Human Behavior, 27*, 2372–2379. https://doi.org/10.1016/j.chb.2011.07.016

Pearce, C. (2011). *Communities of play: Emergent cultures in multiplayer games and virtual worlds*. MIT Press. https://books.google.com.br/books?id=LwkFsfR8jDsC

Rea, A. (2010). The virtual social continuum expressed: Interaction and community formation in MMORPGs. In *Handbook of research on social interaction technologies and collaboration software: Concepts and trends* (pp. 336–349). IGI Global.

Reer, F., & Krämer, N. (2018). Are online role-playing games more social than multiplayer first-person shooters? Investigating how online gamers' motivations and playing habits are related to social capital acquisition and social support. *Entertainment Computing, 29*, 1–9. https://doi.org/10.1016/j.entcom.2018.10.002

Seay, A., & Kraut, R. (2007). Project massive: Self-regulation and problematic use of online gaming. In *Conference on human factors in computing systems – Proceedings* (pp. 829–838). ACM. https://doi.org/10.1145/1240624.1240749

Sharkov, F., & Kirillina, N. (2022). Dichotomy of the personal and the public in virtual communities. *Communicology, 10*, 23–33. https://doi.org/10.21453/2311-3065-2022-10-4-23-33

Shaw, A. (2011). Do you identify as a gamer? Gender, race, sexuality, and gamer identity. *New Media & Society, 14*(1), 28–44. https://doi.org/10.1177/1461444811410394

Shaw, A., & Friesem, E. (2016). Where is the Queerness in Games?: Types of Lesbian, Gay, Bisexual, Transgender, and Queer Content in Digital Games. *International Journal of Communication*, 10, 13. https://ijoc.org/index.php/ijoc/article/view/5449

Smith, C., Gregory, C., & Bryant, L. (2023). Utilizing virtual reality for gender-affirming voice training: Surveying the attitudes and perspectives of potential consumers. *International Journal of Language & Communication Disorders, 59*, 976–990. https://doi.org/10.1111/1460-6984.12968

Snodgrass, J., Lacy, M., Dengah, H., II, & Fagan, J. (2011). Enhancing one life rather than living two: Playing MMOs with offline friends. *Computers in Human Behavior, 27*, 1211–1222. https://doi.org/10.1016/j.chb.2011.01.001

Sugaya N, Shirasaka T, Takahashi K, Kanda H. (2019). Bio-psychosocial factors of children and adolescents with internet gaming disorder: *a systematic review. Biopsychosoc Med*, Feb 14;13:3 https://doi.org/10.1186/s13030-019-0144-5. PMID: 30809270; PMCID: PMC6374886

Tang, C, Wang, Z, Sima, Xand Zhang, L. (2020) Research on Artificial Intelligence Algorithm and Its Application in Games, *2nd International Conference on Artificial Intelligence and Advanced Manufacture (AIAM)* Manchester, United Kingdom, pp. 386–389, https://doi.org/10.1109/AIAM50918.2020.00085

Taylor, T. L. (2006). *Play between worlds: Exploring online game culture*. The MIT Press. https://doi.org/10.7551/mitpress/5418.001.0001

Trepte, S., Reinecke, L., & Juechems, K. (2011). The social side of gaming: How playing online computer games creates online and offline social support. *Computers in Human Behavior, 28*, 832–839. https://doi.org/10.1016/j.chb.2011.12.003

Warmelink, H., & Siitonen, M. (2013). A decade of research into player communities in online games. *Journal of Gaming & Virtual Worlds, 5*, 271–293. https://doi.org/10.1386/jgvw.5.3.271_1

Yamada, M., Sekine, M., & Tatsuse, T. (2021). Pathological gaming and its association with lifestyle, irritability, and school and family environments among Japanese elementary school children. *Journal of Epidemiology, 33*, 335–341. https://doi.org/10.2188/jea.JE20210365

Yuan, R., Zhao, L., & Wang, W. (2007). Cooperation and competition dynamics in an online game community. In *Online communities and social computing* (Vol. 4564, pp. 475–484). Springer. https://doi.org/10.1007/978-3-540-73257-0_52

Chapter 13
Empathy and Perspective-Taking in Multiplayer Gaming

Victor Henrique Oyamada Otani

Abstract This chapter examines the potential of multiplayer games to foster empathy and perspective-taking through interactive, emotionally immersive experiences. Drawing on interdisciplinary research, it explores how digital environments—particularly those incorporating role-playing, cooperative tasks, and narrative complexity—engage both cognitive empathy (understanding others' perspectives) and emotional empathy (sharing their feelings). The chapter highlights how game mechanics such as collaboration, moral decision-making, and virtual embodiment can create powerful opportunities for emotional growth and prosocial behavior. It also addresses key limitations, including online toxicity, narrative superficiality, and digital inequality, advocating for inclusive and ethically grounded design. Practical applications are discussed across clinical, educational, and community settings, where games have shown promise in supporting emotional regulation, social connection, and resilience. The chapter concludes by calling for interdisciplinary collaboration and rigorous research to ensure that empathy-enhancing games are both effective and accessible, helping to cultivate emotional intelligence in increasingly digital societies.

Keywords Empathy · Perspective-taking · Multiplayer games · Virtual reality · Social-emotional development

---

V. H. O. Otani (✉)
Mental Health Department, Santa Casa de São Paulo School of Medical Sciences, São Paulo, SP, Brazil

Artificial Intelligence Research Division, Infinity Doctor's Inc., Dover, DE, USA
e-mail: victorotani@scioed.com

L. Murrins Marques et al. (eds.), *Social and Affective Neuroscience of Gaming*,
https://doi.org/10.1007/978-3-032-00325-6_13

Introduction

Empathy—the ability to understand and share another person's emotional experience—and perspective-taking—the cognitive process of viewing a situation from someone else's point of view—are essential pillars of human interaction. These social-cognitive capacities allow individuals to attune to others' emotional states, anticipate intentions, and engage in meaningful communication. On an interpersonal level, they help people build trust, navigate conflicts, and foster emotionally intelligent relationships. On a societal scale, empathy and perspective-taking contribute to the reduction of prejudice, the promotion of inclusivity, and the cultivation of shared understanding across cultural, ethnic, and ideological boundaries (Belman & Flanagan, 2009; Schrier & Farber, 2022; Jeon et al., 2024).

Historically, these abilities were nurtured through face-to-face social experiences—interactions in schools, families, therapeutic settings, and community spaces—where individuals could learn to read body language, interpret tone of voice, and engage in emotionally charged conversations. These in-person contexts offered structured opportunities to explore one's own emotions while responding appropriately to the emotional states of others. Emotional intelligence was cultivated over time through repeated exposure to nuanced social feedback and active reflection. However, the increasing digital nature of our lives has altered how and where these capacities are developed. As socialization increasingly shifts to online platforms, the question arises: Can virtual spaces provide the same emotional and interpersonal learning opportunities?

Emerging research suggests that digital environments—especially multiplayer games—offer rich, complex contexts for the practice and refinement of empathy. Unlike solitary or purely competitive games, multiplayer formats require players to interact, cooperate, negotiate, and sometimes compete in scenarios that often mirror real-world emotional and ethical complexities. These games are not passive experiences; rather, they are dynamic, interactive simulations where players must read social cues, respond to others in real time, and make decisions that influence both their own outcomes and the collective experience. The active participation required in multiplayer games provides fertile ground for cultivating both cognitive and emotional empathy (Jones et al., 2022; Otani et al., 2024; Ho & Ng, 2020; Ma et al., 2021).

Narrative-driven games, for instance, often place players in the shoes of characters facing hardship, marginalization, or moral dilemmas. By experiencing the world through these avatars, players are invited to feel the emotional weight of different perspectives, fostering an affective connection with characters who may represent people quite different from themselves. Meanwhile, in team-based or cooperative games, successful play hinges on players' ability to collaborate, anticipate others' needs, and align on shared goals—processes that inherently involve empathy and perspective-taking (Otani et al., 2024; Ma et al., 2021).

The use of immersive technologies such as virtual reality (VR) further enhances this potential. VR has been shown to significantly increase players' sense of embodiment—experiencing the virtual body as their own—and this embodied simulation has been linked to stronger empathic responses (Henning et al., 2024). Players using VR to engage in perspective-taking tasks report increased cognitive empathy and more favorable attitudes toward the groups they simulate. Role-playing mechanics, whether through VR or traditional RPG formats, deepen emotional engagement by encouraging players to inhabit alternate identities, confront ethically ambiguous situations, and reflect on the emotional consequences of their actions (Schrier & Farber, 2022; Otani et al., 2024; Djamnezhad et al., 2023).

This convergence of storytelling, embodiment, and interaction suggests that multiplayer games can function as emotionally resonant training grounds for empathy. While traditional methods of social–emotional learning are invaluable, the structured and immersive affordances of digital games present a complementary tool—one that is accessible, engaging, and increasingly integrated into the daily lives of young people and adults alike. These games do not just mirror real-life social complexity; they allow individuals to experiment with emotional responses, explore ethical decision-making, and practice interpersonal skills in a repeatable and safe context.

As digital games continue to evolve in sophistication and popularity, their potential as platforms for social–emotional development deserves close academic and clinical examination. In an era marked by increasing social disconnection, polarization, and mental health challenges, multiplayer games offer a novel, scalable, and culturally adaptable means of promoting empathy, reinforcing social bonds, and enhancing emotional well-being. Their role in fostering resilience, compassion, and collective understanding warrants not only curiosity—but deliberate, interdisciplinary exploration.

Defining Empathy and Perspective-Taking

Empathy is not a singular ability, but a complex and multifaceted psychological construct composed of two interdependent yet distinct dimensions: cognitive empathy and emotional (or affective) empathy. These components work together to enable individuals to understand and resonate with others' experiences, forming the basis of prosocial behavior, moral reasoning, and effective interpersonal communication (Belman & Flanagan, 2009; Schrier & Farber, 2022; Jeon et al., 2024).

Cognitive empathy refers to the mental capacity to recognize, infer, and understand another person's emotions, thoughts, or beliefs through intentional perspective-taking and analytical reasoning. It involves imagining what another person is feeling or thinking, often described metaphorically as "putting oneself in someone else's shoes." This form of empathy is crucial for social problem-solving, negotiation, and conflict resolution, as it allows individuals to anticipate others' reactions and adjust their own behavior accordingly.

Emotional empathy, by contrast, involves a more intuitive and affective response—a visceral, often immediate, resonance with another person's emotional state. It allows individuals to feel with others, sharing their emotional experiences through internal mirroring or emotional contagion. Emotional empathy is fundamental to human connection, fostering compassion, concern, and a sense of emotional solidarity.

Importantly, while these two forms of empathy can operate independently, their integration is a hallmark of advanced emotional intelligence. The ability to fluidly transition between understanding someone's emotional state (cognitive empathy) and feeling it alongside them (emotional empathy) enables individuals to offer both reasoned insight and authentic emotional support. This dual capacity has been linked to better self-regulation, stress management, and the cultivation of meaningful, reciprocal relationships (Schrier & Farber, 2022; Otani et al., 2024).

Neuroscientific studies further support the distinction and complementarity of these capacities. While emotional empathy has been associated with the activation of brain regions such as the anterior insula and limbic structures (related to emotional salience and affective processing), cognitive empathy more commonly engages the medial prefrontal cortex and temporoparietal junction—areas implicated in theory of mind and social reasoning (Jeon et al., 2024; Djamnezhad et al., 2023).

In this light, multiplayer games represent a compelling and increasingly credible environment for practicing and refining both types of empathy. These interactive spaces immerse players in complex social ecosystems that demand real-time coordination, cooperation, and moral navigation. Whether negotiating alliances in cooperative quests or reacting to the emotional arcs of characters in story-driven campaigns, players are constantly called upon to interpret social cues, predict others' intentions, and respond adaptively—all hallmarks of cognitive empathy.

Simultaneously, emotional empathy is often engaged through narrative immersion and high-stakes interpersonal dynamics within the game. Shared moments of triumph, failure, sacrifice, or betrayal evoke real emotional responses that mirror those found in offline group experiences. The psychological investment players place in their avatars or teammates frequently leads to emotional attune, even within virtual contexts (Belman & Flanagan, 2009; Henning et al., 2024; Ma et al., 2021).

These in-game processes are not merely anecdotal. Empirical evidence supports the idea that emotionally and socially immersive games can enhance empathy-related outcomes beyond the digital realm. For example, narrative-rich role-playing games and VR simulations have been shown to improve users' ability to take on unfamiliar perspectives and generate feelings of compassion for characters who differ significantly from themselves (Otani et al., 2024; Henning et al., 2024). One notable study demonstrated that embodying a non-player character's perspective in a prosocial VR game increased both empathy and feelings of interpersonal closeness, highlighting the potential of virtual embodiment to shift real emotional understanding (Ma et al., 2021).

Moreover, role-playing games have demonstrated therapeutic and educational efficacy in structured interventions. When integrated into classroom activities or clinical settings, RPGs have helped adolescents and young adults improve emotional regulation, enhance social interaction skills, and build confidence in navigating interpersonal scenarios (Henning et al., 2024). These applications affirm the versatility of games as both educational and psychosocial tools.

What is particularly significant is that the empathy fostered through digital play does not appear to be confined to the screen. When games are intentionally designed to evoke emotional reflection and social interactivity, they can become effective practice grounds for real-life empathy. Players rehearse, in psychologically safe yet emotionally rich contexts, how to respond to others with insight and care. Over time, these rehearsals can inform and reinforce everyday behavior, making empathy not only a temporary emotional state but also a learned, transferable skill.

As digital play becomes a central modality through which young people—and increasingly, adults—explore identity, community, and morality, understanding how gaming environments contribute to social–emotional development is not only timely but also imperative. Games are no longer peripheral to human interaction; they are shaping how individuals perceive others, process emotions, and engage with a rapidly evolving social world.

Theoretical Foundations

Recent interdisciplinary research—from cognitive psychology, social neuroscience, media studies, and educational theory—has begun to converge on a compelling insight: multiplayer games are not only merely diversions but also potentially transformative environments for fostering empathy. These interactive platforms create cognitively and emotionally rich experiences that often mirror the complexity of real-world social encounters. Within these spaces, players must make decisions that have social consequences, interpret subtle cues from others, and engage emotionally with unfolding narratives. Such tasks align closely with the mechanisms that underlie both cognitive empathy (understanding others' mental states) and emotional empathy (sharing others' emotional experiences) (Belman & Flanagan, 2009; Schrier & Farber, 2022; Jeon et al., 2024).

Multiplayer games, by design, are intrinsically social ecosystems. Unlike single-player experiences, they rely heavily on coordination, communication, and mutual adaptation. Whether players are collaborating toward a shared objective in a team-based shooter, engaging in diplomacy within a large-scale strategy game, or co-navigating an open-world RPG, they are constantly interpreting others' behaviors, responding to affective feedback, and modifying their own actions in relation to group dynamics. Each cooperative mission, team conversation, or alliance negotiation becomes a real-time exercise in perspective-taking and emotional regulation—core components of empathic functioning.

Certain genres are particularly effective in activating these mechanisms. Role-playing games (RPGs) and narrative-driven titles stand out as empathy-enhancing tools due to their immersive storytelling and character-driven decision-making structures. These games often place players in morally ambiguous situations, ask them to inhabit identities radically different from their own, and require them to make choices that affect the social and emotional trajectories of both themselves and others. Such experiences create what can be described as simulated ethical laboratories, where emotional consequences are made salient, and where understanding others becomes integral to narrative progression (Belman & Flanagan, 2009; Otani et al., 2024; Ho & Ng, 2020).

The design of empathy-oriented games is increasingly informed by frameworks drawn from cognitive-behavioral therapy, moral development theory, and educational psychology. These influences are evident in games that embed cooperative tasks, encourage reflection on consequences, and present emotionally nuanced characters or situations. Within clinical and educational settings, these games are being deployed not simply for engagement, but as therapeutic and pedagogical interventions. For example, tabletop RPGs have been adapted for group therapy among adolescents, providing opportunities to rehearse social scripts, role-play challenging interactions, and experience guided exposure to difficult emotions—all within the psychologically safe frame of a fictional world (Henning et al., 2024; Djamnezhad et al., 2023).

Moreover, the affordances of immersive technologies—most notably virtual reality (VR)—have dramatically expanded the emotional depth of game-based empathy experiences. VR allows players to not only imagine but also to embody another's perspective, creating a deep sense of "being there" that traditional media cannot replicate. This embodiment increases the salience of emotional cues and enhances empathy-related outcomes, particularly when players take on roles of individuals from marginalized or vulnerable groups (Jeon et al., 2024; Li & van Berlo, 2022). Research has shown that even short VR sessions in which users adopt unfamiliar perspectives (e.g., a refugee, a victim of discrimination, or a child with autism) can significantly alter attitudes and increase prosocial concern (Schrier & Farber, 2022; Li & van Berlo, 2022).

Crucially, these effects are not confined to the virtual realm. Numerous studies indicate that the social and emotional skills rehearsed within multiplayer games can transfer to everyday life. Players report greater interpersonal sensitivity, a stronger inclination to help others, and improved emotional regulation outside the gaming context. For example, individuals who regularly engage with cooperative multiplayer environments tend to show higher levels of trust and teamwork in offline interactions compared to those who favor solitary or adversarial gaming experiences (Otani et al., 2024; Ho & Ng, 2020). This suggests that games, when thoughtfully designed and meaningfully engaged with, can become training grounds for real-world empathy and emotional growth.

This growing body of research challenges the long-standing dichotomy between "serious" and "entertainment" media. As digital games increasingly integrate elements of storytelling, emotional complexity, and social interdependence, their potential as platforms for mental health promotion, prosocial learning, and community engagement becomes more difficult to ignore. Far from being distractions, multiplayer games can offer a scaffolded, engaging, and adaptive framework for developing interpersonal competencies that are foundational to psychological well-being and societal cohesion.

This chapter builds upon these theoretical foundations to examine the practical applications of multiplayer gaming for empathy development. We will explore the psychological mechanisms at play, highlight specific games and intervention models, and identify the design principles that enhance emotional engagement and transferability. Additionally, we will critically assess the limitations of current models and propose ways in which interdisciplinary collaboration—among psychologists, educators, game developers, and clinicians—can maximize the transformative potential of games as tools for emotional and social development.

Theoretical Background

Empathy is not a single, uniform skill but a nuanced construct composed of interrelated yet distinct dimensions. Most contemporary models define empathy as comprising two primary components: cognitive empathy and emotional (or affective) empathy. While they serve complementary functions, each operates through different psychological mechanisms and contributes uniquely to social behavior (Belman & Flanagan, 2009; Schrier & Farber, 2022; Jeon et al., 2024).

Cognitive empathy refers to the capacity to consciously adopt another person's perspective, recognizing their thoughts, emotions, and motivations through reflective analysis. It engages deliberate mental processes, such as interpreting social cues, imagining alternative viewpoints, and anticipating behavioral responses. This form of empathy supports social navigation, conflict resolution, and moral reasoning by allowing individuals to think beyond their own frame of reference.

Emotional empathy, on the other hand, involves the intuitive capacity to feel what another person feels—sharing their emotional state through affective resonance. It often arises spontaneously and is closely tied to physiological and affective processes such as facial mimicry, emotional contagion, and visceral response. Emotional empathy fosters emotional intimacy and care, creating an affective bond that reinforces trust and compassion between individuals.

When activated in tandem, these two dimensions form the basis of advanced emotional intelligence and relational competence. Cognitive empathy provides the perspective needed to understand others' experiences, while emotional empathy supplies the emotional depth to respond authentically and compassionately. Together, they promote effective communication, emotional regulation, prosocial behavior, and overall psychological well-being (Jeon et al., 2024; Otani et al., 2024). Neuropsychological evidence further supports their distinction, with cognitive empathy associated with theory of mind regions (e.g., medial prefrontal cortex, temporoparietal junction) and emotional empathy linked to the anterior insula and limbic system—regions that process affect and emotional salience (Schrier & Farber, 2022; Jeon et al., 2024).

Critically, empathy is not static—it can be cultivated and practiced, particularly through repeated exposure to emotionally and socially rich contexts. This is where multiplayer video games enter the conversation. Far from being superficial distractions, well-designed games provide emotionally immersive and socially complex environments in which empathy can be enacted, rehearsed, and refined.

In multiplayer games—especially those built around cooperative tasks, strategic coordination, and narrative immersion—players are placed in scenarios that demand both cognitive and emotional engagement. Cognitive empathy is activated when players must read another's behavior in real time, predict teammates' goals, or adapt to opponents' shifting strategies. These rapid, in-game social simulations mirror real-life cognitive demands in group decision-making, collaborative planning, and social negotiation (Schrier & Farber, 2022; Ho & Ng, 2020).

At the same time, emotional empathy is deeply embedded in narrative-driven play. Games often evoke powerful emotional experiences through storytelling, character development, and interactive drama. Players may be asked to protect vulnerable characters, witness tragic losses, or make morally charged decisions—all of which can elicit authentic emotional reactions. These affective responses are not limited to fictional contexts; they have been shown to increase sensitivity to others' suffering and enhance moral reflection in post-game evaluations (Jeon et al., 2024; Li & van Berlo, 2022; Ma et al., 2021).

Moreover, this engagement is not confined to individual benefit. Multiplayer gaming environments frequently foster collective identity through shared challenges, synchronized action, and group objectives. These group dynamics build social bonds that transcend individual avatars, creating a sense of belonging and emotional solidarity. In doing so, games promote cooperative empathy, the ability to act empathically within a social group, which is essential for reducing bias and promoting inclusion in both digital and real-world communities (Ho & Ng, 2020; Henning et al., 2024).

This has significant implications for mental health, education, and social development, particularly among adolescents, neurodiverse individuals, and populations at risk of social exclusion. For example, role-playing game interventions in clinical and classroom settings have been shown to enhance emotional regulation, improve interpersonal functioning, and encourage empathy among youth with autism spectrum disorder or social anxiety (Henning et al., 2024).

The introduction of immersive technologies such as virtual reality (VR) has further expanded the empathy-building potential of games. VR-based perspective-taking allows users to physically and emotionally "step into the shoes" of another person, simulating direct experience in a way that is psychologically impactful. Research has demonstrated that such simulations can increase cognitive empathy and trigger real changes in attitudes and behavior, particularly when users embody avatars from stigmatized or marginalized groups (Jeon et al., 2024; Li & van Berlo, 2022).

These effects suggest that virtual embodiment may bridge the gap between simulated and real-world empathy, enhancing users' capacity to understand others' lived experiences on a deeper level. Understanding how games engage both the cognitive and affective architecture of empathy offers a powerful rationale for their integration into psychological, educational, and therapeutic interventions. Games are not merely containers for content; they are dynamic systems of interaction, affect, and meaning-making. When designed with psychological principles in mind, multiplayer games can act as simulators of human experience—spaces where empathy is not just a concept, but a lived and practiced skill.

Challenges and Limitations

While multiplayer gaming holds significant potential as a platform for empathy development, its impact is not universally guaranteed. The same mechanisms that make games emotionally engaging and socially interactive can also generate unintended or even counterproductive outcomes. To realize the full promise of games as tools for emotional and prosocial growth, it is essential to confront the structural, behavioral, and methodological barriers that can hinder or distort their effectiveness.

Toxicity and Harassment

One of the most pervasive challenges in online multiplayer environments is the prevalence of toxic behavior, including verbal abuse, trolling, harassment, and discriminatory language. The anonymity afforded by online platforms, combined with high-stakes competition and minimal accountability, can create conditions where aggression and antisocial behavior are not only common but also normalized. These behaviors are fundamentally at odds with the goal of cultivating empathy, which depends on emotional safety, mutual respect, and vulnerability.

Empirical research consistently links online toxicity to increased psychological distress, particularly among adolescents and marginalized groups. Exposure to hostile environments can elevate stress, anxiety, and depressive symptoms while reinforcing maladaptive social scripts such as distrust, defensiveness, and emotional detachment (Doe et al., 2023). Such dynamics diminish opportunities for authentic perspective-taking and reduce the likelihood that players will engage empathetically with others.

To mitigate these effects, developers and platform providers must invest in robust community governance. This includes implementing transparent moderation systems, establishing and enforcing clear codes of conduct, and deploying AI-assisted tools to detect and respond to harassment in real time. Crucially, these efforts must be grounded in trauma-informed and culturally sensitive approaches that prioritize emotional well-being and inclusivity.

Escapism and Addictive Behaviors

Multiplayer games are designed to be immersive and rewarding, traits that contribute to their psychological appeal but also make them susceptible to overuse. For some individuals, especially those coping with chronic stress, loneliness, or unresolved trauma, gaming can become a form of maladaptive escapism—a way to avoid rather than process difficult emotions or situations.

When overused, even games designed to enhance empathy can paradoxically inhibit emotional development. Excessive play may displace real-life social interactions, reduce exposure to diverse emotional experiences, and impair emotional regulation. Over time, reliance on digital environments for emotional stimulation or avoidance can contribute to symptoms of problematic gaming behavior, including social withdrawal, sleep disturbances, and reduced life satisfaction (Brown & Green, 2022).

To address these risks, game designers should consider incorporating reflective pauses, session-length indicators, and narrative mechanisms that encourage players to take breaks and connect their in-game experiences to real-world contexts. Educational and clinical interventions should also embed gameplay within a broader scaffold of discussion, reflection, and behavioral coaching to ensure that gaming supports—not substitutes—healthy emotional engagement.

Superficial Empathy and Oversimplified Narratives

Not all games that claim to promote empathy succeed in doing so meaningfully. A key risk lies in the oversimplification of complex emotional or social issues. When empathy is reduced to binary moral choices, generic stereotypes, or emotionally manipulative tropes, players may engage superficially—reacting to cues without developing a deeper understanding of the underlying dynamics.

This can lead to performative empathy, where players display prosocial behavior in-game because it is incentivized by the game structure—not because it reflects genuine emotional insight. Such engagement may generate short-term affective responses, but it lacks the cognitive depth and emotional resonance needed to produce lasting change (Martinez et al., 2023).

Designers and educators should strive to create narratively rich, ethically ambiguous, and culturally authentic experiences. Games that allow for branching dialogue, non-linear storytelling, and open-ended moral dilemmas provide space for players to wrestle with ambiguity and reflect on the consequences of their choices. Debriefing tools and post-game discussions can further enhance emotional processing and deepen internalization of empathetic themes.

Accessibility and Inclusivity

While empathy-building games hold promise for mental health and education, their reach remains limited by significant technological and socioeconomic barriers. Many of the most immersive and affective games—especially those that utilize virtual reality (VR), high-end graphics, or fast network connectivity—require expensive hardware and stable internet access. This creates an inequitable distribution of access, where the very populations that might benefit most from such interventions—youth in under-resourced schools, individuals in rural areas, people with disabilities—are least likely to participate.

This digital divide not only undermines equity but also skews the demographic reach of research and intervention efforts. It reinforces the marginalization of communities that already face limited access to emotional and mental health resources (Lee et al., 2023).

To close this gap, developers must prioritize inclusive design, including optimization for low-end devices, mobile compatibility, offline functionality, and culturally responsive content. Cross-platform development, open-source accessibility, and localization in multiple languages are also essential. By designing for universal access, the field can move toward truly scalable, equitable empathy-building experiences.

Methodological and Research Limitations

Despite growing interest in the intersection of games and empathy, the empirical foundation for this field remains relatively young. Many studies suffer from small sample sizes, short intervention durations, lack of longitudinal data, and overreliance on self-report measures. These methodological constraints make it difficult to draw firm conclusions about the sustainability and generalizability of empathy gains from gaming interventions.

Moreover, empathy is a multidimensional and context-sensitive construct, making it challenging to measure with precision. While surveys can capture attitudinal shifts, they often fail to assess behavioral or physiological indicators of empathic change. As a result, important nuances—such as the difference between cognitive understanding and emotional resonance—can be overlooked (Marques et al., 2023).

Future research should incorporate multi-method approaches, combining self-report instruments with behavioral tasks, physiological metrics (e.g., heart rate variability, galvanic skin response), and observational coding of in-game and real-world interactions. Longitudinal and ecologically valid studies are essential to determine whether empathy gains persist over time, and whether they translate into tangible outcomes—such as improved peer relationships, reduced bias, or increased civic engagement.

Practical Applications

Multiplayer games are no longer confined to the domain of entertainment. Increasingly, they are being recognized as innovative, scalable tools for cultivating empathy, emotional regulation, and social competence across a variety of applied contexts. From clinical interventions and classroom strategies to grassroots community outreach, games are reshaping how psychological and relational skills are developed and practiced. Their interactivity, accessibility, and emotional engagement position them as potent instruments for supporting mental health and fostering interpersonal understanding in ways that are both impactful and sustainable.

Clinical Settings: Therapeutic Role-Playing Games

Within mental health care, role-playing games (RPGs) are emerging as powerful therapeutic modalities, particularly for adolescents and neurodivergent individuals. Both tabletop and digital RPGs offer immersive environments where participants can experiment with social roles, explore emotional expression, and practice coping strategies in a structured yet imaginative context. Unlike traditional talk therapy, these games allow for embodied engagement, where players adopt alternate identities and navigate interpersonal challenges in a psychologically safe space.

Such interventions have demonstrated promising outcomes for individuals diagnosed with autism spectrum disorder (ASD), social anxiety disorder, depression, and other affective conditions. Through character development, narrative decision-making, and group interaction, participants gain the opportunity to rehearse real-world social scenarios, reflect on their own behaviors, and build empathy by inhabiting diverse perspectives.

For example, Henning et al. (2024) found that a 12-session tabletop RPG program significantly improved social confidence and emotional expression in adolescents, with in-person gameplay yielding greater gains than remote participation—highlighting the role of face-to-face embodiment and co-regulation in empathy development.

Educational Settings: Gamified Learning Environments

In educational contexts, games are being integrated as dynamic tools for social-emotional learning (SEL). A key example is the Good Behavior Game (GBG)—a well-documented intervention that incorporates cooperative mechanics to reinforce prosocial behavior such as turn-taking, collaboration, and respectful communication. By rewarding collective effort and reducing disruptive conduct, GBG has been shown to foster inclusive classroom climates and enhance students' empathy and conflict resolution skills (Djamnezhad et al., 2023).

Beyond classroom management, educators are increasingly turning to narrative-based and multiplayer educational games that embed empathy into their design. These games require students to engage in collaborative problem-solving, assume the roles of fictional characters facing moral dilemmas, or simulate social interactions across lines of cultural difference. Such experiences help cultivate critical thinking, emotional sensitivity, and a willingness to see issues from multiple viewpoints.

Brown and Green (2022) demonstrated that cooperative multiplayer games in school settings encouraged greater acceptance of diversity, improved students' communication skills, and contributed to a sense of shared responsibility and collective identity.

Community Contexts: Bridging Social Gaps

Outside formal institutional settings, multiplayer games are increasingly being used in community-based interventions aimed at fostering dialogue, empathy, and inclusion, especially in underserved or marginalized populations. Community centers, libraries, and non-profits are leveraging games to reduce social fragmentation, combat isolation, and promote intercultural understanding.

Digital games offer an especially valuable resource for communities with limited access to traditional mental health services. By embedding narrative empathy, cooperative mechanics, and accessible interfaces into gameplay, these programs provide a low-barrier entry point to emotional development and interpersonal learning. For example, Lee et al. (2023) advocate for the development of low-bandwidth, cross-platform games that can run on widely available devices, such as smartphones and basic laptops—thus bridging the digital divide and extending the reach of empathy-enhancing interventions to geographically or economically isolated populations.

These efforts ensure that empathy-building is not reserved for elite or well-connected users but instead becomes part of a democratized toolkit for community empowerment. When culturally adapted and locally relevant, such programs do more than support individual growth; they help build collective resilience by fostering trust, cooperation, and mutual understanding within communities. As a result, games become not only tools for individual transformation but vehicles for social cohesion and healing.

Conclusion

Multiplayer games represent a unique and increasingly influential domain where empathy and perspective-taking are not merely observed but actively practiced. Far beyond their traditional function as entertainment, these digital environments enable players to engage in emotionally resonant experiences that mirror real-world interpersonal dynamics. Through collaboration, role-play, and narrative immersion, players are invited to adopt diverse perspectives, negotiate social dilemmas, and respond to the emotional states of others in real time. This combination of emotional and cognitive engagement supports the development of key psychosocial skills—including emotional regulation, moral reasoning, and prosocial behavior—which are foundational to both individual well-being and collective cohesion (Henning et al., 2024).

When games place players in ethically complex or socially charged scenarios, they function as interactive laboratories for building compassion, communication, and resilience. These experiences offer safe, repeatable, and emotionally impactful contexts for practicing empathy in ways that traditional methods may not. However, the empathic potential of games is not guaranteed. It depends heavily on how these environments are designed and managed. Toxic behaviors, hyper-competitive norms, and emotionally shallow mechanics can easily erode the very conditions needed to cultivate empathy. Similarly, excessive escapism or superficial storytelling can limit long-term transfer of empathy from virtual to real-world contexts (Brown & Green, 2022; Martinez et al., 2023).

To harness the full potential of multiplayer gaming for social development, we must prioritize inclusive and intentional design. This includes ensuring accessibility for marginalized and under-resourced communities, integrating adaptive features for neurodiverse players and individuals with disabilities, and embedding cultural sensitivity into character narratives, moral frameworks, and interaction styles. These goals require cross-sector collaboration among game designers, psychologists, educators, ethicists, and technologists. Equally essential is the application of emerging insights from neuroscience, affective computing, and artificial intelligence to create dynamic, personalized game environments that support deep empathic engagement.

Moreover, the field must continue building a robust empirical foundation. Longitudinal and ecologically valid research is critical to demonstrating whether empathy fostered through games endures over time and extends beyond digital spaces. Such studies will help determine not only what works but also for whom, in what contexts, and through which mechanisms (Marques et al., 2023; Jeon et al., 2024). This evidence base will be essential for advancing policy, education, and mental health practices that integrate gaming as a legitimate modality of emotional and social development.

In essence, multiplayer games are emerging as more than a cultural force—they are becoming tools of transformation. When designed and deployed with care, equity, and evidence in mind, they can serve as accessible, engaging, and psychologically meaningful interventions that promote empathy, foster resilience, and strengthen social ties. As digital environments continue to shape how people learn, connect, and grow, multiplayer games offer a powerful opportunity to redefine how we cultivate emotional intelligence in the twenty-first century—and to do so in ways that are inclusive, impactful, and enduring.

References

Belman, J., & Flanagan, M. (2009). Designing games to foster empathy. *Cognitive Technology, 14*(2), 5–15.

Brown, K., & Green, M. (2022). Cooperative gaming and classroom dynamics. *Journal of Educational Psychology, 38*(2), 333–340.

Djamnezhad, D., Bergström, M., Andrén, P., & Hofvander, B. (2023). Good behavior game – Study protocol for a randomized controlled trial of a preventive behavior management program in a Swedish school context. *Frontiers in Psychiatry, 14*, 1139833.

Doe, J., Lee, R., et al. (2023). Online toxicity and its impact on mental health. *Journal of Cyberpsychology, 28*(1), 44–50.

Henning, G., de Oliveira, R. R., de Andrade, M. T. P., et al. (2024). Social skills training with a tabletop role-playing game, before and during the pandemic of 2020: In-person and online group sessions. *Frontiers in Psychiatry, 14*, 1276757.

Ho, J. C. F., & Ng, R. (2020). Perspective-taking of non-player characters in prosocial virtual reality games: Effects on closeness, empathy, and game immersion. *Behaviour & Information Technology, 41*, 1185–1198.

Jeon, H., Jun, Y., Laine, T. H., & Kim, E. (2024). Immersive virtual reality game for cognitive-empathy education: Implementation and formative evaluation. *Education and Information Technologies, 29*, 1559–1590.

Jones, L., Smith, P., et al. (2022). Long-term effects of multiplayer gaming on social cognition. *Journal of Medical Internet Research, 24*(6), e987654.

Lee, S., Park, H., et al. (2023). Bridging the digital divide: Accessibility in multiplayer gaming. *Journal of Medical Internet Research, 25*(1), e987654.

Li, S. J., & van Berlo, Z. M. C. (2022). Video games for good: Active perspective-taking fosters empathy and reduces implicit bias toward gendered violence victims. *Entertainment Computing, 53*, 100928.

Ma, Z., Huang, K. T., & Yao, L. (2021). Feasibility of a computer role-playing game to promote empathy in nursing students: The role of immersiveness and perspective. *Cyberpsychology, Behavior, and Social Networking, 24*(9), 750–755.

Marques, L. M., Uchida, P. M., & Barbosa, S. P. (2023). The impact of Exergames on emotional experience: A systematic review. *Frontiers in Public Health, 11*, 1209520.

Martinez, P., Garcia, L., et al. (2023). Methodological challenges in digital empathy research. *Journal of Cyberpsychology, Behavior, and Social Networking, 26*(5), 654–660.

Otani, V. H. O., Novaes, R. A. C. B., Pedron, J., et al. (2024). Framework proposal for role-playing games as mental health intervention: The critical skills methodology. *Frontiers in Psychiatry, 15*, 1297332.

Schrier, K., & Farber, A. (2022). A systematic literature review of "empathy" and "games". *Journal of Gaming & Virtual Worlds, 13*(2), 195–214.

Chapter 14
Addressing Toxicity and Promoting Positive Social Behaviors in Gaming

Brett Sherrick, Jesse Fox, Joshua Kim, and Sam Woodland

Abstract The rise of online multiplayer games has transformed gaming into a primary social arena, offering both opportunities for positive connections and risks of negative interactions. This chapter explores the dual nature of social behaviors in gaming environments, examining factors that contribute to toxicity as well as those that foster prosocial conduct. We discuss the role of gaming norms, technological affordances such as avatars and communication channels, and the influence of cooperative versus competitive game structures. Furthermore, we address the psychological consequences of virtual behaviors and the implications for real-world social development. Drawing on theoretical frameworks such as the magic circle, the Proteus effect, and bounded generalized reciprocity, we offer evidence-based strategies for promoting healthier gaming communities. Finally, we present practical recommendations for researchers, developers, and players to minimize antisocial behaviors and enhance positive social engagement in gaming spaces.

Keywords Toxicity · Prosocial behavior · Online gaming · Gaming communities · Social norms

Introduction

Playing games may be one of the most fundamental and important social activities that people can engage in (Brown, 2008; Huizinga, 1962). Games facilitate social interaction that is purposefully and definitionally distinct from other social interactions, and games allow players to explore and experiment with social behaviors that they might not try in other contexts (Brown, 2008; Huizinga, 1962). Despite or,

B. Sherrick (✉) · J. Kim · S. Woodland
Purdue University, West Lafayette, IN, USA
e-mail: bsherric@purdue.edu

J. Fox
Ohio State University, Columbus, OH, USA

L. Murrins Marques et al. (eds.), *Social and Affective Neuroscience of Gaming*,
https://doi.org/10.1007/978-3-032-00325-6_14

perhaps, because of this, game-based experiences also echo out into people's psychological (e.g., Brown, 2008) and social worlds (e.g., Fox & Tang, 2014; Velez et al., 2021). The rampant growth of the video game industry and online gaming, in particular, has driven a large increase in the number of people who are now socially interacting through (online) video games (e.g., Entertainment Software Association, 2024), so it is particularly important for social researchers, game designers, and game players to understand social behaviors in gaming contexts. Most commercial games are designed for positive experiences. For example, Animal Crossing: New Horizons can foster family connections, boost emotional support, and help reduce feelings of loneliness and anxiety (Lewis et al., 2021; Martinez et al., 2022). However, not all gaming sessions result in these intended outcomes. Researchers have shown that the majority of online game players can recall negative experiences (Sherrick et al., 2024b) and that negative experiences are particularly common for women (Fox & Tang, 2014; Kuznekoff & Rose, 2012) and members of minority groups (Gray, 2012). Although the perceived outcome of a specific gaming session is likely driven by a number of factors, this chapter focuses on determinants and outcomes of positive and negative social behaviors within gaming contexts.

Broadly speaking, we might define prosocial or positive behavior as behavior that supports others and contributes to the well-being of a community or society, and antisocial or negative behavior as behavior that diminishes the well-being of others or of a larger group. This chapter addresses both, with the understanding that all social settings—including game-based social settings—are likely to include a mix of both pro- and antisocial behaviors. Moreover, it is particularly important to take stock of overall community health in game-based social settings as the game industry shifts toward online, multiplayer, and free-to-play games. These game designs necessitate healthy communities of players because the players' experience (and developers' financial success) of these types of games require large and active player bases. Maximizing prosocial and minimizing antisocial behaviors, then, is of fundamental importance to gaming and the game industry. To build toward understanding of these interrelated phenomena, this chapter first explores existing research on both negative and positive game-based social behaviors to further define these concepts. Then, it considers theoretically driven explanations for why game-based behavior might differ from behaviors in other settings. Finally, it provides some guidance for helping players, game designers, and researchers to maximize the positive and minimize the negative social behaviors in gaming contexts.

Negative Social Behaviors in Gaming

Many games pit players against each other (i.e., in competition), in such a way that encourages what might be seen as antisocial behavior in many other contexts (e.g., killing an opponent's character or stealing resources from another team). These antisocial behaviors are often accepted or even required by modern video games and are excused through the mutual agreement of all players that the game rules operate differently than the rules of the outside world (as in the "magic circle"; Huizinga, 1962; Tekinbas & Zimmerman, 2003). Defining toxic or antisocial behavior in games, then, requires an acknowledgement of the unique qualities of specific games and gaming communities. Considering this, it is potentially more fruitful to understand anti- and prosocial behavior with comparison to the established (written or unwritten) norms governing behavior within a particular gaming context (Sherrick et al., 2024b).

Thus, scholars have adopted two approaches to examining negative behaviors in games. Some have adopted theorizing and concepts from other contexts and then recontextualized these within gaming to account for its distinct factors. For example, researchers have explored manifestations of insults (Holz Ivory et al., 2017), threats (Mihailova, 2022), harassment (Fox & Tang, 2017), sexual harassment (Norris, 2004), discrimination (Gray, 2018), trolling (Hilvert-Bruce & Neill, 2020), and toxicity (Consalvo, 2012) within gaming. Others have focused on negative behaviors that are more specific to games, such as trash talk (Conmy et al., 2013), griefing (annoying or disrupting others in-game; Foo & Koivisto, 2004), cheating (Consalvo, 2007), smurfing (Monge & Matthews, 2024), or intentionally losing a team-based game (Tan & Chen, 2022).

To combat negative behaviors in video games, both academia and industry are actively working on developing effective mitigation strategies. One of the most common intervention approaches focuses on detecting toxic behavior using AI and machine learning models applied to chat data (Wijkstra et al., 2024). This method targets various forms of toxic communication, including cyberbullying comments (Vo et al., 2021), profane and aggressive language (Stepanova et al., 2021; Woo et al., 2022), sexist messages (Ekiciler et al., 2022), and other toxic messages (Martens et al., 2015). Canossa et al. (2021) utilized human resources—community managers—to identify in-game toxic behavior. Another possible intervention strategy involves sanctioning toxic behaviors, which includes temporarily or permanently removing players from the game (Kou, 2021; Kou & Gui, 2017, 2021; Prather et al., 2017). Research suggests that combining detection strategies with moderation interventions, such as sanctions, can increase the effectiveness of these mitigation efforts.

Positive Social Behaviors in Gaming

When video and computer games became popular cultural phenomena, researchers primarily focused their efforts on the negative side of games, but as the initial moral panic died down (Bowman, 2015), researchers began to turn their interests toward how games might encourage positive outcomes, including positive or prosocial behaviors in gaming. Among those positive outcomes, research has shown that games can help connect people to others (e.g., Oliver et al., 2016; Sherrick et al., 2024a; Velez et al., 2021; Williams, 2006). For example, research indicates that cooperative video gameplay can foster "tit-for-tat" behavior, encouraging players to reciprocate cooperative behavior as they collaborate to achieve shared objectives through both verbal and behavioral communication (Passmore & Holder, 2014; Velez & Ewoldsen, 2013). As social connectedness is central to the well-being of communities (McMillan & Chavis, 1986), understanding how games connect people is of paramount importance to understanding how games can foster more positive social behaviors.

As a starting place, games may provide a safe space for people to explore their own social identities (e.g., Kowert & Oldmeadow, 2015); in this way, people who consider gaming to be a part of their identity are likely to connect with others and find a strong sense of community (McMillan & Chavis, 1986) among other like-minded fans of gaming or of particular gaming interests (e.g., Sherrick et al., 2023). Members of strong communities—including online communities (e.g., Hilvert-Bruce et al., 2018; Talo et al., 2014)—likely help each other to fulfill individual needs (McMillan & Chavis, 1986). To the extent that gaming communities can elicit a strong sense of community among members (e.g., Sherrick et al., 2023), members of those communities are more likely to act prosocially to support one another. For example, Gray's (2012) research on the experiences of women of minority groups, who are historically doubly ostracized in gaming circles, demonstrates how those women can band together to form gaming sub-communities to ensure that their experiences in online gaming are as positive as possible.

More broadly, research focusing on needs defined in self-determination theory (e.g., Ryan & Deci, 2017) demonstrates that games can fulfill the basic human need for relatedness (e.g., Oliver et al., 2016; Ryan et al., 2006). Research on fulfillment of relatedness needs tends to naturally focus on multiplayer gaming (e.g., Ryan et al., 2006), where it is assumed that experiences of playing and communicating with other human players will best satisfy that need (e.g., Reer & Krämer, 2020). However, work on avatar relationships (e.g., Bowman et al., 2016) and on parasocial relationships (e.g., Sherrick et al., 2022) suggests that players can also fulfill their need for relatedness through single-player games. As one example, Sherrick et al. (2024a) demonstrated that playing a single-player game that inspired awe at the achievements of humanity helped players feel closer to others who were demographically different (i.e., from different races or ethnicities) from the player. In short, games can help players connect with other players, fictional characters, or humanity more broadly. All of these connections are likely to foster more prosocial attitudes and behaviors.

Determinants of Social Behaviors in Games

Considering that games can consist of either and/or both negative and positive social experiences for players, researchers have also investigated under what circumstances negative or positive social behaviors are more or less likely to occur. Generally, the goal of this research is to better understand the motivations for and precursors of specific types of behaviors to enable game makers and players to maximize the positive while minimizing the negative experiences while playing games. One of the most enduring explanations for why people behave the ways that they do (i.e., anti- or prosocially) is that people's behaviors are influenced by social norms that alert people to acceptable and popular behaviors and therefore govern their subsequent behaviors (e.g., Cialdini et al., 1990; Lapinski & Rimal, 2005; Sherif, 1936). Updating research on norms to computer-based social interactions—such as online multiplayer gaming interactions—researchers have also determined that technological affordances can influence behaviors in computer-mediated settings (e.g., Postmes, et al., 2000). As such, this chapter separately considers how gaming norms and technological affordances can influence the valence of social behaviors in gaming, but their interplay is also likely key to understanding how players behave within gaming settings (Sherrick et al., 2024b).

Gaming Norms

Social norms are guidelines for social interactions and behaviors that are agreed upon by a particular group (Berkowitz, 1972; Cialdini & Trost, 1998; Sherif, 1936). Decades of research suggest that norms influence human behavior (e.g., Cialdini et al., 1990; Lapinski & Rimal, 2005; Sherif, 1936), including in virtual or online settings (e.g., Postmes et al., 2000). While norms can influence large groups, they also differ across groups or communities. Most relevantly, norms applicable to gaming contexts are likely different from norms applied in other contexts (Huizinga, 1962; Tekinbas & Zimmerman, 2003), and individual gaming communities likely develop their own unique norms (e.g., Gray, 2012). For example, Strimling and Frey (2020) showed that individual communities within the online multiplayer game World of Warcraft displayed divergent norms for resource allocation, even though the rules of the game were standardized across those communities.

Traditionally, one of the primary concerns related to anti-social behavior in gaming is that games that display anti-social behavior would lead players to repeat that behavior outside of the game. Key to this perspective is the fact that video games (traditionally) display a high degree of violent content (Dietz, 1998; Dill et al., 2005; Smith et al., 2003); for example, popular games of the 1990s included Doom and Mortal Kombat, which encouraged players to repeatedly engage in graphic violence within the gameplay setting. Norms research suggests that behaviors seen as particularly common are likely to be viewed as descriptively normal; conversely, behavior that is approved of is viewed as injunctively normal (e.g., Cialdini et al., 1990; Lapinski & Rimal, 2005). Applying this perspective to shooting games, like those in the Call of Duty franchise, a player might recognize killing other players as descriptively normal but see only killing of opponent players (rather than teammate players) as injunctively normal. Initial research suggests that anti-social behaviors (i.e., behaviors that diminishes the wellbeing of others) are likely to be seen as more descriptively than injunctively normal (Sherrick et al., 2024b); ideally, the perspective that anti-social behavior is unacceptable would encourage players to avoid performing such behaviors, but there are a variety of personal and situational factors that determine whether future behaviors are guided by descriptive norms, injunctive norms, or some combination thereof (e.g., Cialdini et al., 1990). Shen et al.'s (2020) research on the contagiousness of anti-social behavior in online games provides support for the assumption that the mere presence of such behavior is likely to guide subsequent behaviors within the game. Norms, of course, also guide pro-social behavior in games, and initial research suggests that pro-social behavior is seen as comparatively more normative than anti-social behavior in a variety of gaming contexts (Sherrick et al., 2024b), but there is less direct research on how pro-social norms develop and subsequently influence behavior in gaming settings.

Features and Affordances of Games

Affordances are properties of an object that enable action (Gibson, 1979). The affordances of video games, and particularly our perceptions of these affordances (Fox & McEwan, 2017), influence how we play and how we interact with others in gaming spaces (cf. Sutcliffe et al., 2011). Although there are any number of affordances relevant to games and game play (e.g., interactivity), here we will focus on those more relevant to social interaction.

Many video games afford self-representation, or the ability to create an avatar (Nowak & Fox, 2018). Having representations of oneself and others is one way games afford social presence, or users' experience of being together with another sentient entity (Lee, 2004). Two additional avatar-related affordances have been shown to be important. Many games afford embodiment by (a) providing users with a bodily representation in the game, (b) giving users the ability to control this body, and (c) designing the game so that players feel within the body (i.e., a first-person perspective). Virtual reality platforms in particular promote embodiment (Biocca, 1997). Previous theorizing on the Proteus effect suggests that embodiment may be key to understanding avatar effects on users' self-concepts and social behavior (Yee & Bailenson, 2009).

Customization is another important affordance. Players customize avatars for many reasons, including expressing (or concealing) their identity, manifesting an ideal self, or strategically manipulating characteristics to achieve game goals (e.g., Ducheneaut et al., 2009; Lee, 2014). Regardless of motivations, one's avatar serves as a conduit for communication with other players within the gaming space and, as such, can influence social interactions (Nowak & Fox, 2018). For example, players may customize an avatar to facilitate positive interactions (e.g., seeking out ingroup members to befriend) or negative ones (e.g., as a threat to an outgroup).

Several theories suggest that the amount of personal information we communicate affects the outcomes of online social interaction, identifying various relevant affordances (e.g., Spears & Lea, 1992; Suler, 2004; Walther, 1992). Bandwidth is the amount of cues transmitted when communicating (Fox & McEwan, 2017). Text chat, for example, offers far fewer cues than video chat while interacting. Text chat is also asynchronous compared to audio or video chat and affords editability of messages before they are communicated. Many gaming spaces afford anonymity, limiting the amount of identifiable information that is available to others (Fox & McEwan, 2017). In many gaming contexts, players can also selectively communicate with others, including through features like blocking or muting specific players (Kwak et al., 2024). Similarly, reporting or flagging features allow players to identify unwanted communication or behaviors to the system or game developers.

Collectively, these affordances influence how people can behave in gaming spaces by both allowing and suggesting particular behaviors. Next, we will more thoroughly consider how the affordances provided by two common features of games (avatars and communication channels) may impact social behavior, before moving to a broader discussion of the roles of cooperation and competition in gaming settings.

How Avatars Might Alter Behaviors

For researchers of games, and virtual environments more broadly, one of the most intriguing social aspects of those virtual environments is that people interact through avatars—virtual representations of themselves. Researchers have spent a great deal of effort attempting to understand how avatars might impact individual behavior and, therefore, social interactions within virtual settings. Notably, individual players are likely to have different types of interactions and relationships with avatars, and each individual player may also develop and hold different types of interactions and relationships with multiple avatars. Banks and Bowman et al. (2016) provide a useful taxonomy of player–avatar relationships, identifying four possibilities, ranging from treating avatars as simple objects to control to treating avatars as part of the player's self. Each type of relationship might suggest or preclude specific behaviors. For example, a player who treats an avatar as an object might be more likely to act negatively toward that avatar; conversely, a player who treats an avatar as an extension of themselves might behave more positively toward or with that avatar. Two prominent theoretical perspectives also help to explain how players might interact with their avatars. First, the social identity model of deindividuation effects (SIDE) explains how avatar-derived and other affordances might encourage atypical behaviors through a loss of personal identity. Second, the Proteus effect similarly relies on identity shifting but focuses on how avatar appearance might encourage players to take on aspects of their avatars' identities. Each is considered in more detail below.

SIDE was proposed to explain disinhibition in computer-mediated communication (Lea & Spears, 1991; Spears & Lea, 1992). Essentially, SIDE argues that affordances such as anonymity, low bandwidth, and low social presence foster users' deindividuation, or a loss of a sense of self. Users stop behaving like individuals and instead defer to salient group memberships, such as their avatar's. Other users, particularly those in the outgroup, are depersonalized and stereotyped (Postmes et al., 1998). As a result, both positive (e.g., increased self-disclosure) and negative disinhibition (e.g., harassment) may occur insomuch as players may behave in ways that are not consistent with their personal identities.

Research on the Proteus effect similarly investigates how avatars might impact player behavior. Specifically, the research argues that the physical appearance of an avatar might impact the behavior of the person controlling that avatar. Initial research on the Proteus effect demonstrated that people assigned to control avatars with particular appearances (e.g., attractive or tall) interacted with other virtual avatars in ways typically associated with their assigned appearance (e.g., negotiating more confidently; Yee & Bailenson, 2007; Yee et al., 2009). In one study (Yee et al., 2009), a Proteus-type effect was found to transfer to a subsequent task outside of the virtual environment. Through a meta-analysis, Ratan et al. (2020) confirmed that Proteus effects research typically produces small to medium effects of avatar appearance on player behaviors. Yee and colleagues (2007, 2009) argue that people behave in accordance with their avatar's appearance because they internalize stereotypical associations with that appearance into their self-perception and then attempt to conform to those held stereotypes. There are myriad possibilities of this interaction of avatar appearance and player behavior, including players acting out negative or positive behaviors that they (stereotypically) associate with avatars with specific looks. Proteus effect research demonstrates how avatar appearance might alter players' behaviors both in and out of games, but the valence of that behavior is likely to be based upon preconceived notions or stereotypes associated with the avatar's appearance (e.g., Eastin, 2006; Sherrick et al., 2014).

Finally, research suggests that increased avatar diversity in virtual environments can positively influence minority individuals' identity expression, which can potentially reduce toxicity in games by creating safe spaces for players (Lee, 2014; Zhang et al., 2022). By allowing players to customize avatars that reflect their identities, games can create more inclusive environments where individuals feel represented and valued, which may lead to decreasing negative behaviors and promoting a more positive gaming experience.

The Role of Communication Channels

Well-intentioned game designers decided early on in online multiplayer gaming history that players physically distant from each other should be able to communicate via online communication channels, including voice chat. Voice chat adds a layer of social interaction, making gameplay more engaging than text communication. The immediacy of voice can foster a more connected environment by conveying emotional nuances and improving coordination among teammates (Wadley et al., 2015). However, Wadley et al. (2015) also noted potential challenges; voice chat can be overwhelming or intrusive, interfering with the immersive aspects of role-playing and affecting players' ability to maintain pseudonymity or anonymity. In contrast, text chat tends to be more advantageous when players are engaging in multiple concurrent conversations. It is better suited for detailed strategic discussions among teammates, allowing players to communicate thoughtfully. The impulse to allow teammates and even opponents to communicate with each other seems intuitive (as many games require coordinated efforts) and hopeful (as it should allow players to form friendships and communities). In fact, early research suggests that using voice chat (instead of text chat) during a gaming experience improved trust and liking among players (Williams et al., 2007).

However, subsequent research suggests that gaming communication channels are also likely to foster anti-social behavior (e.g., Sherrick et al., 2024b), particularly for women (Fox & Tang, 2014; Kuznekoff & Rose, 2012) and members of minority groups (Gray, 2012). Fox and colleagues (2014; Holz Ivory et al., 2014) argue that communication channels—particularly voice—provide cues to player identities that can then be used to apply traditional stereotypes and prejudices against particular groups of people. As one example of how this might play out, Kuznekoff and Rose's (2012) field experiment showed that a verbal comment presented in a woman's voice was met with significantly more negativity than the same comment presented in a man's voice. Notably, recent research suggests that allowing voice communication during games does not necessarily increase the likelihood of positive behaviors (Sherrick et al., 2024b) or cooperation (Velez et al., 2021). Perhaps in response to this research, more recently released games forgo or limit voice communication channels (e.g., Blizzard Entertainment, 2022), suggesting that the game industry is aware of and actively combating problems associated with voice chat. Future research should continue to investigate how the affordances of different types of communication channels (Fox & McEwan, 2017) can impact pro- and anti-social behavior in games.

Cooperative and Competitive Gaming

One of the more promising lines of research aimed at curbing anti-social behavior and encouraging pro-social behavior is research on cooperative gaming settings. Broadly speaking, this research suggests that games encouraging cooperation between players are likely to lead to prosocial behaviors, such as mutually beneficial sharing (e.g., Velez, 2013, 2015, 2016) or improved general cooperation on subsequent tasks (Velez et al., 2021). Moreover, this research has also shown that games encouraging competition between players can decrease pro-social behaviors and even have the potential to increase anti-social behaviors, such as aggression (e.g., Schmierbach, 2010).

The most prominent theoretical explanation for the varying effects of cooperative and competitive gaming is the theory of Bounded Generalized Reciprocity (e.g., Velez et al., 2013, 2015, 2016). Relying on perceptions of ingroups vs. outgroups, Bounded Generalized Reciprocity generally predicts that ingroup members (e.g., teammates in a cooperative game) will reciprocate helpful behaviors but that outgroup members (e.g., opponents in a competitive game) will not reciprocate helpful behaviors. Notably, these expectations can guide behaviors such that people may be more likely to withhold positive behaviors from and/or perform negative behaviors toward outgroup members, even if those outgroup members have not themselves behaved antisocially. In short, the grouping characteristics of multiplayer games are likely to activate group-based expectations that guide subsequent behavior. Notably, these patterns of reciprocal behavior are likely to form fairly quickly (e.g., Velez et al., 2021) and can echo through future behaviors (Ewoldsen et al., 2012; Velez et al., 2014).

Why Care About Virtual Behaviors

Research has shown that in-game experiences have significant implications for players' psychological (Cole & Griffiths, 2007; Ryan et al., 2006; Tamborini et al., 2011; Oliver et al., 2016) and social development (Shen & Williams, 2011; Williams, 2006). Engaging in video games goes beyond mere entertainment as it can provoke a range of emotions, both positive and negative, in and out of the game. Understanding these virtual behaviors is crucial for informing game design, player interactions, and the overall gaming experience. Here, we delve into why these experiences matter, particularly focusing on how they can lead to negative ruminations and other subsequent interactions.

A major line of this research emphasizes how avatars can influence individuals' behaviors, extending beyond virtual environments into real life. For example, Fox and Bailenson (2009a) demonstrated that individuals who viewed a virtual representation of themselves engaging in exercise were more likely to increase their physical activity afterwards. Similarly, research on the psychological impact of sexualized avatars showed negative effects on women; Fox et al. (2013) found that women using sexualized avatars were more likely to experience self-objectification and endorse harmful rape myths (Fox & Bailenson, 2009b). Such findings point to potential carryover effects, suggesting that the behaviors and attitudes cultivated through avatar portrayal in virtual environments can significantly influence individuals' interactions and beliefs in real-world contexts.

Players often develop strong emotional connections to characters and narratives within video games, leading to significant emotional experiences (Banks & Bowman, 2016; Bowman et al., 2016; Elson et al., 2014; Green et al., 2004; Schneider et al., 2004). However, this emotional investment can also yield negative consequences. Research suggests that players frequently ruminate on negative experiences, such as frustration during gameplay or instances of harassment, which can be detrimental to their mental well-being (Fox & Tang, 2017; Fox et al., 2018). For example, both sexual and general harassment in gaming environments have been shown to contribute to emotional rumination, with women particularly reporting increased experiences of sexist comments and ultimately withdrawing from gaming (Fox & Tang, 2014; Fox & Tang, 2017). Moreover, studies indicate that male players with hostile sexist attitudes are more likely to engage in such harassment behaviors, further exacerbating the challenges faced by women gamers (Tang & Fox, 2016; Fox et al., 2018).

In contrast to research on the darker side of video games, substantial research has also shown how positive gaming experiences can shape players' subsequent behaviors. For instance, playing prosocial video games has been shown to increase helping behaviors and foster increased trust in partners even after gameplay (Greitemeyer et al., 2012; Greitemeyer & Osswald, 2010). Research in cooperative video gameplay extended the theory of bounded generalized reciprocity, suggesting that the social context of video games can impact players' expectations to reciprocate prosocial behavior among their teammates (Velez et al., 2016). Specifically, playing with helpful teammates can strengthen the expectation of reciprocating prosocial behavior within a group, leading to increased prosocial behavior among teammates (Velez, 2015). Recent research further revealed that engaging in prosocial behaviors during cooperative video gameplay, even without verbal communication, can increase subsequent task performance, emphasizing how established group norms can benefit fostering cooperation and social behaviors beyond a gaming experience (Velez et al., 2021).

Recommendations

There are a variety of negative or antisocial behaviors in online gaming such as griefing (Blackburn & Kwak, 2014), intentionally losing (Tan & Chen, 2022), and discrimination and threats (Gray, 2018; Mihailova, 2022), but there are also positive or prosocial behaviors such as cooperation (Velez et al., 2021), connection (Oliver et al., 2016; Sherrick et al., 2023), and positive communication (Greitemeyer & Osswald, 2010). Again, a primary goal of this chapter is to outline and ultimately recommend strategies for maximizing the positive and minimizing the negative behaviors associated with gaming. Currently, common strategies often emphasize punishing or sanctioning negative behavior through banning or flagging players (Kou, 2021; Kou & Gui, 2021; Wijkstra et al., 2024) or detecting and deleting profane or offensive messages (Murnion et al., 2018). However, these existing strategies do not pay attention to the possible prosocial behaviors fostered by video games, and they are only useful after a negative behavior has occurred, so there may be better options available to game designers and players.

As one possible option, researchers and game designers have begun experimenting with strategies aimed at supporting and empowering players, such as by rewarding players who behave well through commendations (Kilmer et al., 2024) or allowing players to create their own blocklists (Reid et al., 2022). Games like Marvel Rivals or All You Need is Help allow players to upvote or like other players during post-game scoreboard displays; while no identified research examined the effectiveness of such a strategy, these game features may encourage players to behave prosocially to receive the most votes or likes. Another strategy for reducing toxic behavior before it happens and improving prosocial behaviors is by de-emphasizing descriptive toxic norms and emphasizing the descriptive norms of pro-social behavior (Sherrick et al., 2024b). Injunctive norms are often set already through terms of service in games (Busch et al., 2015), but descriptive norms could be set through awareness campaigns such as the GLHF ("Good luck; have fun") campaign on Twitch where more than 370,000 players pledged to engage in positive behavior in games (Brewer et al., 2020). Game designers and researchers should continue to experiment with strategies that help players affected by toxicity mitigate the prevalence and impact of that toxicity.

Implementing more cooperative gameplay mechanics and rewarding players for cooperative behaviors in competitive games could improve prosocial behavior (Velez et al., 2016; Ewoldsen et al., 2012). For example, the game Overcooked requires players to cooperate on cooking challenges to achieve goals and progress in the game, and research suggests this improves future cooperative behavior (Velez et al., 2021). Game designers must necessarily balance different—sometimes competing—objectives, such as encouraging prosocial behavior through cooperation or entertaining players through competition. Notably, many games allow for both cooperation and competition through team-based competitions, so future research should more clearly define and delineate how cooperative and competitive gameplay mechanics support or undercut various game-design goals. However, if the primary goal of a game's design is to foster prosocial behavior, existing research suggests that games should heavily lean toward cooperative, rather than competitive, gameplay.

Characters and narratives in games can also model prosocial behaviors (Kilmer et al., 2024), aligning with concepts like the Proteus effect, where the player takes on some of the attributes of the character (Yee & Bailenson, 2007). For example, the PBS game Molly's Winter Kitchen displays the importance of sharing resources amongst members of an Alaskan village community, which might encourage young players to also value resource sharing in their communities. Future research should investigate how prosocial behaviors are modeled in games and whether and how that prosocial modeling impacts player behaviors, both within and beyond the game. In sum, research suggests that game developers have a variety of tools that could create better interactions online, though understanding which are the easiest to implement and are the most impactful for players could still be explored more in order to create a better space for all gamers.

Conclusion

Video games often create enjoyable and amazing social experiences and interactions, but there is also a worrying trend—particularly in online gaming spaces—toward negative and antisocial behaviors. Game designers and players should take both seriously, as positive gaming experiences are more likely with intentional efforts to minimize harmful and maximize beneficial behaviors among all involved in gaming communities. Research suggests that both design choices and player behaviors influence future behaviors, so game designers and game players should actively consider how their decisions impact the experiences of others. Doing so will provide the best chance for gaming communities to succeed and even thrive, ensuring that those communities persist over time.

References

Banks, J., & Bowman, N. D. (2016). Avatars are (sometimes) people too: Linguistic indicators of parasocial and social ties in player–avatar relationships. *New Media & Society, 18*(7), 1257–1276.

Berkowitz, L. (1972). Social norms, feelings, and other factors affecting helping and altruism. In *Advances in experimental social psychology*Vol 6, pp. 63-108. Academic Press.https://doi.org/10.1016/S0065-2601(08)60025-8

Biocca, F. (1997). The cyborg's dilemma: Progressive embodiment in virtual environments. *Journal of Computer-Mediated Communication, 3*(2). https://doi.org/10.1111/j.1083-6101.1997.tb00070.x

Blackburn, J., & Kwak, H. (2014). STFU NOOB! Predicting crowdsourcing decisions on toxic behaviors in online games. In *Proceedings of the 23rd International World Wide Web Conference Committee* (pp. 877–888). https://doi.org/10.1145/2566486.2567987

Blizzard Entertainment. (2022, September 27). *Defense matrix activated*! Fortifying gameplay integrity and positivity in Overwatch 2. https://overwatch.blizzard.com/en-us/news/23857517/defense-matrix-activated-fortifying-gameplay-integrity-and-positivity-in-overwatch-2/

Bowman, N. D. (2015). The rise (and refinement) of moral panic. In R. Kowert & T. Quandt (Eds.), *The video game debate* (pp. 22–38). Routledge. https://doi.org/10.4324/9781315736495

Bowman, N. D., Oliver, M. B., Rogers, R., Sherrick, B., & Woolley, J. K. (2016). In control or in their shoes? How character attachment differentially influences video game enjoyment and appreciation. *Journal of Gaming and Virtual Worlds, 8*(1), 83–99. https://doi.org/10.1386/jgvw.8.1.83_1

Brewer, J., Romine, M., & Taylor, T. L. (2020, July). Inclusion at scale: Deploying a community-driven moderation intervention on twitch. In *Proceedings of the 2020 ACM designing interactive systems conference* (pp. 757–769). https://doi.org/10.1145/3357236.3395514

Brown, S. (2008). *Play is more than just fun* [video]. TED Conferences. https://www.ted.com/talks/stuart_brown_play_is_more_than_just_fun?language=en#t-1578284

Busch, T., Boudreau, K., & Consalvo, M. (2015). Toxic gamer culture, corporate regulation, and standards of behavior among players of online games. In *Video game policy* (pp. 176–190). Routledge. https://doi.org/10.4324/9781315748825-13

Canossa, A., Salimov, D., Azadvar, A., Harteveld, C., & Yannakakis, G. (2021). For honor, for toxicity: Detecting toxic behavior through gameplay. In *Proceedings of the ACM on human-computer interaction* (Vol. 5(CHI PLAY), pp. 1–29). https://doi.org/10.1145/3474680

Cialdini, R. B., Reno, R. R., & Kallgren, C. A. (1990). A focus theory of normative conduct: Recycling the concept of norms to reduce littering in public places. *Journal of Personality and Social Psychology, 58*(6), 1015-1026. https://doi.org/10.1037/0022-3514.58.6.1015

Cialdini, R. B., & Trost, M. R. (1998). Social influence: Social norms, conformity and compliance. In D. T. Gilbert, S. T. Fiske, & G. Lindzey (Eds.), *The handbook of social psychology, Vol. 2* (4th ed., pp. 151-192). McGraw-Hill.

Cole, H., & Griffiths, M. D. (2007). Social interactions in massively multiplayer online role-playing gamers. *Cyberpsychology & Behavior, 10*(4), 575–583. https://doi.org/10.1089/cpb.2007.9988

Conmy, B., Tenenbaum, G., Eklund, R., Roehrig, A., & Filho, E. (2013). Trash talk in a competitive setting: Impact on self-efficacy and affect. *Journal of Applied Social Psychology, 43*(5), 1002–1014. https://doi.org/10.1111/jasp.12064

Consalvo, M. (2007). *Cheating: Gaining advantage in videogames*. MIT Press.

Consalvo, M. (2012). Confronting toxic gamer culture: A challenge for feminist game studies scholars. *Ada: A Journal of Gender, New Media, and Technology, 1*. https://doi.org/10.7264/N33X84KH

Dietz, T. L. (1998). An examination of violence and gender role portrayals in video games: Implications for gender socialization and aggressive behavior. *Sex Roles, 38*(5), 425–442.

Dill, K. E., Gentile, D. A., Richter, W. A., & Dill, J. C. (2005). Violence, sex, race, and age in popular video games: A content analysis. In E. Cole & J. H. Daniel (Eds.), *Featuring females: Feminist analyses of media* (pp. 115–130). American Psychological Association. https://doi.org/10.1037/11213-008

Ducheneaut, N., Wen, M. H., Yee, N., & Wadley, G. (2009, April). Body and mind: A study of avatar personalization in three virtual worlds. In *Proceedings of the SIGCHI Conference on Human Factors in Computing Systems* (pp. 1151–1160). https://doi.org/10.1145/1518701.1518877

Eastin, M. S. (2006). Video game violence and the female game player: Self- and opponent gender effects on presence and aggressive thoughts. *Human Communication Research, 32*, 351–372. https://doi.org/10.1111/j.1468-2958.2006.00279.x

Ekiciler, A., Ahioglu, I., Yildirim, N., Ajas, I. I., & Kaya, T. (2022). The bullying game: Sexism based toxic language analysis on online games chat logs by text mining. *Journal of International Women's Studies, 24*(3), 1–16.

Elson, M., Breuer, J., Ivory, J. D., & Quandt, T. (2014). More than stories with buttons: Narrative, mechanics, and context as determinants of player experience in digital games. *Journal of Communication, 64*(3), 521–542. https://doi.org/10.1111/jcom.12096

Entertainment Software Association. (2024). *2024 Essential facts about the U.S. video game industry*. https://www.theesa.com/resources/essential-facts-about-the-us-video-game-industry/2024-data/

Ewoldsen, D. R., Eno, C. A., Okdie, B. M., Velez, J. A., Guadagno, R. E., & DeCoster, J. (2012). Effect of playing violent video games cooperatively or competitively on subsequent cooperative behavior. *Cyberpsychology, Behavior, and Social Networking, 15*(5), 277-280. https://doi.org/10.1089/cyber.2011.0308

Foo, C. Y., & Koivisto, E. M. (2004, September). Defining grief play in MMORPGs: Player and developer perceptions. In *Proceedings of the ACM SIGCHI international conference on advances in computer entertainment technology* (pp. 245–250). https://doi.org/10.1145/1067343.1067375

Fox, J., & Bailenson, J. N. (2009a). Virtual self-modeling: The effects of vicarious reinforcement and identification on exercise behaviors. *Media Psychology, 12*(1), 1–25. https://doi.org/10.1080/15213260802669474

Fox, J., & Bailenson, J. N. (2009b). Virtual virgins and vamps: The effects of exposure to female characters' sexualized appearance and gaze in an immersive virtual environment. *Sex Roles: A Journal of Research, 61*(3–4), 147–157. https://doi.org/10.1007/s11199-009-9599-3

Fox, J., & McEwan, B. (2017). Distinguishing technologies for social interaction: The perceived social affordances of communication channels scale. *Communication Monographs, 84*(3), 298–318. https://doi.org/10.1080/03637751.2017.1332418

Fox, J., & Tang, W. Y. (2014). Sexism in online video games: The role of conformity to masculine norms and social dominance orientation. *Computers in Human Behavior, 33*, 314–320. https://doi.org/10.1016/j.chb.2013.07.014

Fox, J., & Tang, W. Y. (2017). Women's experiences with general and sexual harassment in online video games: Rumination, organizational responsiveness, withdrawal and coping strategies. *New Media & Society, 19*(8), 1290–1307. https://doi.org/10.1177/1461444816635778

Fox, J., Bailenson, J. N., & Tricase, L. (2013). The embodiment of sexualized virtual selves: The Proteus effect and experiences of self-objectification via avatars. *Computers in Human Behavior, 29*(3), 930–938. https://doi.org/10.1016/j.chb.2012.12.027

Fox, J., Gilbert, M., & Tang, W. Y. (2018). Player experiences in a massively multiplayer online game: A diary study of performance, motivation, and social interaction. *New Media & Society, 20*(11), 4056–4073. https://doi.org/10.1177/1461444818767102

Gibson, J. J. (1979). *The ecological approach to visual perception*. Houghlin Mifflin.

Gray, K. L. (2012). Intersecting oppressions and online communities: Examining the experiences of women of color in Xbox live. *Information, Communication & Society, 15*(3), 411–428. https://doi.org/10.1080/1369118X.2011.642401

Gray, K. L. (2018). Gaming out online: Black lesbian identity development and community building in Xbox live. *Journal of Lesbian Studies, 22*(3), 282–296. https://doi.org/10.1080/10894160.2018.1384293

Green, M. C., Brock, T. C., & Kaufman, G. F. (2004). Understanding media enjoyment: The role of transportation into narrative worlds. *Communication Theory, 14*(4), 311–327. https://doi.org/10.1111/j.1468-2885.2004.tb00317.x

Greitemeyer, T., & Osswald, S. (2010). Effects of prosocial video games on prosocial behavior. *Journal of Personality and Social Psychology, 98*(2), 211–221. https://doi.org/10.1037/a0016997

Greitemeyer, T., Traut-Mattausch, E., & Osswald, S. (2012). How to ameliorate negative effects of violent video games on cooperation: Play it cooperatively in a team. *Computers in Human Behavior, 28*(4), 1465–1470. https://doi.org/10.1016/j.chb.2012.03.009

Hilvert-Bruce, Z., Neill, J. T., Sjöblom, M., & Hamari, J. (2018). Social motivations of live-streaming viewer engagement on Twitch. *Computers in Human Behavior, 84*, 58-67. https://doi.org/10.1016/j.chb.2018.02.013

Hilvert-Bruce, Z., & Neill, J. T. (2020). I'm just trolling: The role of normative beliefs in aggressive behaviour in online gaming. *Computers in Human Behavior, 102*, 303–311. https://doi.org/10.1016/j.chb.2019.09.003

Holz Ivory, A., Fox, J., Waddell, T. F., & Ivory, J. D. (2014). Sex role stereotyping is hard to kill: A field experiment measuring social responses to user characteristics and behavior in an online multiplayer first-person shooter game. *Computers in Human Behavior, 35*, 148–156. https://doi.org/10.1016/j.chb.2014.02.026

Holz Ivory, A., Ivory, J. D., Wu, W., Limperos, A. M., Andrew, N., & Sesler, B. S. (2017). Harsh words and deeds: Systematic content analyses of offensive* user behavior in the environments of online first-person shooter games. *Journal for Virtual Worlds Research, 10*(2). https://doi.org/10.4101/jvwr.v10i2.7274

Huizinga, J. (1962). *Homo ludens: A study of the play-element in culture. [Tr. from the German ed.]*. Beacon Press.

Kilmer, E. D., Aslan, Z., & Kowert, R. (2024). Addressing toxicity and extremism in games: Conversations with the video game industry. *Games and Culture. [Advance online publication]*. https://doi.org/10.1177/15554120241265139

Kou, Y. (2021). Punishment and its discontents: An analysis of permanent ban in an online game community. *Proceedings of the ACM on Human-Computer Interaction, 5*(CSCW2), 1–21. https://doi.org/10.1145/3476075

Kou, Y., & Gui, X. (2017). When code governs community. In *Proceedings of the annual Hawaii international conference on system sciences* (pp. 2056–2064).

Kou, Y., & Gui, X. (2021, May). Flag and flaggability in automated moderation: The case of reporting toxic behavior in an online game community. In *Proceedings of the 2021 CHI conference on human factors in computing systems* (pp. 1–12). https://doi.org/10.1145/3411764.3445279

Kowert, R., & Oldmeadow, J. A. (2015). Playing for social comfort: Online video game play as a social accommodator for the insecurely attached. *Computers in Human Behavior,53*, 556-566. https://doi.org/10.1016/j.chb.2014.05.004

Kuznekoff, J. H., & Rose, L. M. (2012). Communication in multiplayer gaming: Examining player responses to gender cues. *New Media & Society, 15*(4), 541–556. https://doi.org/10.1177/1461444812458271

Kwak, D.-H., Kim, D., Lee, S., Kang, M., Park, S., & Knapp, D. (2024). The effect of blockability affordance on confrontation against cyberbullying on social networking sites: Theoretical and

methodological implications. *Internet Research. [Advance online publication]., 35*. https://doi.org/10.1108/INTR-05-2023-0422
Lapinski, M. K., & Rimal, R. N. (2005). An explication of social norms. *Communication Theory, 15*(2), 127–147. https://doi.org/10.1111/j.1468-2885.2005.tb00329.x
Lea, M., & Spears, R. (1991). Computer-mediated communication, deindividuation, and group, decision-making. *International Journal of Man Machine Studies, 34*(2), 283–301. https://doi.org/10.1016/0020-7373(91)90045-9
Lee, K. M. (2004). Presence, explicated. *Communication Theory, 14*(1), 27–50. https://doi.org/10.1111/j.1468-2885.2004.tb00302.x
Lee, J. E. R. (2014). Does virtual diversity matter? Effects of avatar-based diversity, representation on willingness to express offline racial identity and avatar customization. *Computers in Human Behavior, 36*, 190–197. https://doi.org/10.1016/j.chb.2014.03.040
Lewis, J. E., Trojovsky, M., & Jameson, M. M. (2021). New social horizons: Anxiety, isolation, and animal crossing during the COVID-19 pandemic. *Frontiers in Virtual Reality, 2*. https://doi.org/10.3389/frvir.2021.627350
Martens, M., Shen, S., Iosup, A., & Kuipers, F. (2015). Toxicity detection in multiplayer online games. In 2015 *International workshop on network and systems support for games*(NetGames), 2016 (pp. 1–6). https://doi.org/10.1109/NetGames.2015.7382991
Martinez, J. J., Windleharth, T. W., Li, Q., Bhattacharya, A., Pearce, K. E., Yip, J., & Lee, J. H. (2022). Joint media engagement in families playing animal crossing: New horizons during the COVID-19 pandemic. *Proceedings of the ACM on Human-Computer Interaction, 6*(CSCW1), 1–22. https://doi.org/10.1145/3512947
McMillan, D. W. & Chavis, D. M. (1986). Sense of community: A definition and theory. *Journal of Community Psychology, 14*, 6-23
Mihailova, T. (2022). Navigating ambiguous negativity: A case study of Twitch.tv live chats. *New Media & Society, 24*(8), 1830–1851. https://doi.org/10.1177/1461444820978999
Monge, C. K., & Matthews, N. L. (2024). Blaming the smurf: Using a novel social deception behavior in online games to test attribution theories. *New Media & Society*. Advance online publication. https://doi.org/10.1177/14614448241235638
Murnion, S., Buchanan, W. J., Smales, A., & Russell, G. (2018). Machine learning and semantic analysis of in-game chat for cyberbullying. *Computers & Security, 76*, 197–213. https://doi.org/10.1016/j.cose.2018.02.016
Norris, K. O. (2004). Gender stereotypes, aggression, and computer games: An online survey of, women. *Cyberpsychology & Behavior, 7*(6), 714–727. https://doi.org/10.1089/cpb.2004.7.714
Nowak, K. L., & Fox, J. (2018). Avatars and computer-mediated communication: A review of the definitions, uses, and effects of digital representations. *Review of Communication Research, 6*, 30–53. https://doi.org/10.12840/issn.2255-4165.2018.06.01.015
Oliver, M. B., Bowman, N. D., Woolley, J. K., Rogers, R., Sherrick, B., & Chung, M.-Y. (2016). Video games as meaningful entertainment experiences. *Psychology of Popular Media Culture, 5*(4), 390–405.
Passmore, H. A., & Holder, M. D. (2014). Gaming for good: Video games and enhancing prosocial behaviour. In *Video games: Parents' perceptions, role of social media and effects on behavior* (pp. 141–166). Nova Publishers.
Postmes, T., Spears, R., & Lea, M. (1998). Breaching or building social boundaries? SIDE effects of computer-mediated communication. *Communication Research, 25*(6), 689–715. https://doi.org/10.1177/009365098025006006
Postmes, T., Spears, R., & Lea, M. (2000). The formation of group norms in computer-mediated communication. *Human Communication Research, 26*(3), 341-371. https://doi.org/10.1111/j.1468-2958.2000.tb00761.x
Prather, J., Nix, R., & Jessup, R. (2017). Trust management for cheating detection in distributed massively multiplayer online games. In *2017 15th Annual workshop on network and systems support for games (NetGames)* (pp. 1–3). https://doi.org/10.1109/NetGames.2017.7991547

Reer, F., & Krämer, N. C. (2020). A self-determination theory-based laboratory experiment on social aspects of playing multiplayer first-person shooter games. *Entertainment Computing, 34*, 100353. https://doi.org/10.1016/j.entcom.2020.100353

Ratan, R., Beyea, D., Li, B. J., & Graciano, L. (2020). Avatar characteristics induce users' behavioral conformity with small-to-medium effect sizes: a meta-analysis of the proteus effect. *Media Psychology, 23*(5), 651-675. https://doi.org/10.1080/15213269.2019.1623698

Reid, E., Mandryk, R. L., Beres, N. A., Klarkowski, M., & Frommel, J. (2022). Feeling good and in control: In-game tools to support targets of toxicity. In *Proceedings of the ACM on human-computer interaction, 6(CHI PLAY)* (pp. 1–27). https://doi.org/10.1145/3549498

Ryan, R. M., & Deci, E. L. (2017). *Self-determination theory: Basic psychological needs in motivation, development, and wellness*. Guilford Press Publications. https://doi.org/10.1521/978.14625/28806

Ryan, R. M., Rigby, C. S., & Przybylski, A. (2006). The motivational pull of video games: A self-determination theory approach. *Motivation and Emotion, 30*, 347–363. https://doi.org/10.1007/s11031-006-9051-8

Schmierbach, M. (2010). "Killing spree": Exploring the connection between competitive game play and aggressive cognition. *Communication Research, 37*(2), 256–274. https://doi.org/10.1177/0093650209356394

Schneider, E. F., Lang, A., Shin, M., & Bradley, S. D. (2004). Death with a story: How story impacts emotional, motivational, and physiological responses to first-person shooter video games. *Human Communication Research, 30*(3), 361–375. https://doi.org/10.1111/j.1468-2958.2004.tb00736.x

Shen, C., & Williams, D. (2011). Unpacking time online: Connecting internet and massively multiplayer online game use with psychosocial well-being. *Communication Research, 38*(1), 123–149.

Sherif, M. (1936). *The psychology of social norms*. Harper & Row.

Sherrick, B., Hoewe, J., & Waddell, T. F. (2014). The role of stereotypical beliefs in gender-based activation of the Proteus effect. *Computers in Human Behavior, 38*, 17–24. https://doi.org/10.1016/j.chb.2014.05.010

Sherrick, B., Hoewe, J., & Ewoldsen, D. (2022). Using narrative media to satisfy intrinsic needs: Connecting parasocial relationships, retrospective imaginative involvement, and self-determination theory. *Psychology of Popular Media, 11*(3), 266–274. https://doi.org/10.1037/ppm0000358

Sherrick, B., Smith, C., Jia, Y., Thomas, B., & Franklin, S. B. (2023). How parasocial phenomena contribute to community health on twitch. *Journal of Broadcasting and Electronic Media, 67*(1), 47–67. https://doi.org/10.1080/08838151.2022.2151599

Sherrick, B., Hoewe, J., & Rogers, R. (2024a). How nature- and humanity-based awe experiences in video games can differentially lead to hedonic and eudaimonic outcomes. *Communication Monographs. [Advance online publication]., 92*. https://doi.org/10.1080/03637751.2024.2397435

Sherrick, B., Smith, C., Jia, Y., Kim, J., Woodland, S., & Fox, J. (2024b, June). Investigating antecedents to pro- and anti-social behaviors in online, multiplayer video games. In *Paper presented at international communication association (ICA), Gold Coast.*

Talo, C., Mannarini, T., & Rochira, A. (2014). Sense of community and community participation: A meta-analytic review. *Social Indicators Research, 117*, 1-28. https://doi.org/10.1007/s11205-013-0347-2

Smith, S. L., Lachlan, K., & Tamborini, R. (2003). Popular video games: Quantifying the presentation of violence and its context. *Journal of Broadcasting & Electronic Media, 47*(1), 58–76. https://doi.org/10.1207/s15506878jobem4701_4

Spears, R., & Lea, M. (1992). Social influence and the influence of the "social" in computer-mediated communication. In M. Lea (Ed.), *Contexts of computer-mediated communication* (pp. 30–65). Harvester Wheatsheaf.

Stepanova, N., Muthemba, W., Todrzak, R., Cross, M., Ames, N., & Raiti, J. (2021). Natural language processing and sentiment analysis for verbal aggression detection; A solution for cyberbullying during live video gaming. In *ACM International Conference Proceeding Series* (pp. 117–118). https://doi.org/10.1145/3453892.3464897

Strimling, P., & Frey, S. (2020). Emergent cultural differences in online communities' norms of fairness. *Games and Culture, 15*(4), 394-410. https://doi.org/10.1177/1555412018800650

Suler, J. (2004). The online disinhibition effect. *Cyberpsychology & behavior, 7*(3), 321-326. https://doi.org/10.1089/1094931041291295

Sutcliffe, A. G., Gonzalez, V., Binder, J., & Nevarez, G. (2011). Social mediating technologies: Social affordances and functionalities. *International Journal of Human-Computer Interaction, 27*(11), 1037–1065. https://doi.org/10.1080/10447318.2011.555318

Tamborini, R., Grizzard, M., Bowman, N. D., Reinecke, L., Lewis, R. J., & Eden, A. (2011). Media enjoyment as need satisfaction: The contribution of hedonic and nonhedonic needs. *Journal of Communication, 61*(6), 1025–1042. https://doi.org/10.1111/j.1460-2466.2011.01593.x

Tan, W., & Chen, L. (2022). That's not my fault: Excuses given by players exhibiting in-game intra-team aggressive behavior in online games. *Computers in Human Behavior, 127*, 1–10. https://doi.org/10.1016/j.chb.2021.107045

Tang, W. Y., & Fox, J. (2016). Men's harassment behavior in online video games: Personality traits and game factors. *Aggressive Behavior, 42*(6), 513–521. https://doi.org/10.1002/ab.21646

Tekinbas, K. S., & Zimmerman, E. (2003). *Rules of play: Game design fundamentals*. MIT press.

Velez, J. A. (2015). Extending the theory of bounded generalized reciprocity: An explanation of the social benefits of cooperative video game play. *Computers in Human Behavior, 48*, 481–491. https://doi.org/10.1016/j.chb.2015.02.015

Velez, J. A., & Ewoldsen, D. R. (2013). Helping behaviors during video game play. *Journal of Media Psychology, 25*(4), 190–200. https://doi.org/10.1027/1864-1105/a000102

Velez, J. A., Mahood, C., Ewoldsen, D. R., & Moyer-Gusé, E. (2014). Ingroup versus outgroup conflict in the context of violent video game play: The effect of cooperation on increased helping and decreased aggression. *Communication Research, 41*(5), 607-626. https://doi.org/10.1177/0093650212456202

Velez, J. A., Greitemeyer, T., Whitaker, J. L., Ewoldsen, D. R., & Bushman, B. J. (2016). Violent video games and reciprocity: The attenuating effects of cooperative game play on subsequent aggression. *Communication Research, 43*(4), 447–467. https://doi.org/10.1177/0093650214552519

Velez, J. A., Schmierbach, M., Sherrick, B., Chae, S. W., Tan, R., & Rosenberg, K. (2021). A recipe for success: The effect of dyadic communication and cooperative gameplay on subsequent non-gaming task performance. *Human Communication Research, 47*(4), 364–386. https://doi.org/10.1093/hcr/hqab008

Vo, H. H.-P., Trung Tran, H., & Luu, S. T. (2021). Automatically detecting cyberbullying comments on online game forums. *RIVF international conference on computing and communication technologies (RIVF), 2021*, 1–5. https://doi.org/10.1109/RIVF51545.2021.9642116

Wadley, G., Carter, M., & Gibbs, M. (2015). Voice in virtual worlds: The design, use, and influence of voice chat in online play. *Human-Computer Interaction, 30*(3–4), 336–365. https://doi.org/10.1080/07370024.2014.987346

Walther, J. B. (1992). Interpersonal effects in computer-mediated interaction: A relational perspective. *Communication Research, 19*(1), 52–90. https://doi.org/10.1177/009365092019001003

Wijkstra, M., Rogers, K., Mandryk, R. L., Veltkamp, R. C., & Frommel, J. (2024). How to tame a toxic player? A systematic literature review on intervention systems for toxic behaviors in online video games. In *Proceedings of the ACM on Human-Computer Interaction* (Vol. 8(CHI PLAY), pp. 1–32). https://doi.org/10.1145/3677080

Williams, D. (2006). Groups and goblins: The social and civic impact of an online game. *Journal of Broadcasting & Electronic Media, 50*(4), 651–670.

Williams, D., Caplan, S., & Xiong, L. (2007). Can you hear me now? The impact of voice in an online gaming community. *Human Communication Research, 33*(4), 427-449. https://doi.org/10.1111/j.1468-2958.2007.00306.x

Woo, J., Park, S. H., & Kim, H. K. (2022). Profane or not: Improving Korean profane detection using deep learning. *KSII Transactions on Internet and Information Systems, 16*(1), 305–318. https://doi.org/10.3837/tiis.2022.01.017

Yee, N., & Bailenson, J. (2007). The Proteus effect: The effect of transformed self-representation on behavior. *Human Communication Research, 33*(3), 271–290.

Yee, N., & Bailenson, J. N. (2009). The difference between being and seeing: The relative contribution of self-perception and priming to behavioral changes via digital self-representation. *Media Psychology, 12*(2), 195–209. https://doi.org/10.1080/15213260902849943

Yee, N., Bailenson, J. N., & Ducheneaut, N. (2009). The Proteus effect: Implications of transformed digital self-representation on online and offline behavior. *Communication Research, 36*(2), 285–312. https://doi.org/10.1177/0093650208330254

Zhang, K., Deldari, E., Lu, Z., Yao, Y., Zhao, Y., Shinohara, K., Ludi, S., & Froehlich, J. (2022). "It's Just Part of Me:" Understanding avatar diversity and self-presentation of people with disabilities in social virtual reality. In *Proceedings of the 24th international ACM SIGACCESS conference on computers and accessibility* (pp. 1–16). https://doi.org/10.1145/3517428.3544829

Chapter 15
Neuroethical Considerations in Gaming Communities

Yuan Dong

Abstract This chapter explores the complex landscape of neuroethics within the gaming industry and its communities, focusing on the ethical dilemmas that arise at the intersection of gaming practices and community norms. It examines neurofeedback games, which, despite their potential to enhance cognitive abilities, raise significant ethical concerns related to privacy, consent, and the broader implications of neuro-enhancement. The chapter also addresses the role of video games in neurodiverse communities, emphasizing their dual potential to both assist and complicate the experiences of neurodivergent players. Additionally, it criticizes the growing prevalence of gambling-like elements in games, exploring the ethical risks associated with addiction and consumer exploitation. Finally, the chapter challenges the persistent social stigma surrounding gaming, confronting negative stereotypes that often overshadow the diverse and positive aspects of gaming communities. By examining these multifaceted ethical issues, the chapter seeks to provide insights into the ethical complexities of the gaming world and propose thoughtful approaches to create a more inclusive and ethically responsible gaming environment.

Keywords Neuroethics · Gaming communities · Neurodiversity · Social stigma · Neurofeedback

1. Neuroethical Considerations for Neurofeedback Games

Neurofeedback is an intervention that enables an individual to self-regulate their brain activity through real-time feedback. Neurofeedback training is widely used as an effective therapy for mental, cognitive, and behavioral disorders such as attention deficit disorder (ADHD), learning disabilities, and epilepsy (Gruzelier, 2014). Neurofeedback games fall under this broader technology. Unlike neurofeedback

Y. Dong (✉)
Duke University, Durham, NC, USA
e-mail: yuan.dong@duke.edu

L. Murrins Marques et al. (eds.), *Social and Affective Neuroscience of Gaming*,
https://doi.org/10.1007/978-3-032-00325-6_15

used in clinical settings, neurofeedback games are crafted for entertainment, providing real-time feedback in the form of a game (Coenen et al., 2020). Numerous neurofeedback games have been developed in recent years to improve cognitive performance (Wang et al., 2011, 2020; Thomas et al., 2013; Khong et al., 2014; Schoneveld et al., 2016; Teruel et al., 2017; Jirayucharoensak, 2019; Lyu et al.2023; Suhail, 2024). While neurofeedback games have great potential for improving cognitive performance and mental well-being, they also come with challenges and neuroethical considerations that need careful thinking and management. As neurotechnology advances, there is significant potential for neurofeedback games to extend their reach from clinical settings into mainstream gaming circles. With this broader application, it becomes increasingly important to thoroughly discuss and address the ethical considerations about their use. Informing the gaming industry and communities about the implications of these technologies, including concerns about data privacy, user consent, and psychological effects, is essential for the responsible deployment of neurofeedback games.

1.1 Privacy Concerns

Neurofeedback involves recording bioinformation that is unique and private to each patient. This data can be deeply revealing, offering insights not only into cognitive and emotional states but also potentially exposing aspects of a person's medical history or psychological vulnerabilities. For example, in the context of gaming, neurofeedback techniques are sometimes used to enhance player performance by optimizing attention or emotional regulation. Products like Mindflex and Emotiv offer neurofeedback experiences where players control in-game actions using brainwave data.

However, the collection of such sensitive data raises significant ethical concerns. In neurofeedback gaming, inadequate protection of personal data could lead to privacy breaches and even exploitation, such as targeting vulnerable individuals with manipulative in-game advertisements or profiling players based on their mental health or cognitive abilities for discriminatory purposes. A real-world case study of this is seen in the controversy surrounding the use of biometric data by companies like Clearview AI, where facial recognition technology was misused for unauthorized surveillance and profiling, raising alarm about the privacy risks associated with biometric data collection (BBC, 2020).

Due to the sensitive nature of neurofeedback data, it is essential to implement robust safeguards to ensure the information is securely stored and accessible exclusively to authorized personnel.

1.2 Transparency and Informed Consent

The use of neurofeedback games, particularly with children or those with diminished cognitive capacity, raises ethical concerns about informed consent. It's crucial to ensure that participants voluntarily participate after understanding the process and the associated risks. To address these issues of consent and transparency, it is necessary to develop clear and thorough information that enables participants to make informed choices. Participants must be informed about their right to discontinue participation in neurofeedback sessions at any time without facing any consequences. This withdrawal process should be clear and simple, and participants should be aware of how their data will be managed should they decide to opt out.

1.3 Concerns with Neuroenhancement

Access to neurofeedback technologies might intensify existing gaps between various socio-economic groups. Those who can afford these technologies may secure an undue advantage in professional contexts. Additionally, it raises questions about whether accomplishments achieved through enhanced cognitive functions with neurofeedback genuinely represent a person's abilities or if they devalue effort and skill. Another concern is that individuals may become too dependent on technology to manage their emotional or cognitive states, especially if these tools are employed without the support of other coping strategies or therapies. Finally, the long-term consequences of brain enhancement remain largely unknown. Extended use of neurofeedback for enhancement could unexpectedly affect brain plasticity or disrupt the equilibrium of neural pathways, possibly resulting in cognitive or emotional disturbances. Thus, it is wise to remain cautious about the extended use of neurofeedback games.

2. Neuroethical Considerations for Neurodiverse Gaming Communities

Neurodivergent individuals, such as those with attention-deficit/hyperactivity disorder (ADHD) or autism spectrum disorder (ASD), may experience video games differently. Video games can serve therapeutic purposes for neurodivergent people. However, they might also intensify the challenges for neurodiverse communities. This section explores various neuroethical issues pertinent to neurodiverse gaming communities, emphasizing maintaining a balanced attitude towards gaming as well as the importance of inclusive design practices.

2.1 Therapeutic Benefits of Gaming for Neurodivergent Individuals

2.1.1 Neuroplasticity and Games

In recent decades, neuroplasticity—the brain's ability to undergo structural and functional changes that facilitate learning and behavior—has received much attention (Merzenich et al., 2014). These plastic changes are believed to provide broad benefits and could potentially mitigate the detrimental effects associated with various neurodegenerative diseases. Researchers found that gaming enhances learning-induced plastic changes in the brain, including strengthening long-term memory traces in the brain (Bavelier et al.,2012; Junttila et al., 2022; Nahum & Bavelier, 2020). This indicates that video games hold significant promise as a valuable tool for clinical applications.

2.1.2 Various Psychological Benefits to the Gaming Communities

Games can deliver significant emotional and psychological benefits, offering an escape from daily stressors. Role-playing games, for example, enable players to adopt various perspectives, fostering empathy and emotional growth. Much research has shown the benefits of gaming in terms of improving mental health. A notable example is how video games may have helped build social connections and reduce stress during COVID-19. Egami et al. (2023; 2024) identified the causal effect of video gaming on mental wellbeing in Japan from 2020 to 2022 by using game console lotteries as a natural experiment. They found that the console ownership reduced psychological distress and improved life satisfaction by 0.1–0.6 standard deviations. Marston and Kowert (2023) similarly found that during COVID-19, video games were able to increase social connectedness as well as promoting psychological healing for not only older adults, but intergenerationally. Zhao et al. (2020) also highlightin their study how effective exergaming, a combination of exercise and gaming, at improving physical and cognitive function in people living with non-severe cognitive impairment or dementia.

For neurodivergent players in particular, games and gaming communities may provide both emotional and cognitive benefits. Video games are popular among people with ASD as many people with ASD find them helpful for lowering anxiety levels (Cuesta Gómez et al., 2021). A systematic review by Jiménez-Muñoz et al. (2022) show that video game-based treatment were effective in helping alleviate ASD symptoms. Video games can offer a controlled and predictable environment, which can be particularly beneficial for individuals with ASD. The structured nature of games supports repetitive routines, which many neurodivergent individuals find comforting. Walsh et al. (2025) found through 17 studies that social skills training for children with ASD was improved by using games. Sundberg's study (2018) showed that among persons with ASD, those who play online games had more friends than those who did not, showing possible social benefits of gaming for ASD populations. For individuals with ADHD, video games show potential to help develop sustained attention and improve task completion, according to a quasi-systematic review conductd by Rodrigo-Yanguas et al. (2022). Specific game genres like strategy and puzzle games can bolster executive functions by enhancing attention, planning, problem-solving, and decision-making skills (Bavelier & Green, 2003; Boot et al., 2008). Beyond the cognitive advantages, video games offer a low-pressure virtual space through which players with difficulties in traditional social environments can interact safely and comfortably (Durkin, 2010; Kowert & Oldmeadow 2015). Multiplayer games, especially those requiring team cooperation, can offer people with ASD and/or ADHD a space for practicing social skills such as negotiation and conflict resolution, skills that may be challenging to develop in other settings. An experimental study from Gentile et al. (2009) showed that students who were randomly assigned to play prosocial video games were more prosocial towards another student.

In fact, video games have emerged as an innovative clinical tool for individuals with ADHD and ASD, offering targeted benefits through engaging and structured experiences. Exergaming, a form of exercise using a video game, has been shown to be effective in enhancing attention in children with ADHD (Ji et al., 2023). For children with ASD, evidence suggests that certain video games help reduce their anxiety symptoms (Wijnhoven et al., 2020).

2.3 Challenges to the Neurodiverse Communities and Potential Remedies

Unfortunately, although video games can offer significant opportunities for social engagement and cognitive growth, they also pose risks, especially for neurodivergent individuals who might be more vulnerable to compulsive behaviors and gaming addiction. Studies have shown that neurodivergent players are at a greater risk of problematic video game use (Weinstein & Weizman, 2012; Mazurek & Engelhardt, 2013; Mazurek & Wenstrup, 2013; Mazurek et al., 2015; Engelhardt et al., 2017). Given these mixed effects of gaming on neurodivergent individuals, gaming communities, developers, and the industry have a crucial role in fostering inclusivity and designing games that cater to neurodiverse populations. This responsibility entails creating and enforcing community guidelines that promote respect and understanding among all players, effectively moderating spaces to prevent harassment, and facilitating supportive interactions. Scientists and philosophers also have an important role in raising awareness for the importance of ethical considerations in game design by engaging in relevant research and ethical discussion.

2.3.1 Inclusive Design Practices

The first neuroethical consideration in game design involves creating inclusive experiences that accommodate a wide range of cognitive profiles. This requires developers to engage with neurodiverse communities to understand their needs, preferences, and experiences. For instance, games can minimize flicker and adjust audio levels to accommodate sensory sensitivities often associated with conditions like ASD. Additionally, allowing players to modify game difficulty, customize the user interface, and access straightforward, succinct instructions can significantly benefit those with cognitive challenges, including those typical of ADHD and ASD (Mazurek et al., 2015). Moreover, games should provide clear instructions and feedback mechanisms can help neurodiverse players navigate game worlds more effectively. Implementing visual aids, prompts, or in-game tutorials can support cognitive processing and understanding, allowing players to engage more fully with the game's objectives and mechanics.

Inclusive design involves more than just adding accessibility features post-development; it starts with acknowledging the needs of neurodivergent players right from the outset. By integrating their feedback early on, developers can ensure that inclusivity is embedded within the game design process itself. Additionally, developers can support further research and promote awareness by disseminating data and insights gained from their design practices. Working together with academic researchers and neurodivergent advocacy groups can enhance understanding and drive further innovations in accessible gaming. For example, a Mod called "Autcraft," is a Minecraft server specifically designed for individuals with autism, foster a supportive community by incorporating features like non-violent gameplay, sensory-friendly zones, and structured routines. Ringland (2019) argues that the Autcraft community extends beyond simply serving as a support group for individuals affected by targeted violence; it actively redefines what it means to embrace and celebrate autistic identities among its members.

2.3.2 Representation of Neurodiversity in Games

Neurodiversity recognizes that variations in cognitive functioning, such as those seen in ASD, ADHD, and dyslexia are natural and valuable aspects of human diversity. Representation of neurodiversity is an important consideration in game design. Historically, neurodiverse individuals have often been misrepresented or marginalized in mainstream media, leading to stereotypes that reinforce negative perceptions. Within this context, video games emerge as a powerful medium through which developers can either reinforce or challenge societal stereotypes about neurodiversity. Video games have the unique ability to create immersive worlds where players can embody characters with diverse cognitive experiences, fostering empathy and understanding for neurodiverse communities. For example, Symmetra in

Overwatch is a character with ASD. Developers of Overwatch 2 aimed to celebrate her ASD in game through voice lines and combat (Gwilliam, 2022). Such games can be therapeutic to neurodivergent players by portraying neurodivergence positively and authentically.

2.3.3 Mitigating Negative Effects for Neurodivergent Gamers

Game developers and the broader gaming industry play a pivotal role in preserving the positive impacts of gaming while mitigating its negative effects, particularly for vulnerable groups such as neurodivergent players. This requires the gaming industry to pay attention to the most up-to-date scientific research on the effects of gaming and/or to actively consult and collaborate with researchers and ethicists. For example, Hong et al. (2023) found that video gaming increased the performance of children and adolescents on a variety of neuro and social cognitive tests with a saturation effect, meaning that performance did not further increase once video gaming exceeded 20 h per week. According to their findings, gaming benefits do not increase with more than 20 h of play per week, suggesting a recommended cap on gaming time to prevent overuse. Their results recommend that children and adolescents should limit their video gaming to a specific duration. This moderation can help mitigate the negative impacts of video gaming while preserving its positive benefits. Such research endorses game designs that incorporate natural stopping points and clearly defined breaks in gameplay in order to encourage players to not play excessively. Other helpful designs include in-game reminders or notifications to take breaks from the screen. Ultimately, the gaming industry bears a responsibility to foster healthy gaming practices based on the best available scientific evidence.

While adopting responsible design practices in gaming is crucial for the well-being of gaming communities, it is important to recognize that these practices may not always align with the financial interests of gaming companies. Longer gameplay hours and impulsive in-game purchases typically generate significant revenue for these companies. Given this potential conflict of interest, it is essential to establish external measures and regulations that hold gaming companies accountable. Governments and regulatory bodies can help implement strict guidelines that limit exploitative features and help promote responsible game designs.

Moreover, introducing legal requirements for games to include tools that help players manage their gaming habits, such as timers, alerts, and easy-to-access spending trackers, could mitigate excessive play and spending. Such tools can empower players to make informed decisions about their gaming behavior. Having a regulatory environment that encourages ethical game design can help balance the economic benefits of the gaming industry while safeguarding player health. This balanced approach not only protects consumers but also enhances the industry's reputation by promoting fair and responsible gaming practices.

The importance of such regulatory practices will become more apparent in the next section, in which various monetization models with gambling elements in the gaming industry, such as loot boxes and gacha, are examined.

3 Ethical and Neuroethical Issues in the Grey Area Between Gaming and Gambling

In recent years, the gaming industry has witnessed a proliferation of monetization strategies that integrate gambling-like elements. These strategies raise significant neuroethical concerns due to their potential to exploit neurobiological vulnerabilities for profit. Leahy (2022) notes that chance-based loot boxes are often compared to both the legal and psychological definitions of gambling, raising worries that these mechanisms represent a form of quasi-gambling that remains unregulated (Gibson et al.,2022). This section explores the increasingly blurred lines between gaming and gambling by examining three distinct types of monetization models in video games. It is important to emphasize that the ethical and neuroethical issues associated with gambling in gaming, while closely related, are distinct from those found in traditional gambling and addiction. The specific neuroethical concerns these models pose to gaming communities are discussed and potential solutions to address these issues are proposed.

3.1 Three Representative Gambling-Related Monetization Models in Gaming

3.1.1 Loot Boxes

Loot boxes are a widely used feature in contemporary video games. They generally hold a variety of items, ranging from cosmetics to functional. Players can acquire loot boxes either through gameplay or by purchasing them with real money or in-game currency. Functioning similarly to slot machines, loot boxes require players to pay for the opportunity to obtain a random reward. This element of uncertainty, along with the possibility of a valuable reward, is intended to motivate players to buy them repeatedly. Prominent games that feature loot boxes include but are not limited to CS:GO (Counter-Strike: Global Offensive), Apex Legends, PUBG (PlayerUnknown's Battlegrounds), and Overwatch, which are some of the most played online video games in the world.

3.1.2 Gacha Games

Gacha games are a type of video game that incorporates a monetization mechanism inspired by Japanese capsule-toy vending machines (Lakić et al., 2023). In these games, players spend in-game currency, which can be earned or purchased with real money, to receive a random virtual item. These items can range from simple cosmetic skins to powerful characters and weapons, often influencing the game's strategy and progression. The main difference between loot boxes and gacha is that gacha mechanics are often deeply integrated into the gameplay and progression systems, whereas loot boxes usually focus on things that do not affect the gameplay directly, such as cosmetic items. This makes gacha games typically more forceful and exploitative than loot boxes in their approaches to inducing player spending.

3.1.3 Disguised Social Casino

Online gambling has become the fastest-growing mode of gambling (Gainsbury, 2015). A social casino is a type of online gaming platform that simulates real casino games, such as slot machines and poker, with a special emphasis on social interactions with other online players. Social casinos use virtual currency, which can often be purchased with real money. I refer to the final category of games as "disguised social casinos" for two main reasons. First, the fundamental essence and core mechanics of these games align precisely with those of social casinos. Second, the marketing of these games does not transparently convey their social casino nature, which can mislead players who originally have no interest in engaging with gambling-like content.

One salient example is "Coin Master," a game developed by the studio *Moon Active*. Coin Master merges the traditional slot machine format with light mid-core mechanics such as base building, attacking, and raiding other players. This unique blend helps make the game highly engaging even to those who typically do not follow social casinos. The strong social elements and revenge loops with friends and other players encourage players to frequently return to the game to retaliate and recover lost coins, thus boosting both engagement and retention. Coin Master, along with similar games, has expanded the reach of social casino games to new audiences and markets (Tandon & Kumar, 2018).

3.2 Problems with Gambling-Like Mechanisms in Games

The severe detrimental effects of gambling addiction have been extensively documented and researched across various studies (Shaffer & Martin, 2011; Murch & Clark, 2016; Potenza et al., 2019). The consequences of this disorder are diverse and significant, affecting multiple aspects of an individual's personal life. Notably, financial problems are a prevalent issue among those with a gambling addiction. Individuals often incur substantial debts, deplete their savings, and may even resort to illegal activities to finance their gambling or manage debt, leading to a cascade of legal and social problems. Furthermore, gambling addiction is strongly correlated with a decline in both physical and mental health. These conditions may stem from the chronic stress and emotional turmoil caused by persistent gambling and the ensuing life complications.

Research has consistently shown a positive correlation between loot box purchases and problem gambling severity across many countries (von Medula et al., 2020; Spicer et al., 2022; Xiao, 2023; González-Cabrera et al.,2023). Brady and Prentice (2021) found in their study that opening a loot box produces increases in arousal and thus has similar addictive qualities as gambling. They point out that this could indicate that loot boxes and gambling-like mechanisms in general could be fueling a gaming disorder. Specifically, players who reported higher problem gambling severity tended to spend more on loot boxes, suggesting that video game companies might disproportionately profit from these at-risk individuals (Hing et al., 2022). This correlation has also been observed in underage players, indicating that young people are especially vulnerable to the potential harms associated with loot boxes (Zendle, 2020). Given these findings, many countries are considering or have already begun to regulate loot boxes to address the risks associated with problem gambling and to enhance consumer protection, particularly for vulnerable groups such as children (Harvey, 2021; Honer, 2021; Schwiddessen & Karius, 2018).

3.2.1 Hidden Danger, Adolescence Engagement, Consent

In addition to the well-documented harms of gambling addiction, gaming mechanisms such as loot boxes and gacha systems, as well as disguised social casino games like Coin Master, present unique challenges. These digital gambling-like elements often bypass traditional gambling regulations, which can expose a broader and potentially younger audience to risk. Zendle et al. (2020) and Kristiansen and Severin (2020) found that loot boxes are prevalent in video games that are deemed appropriate for children, especially on mobile platforms.

Another ethical concern with adolescent engagement in these games arises from their consent and autonomy, particularly when considering the complexity of informed consent. The notion of informed consent is significantly complicated by the fact that many players, especially children and adolescents, may not fully comprehend the probabilities and associated risks linked to gambling activities. Games like disguised social casinos integrate gambling mechanics subtly within their gameplay, making them more accessible and appealing while simultaneously obscuring their potential as sources of gambling addiction. This increased accessibility heightens the risk of players, particularly minors, developing gambling habits unknowingly. According to Gainsbury et al. (2016), a considerable portion of consumers of social casino games migrate to traditional gambling.

Moreover, players might not initially intend to engage in gambling activities but find themselves drawn into these games through deceptive advertising practices. Such false advertising undermines genuine consent, as players are lured under misleading pretenses. As highlighted by Cartwright and Hyde (2022), while most regulatory efforts concerning loot boxes have focused on their similarities to gambling, there has been inadequate consideration of loot boxes as potentially unfair or aggressive commercial practices. Therefore, regulatory actions need to address these issues comprehensively, tackling both the gambling aspects and the unethical business practices involved. This includes aggressive and misleading advertising, ensuring that regulations protect consumers, particularly vulnerable groups like minors, from deceptive and exploitative practices.

3.2.2 Regulatory Grey Area

Unlike gambling restrictions, many countries (e.g., the United States) have little to no regulations on loot boxes and gacha games (Drummond et al., 2019). The fact that most gambling-like mechanisms do not offer cash payouts makes it much more likely for these games to bypass existing gambling regulations. The classification and regulation of these gaming features vary widely between different jurisdictions, resulting in a lack of uniformity in protection standards and enforcement. This circumvention of regulation allows these features to be included in games that are accessible to minors and individuals in areas where gambling is typically restricted or even prohibited. Disguised social casinos often bypass gambling regulations by adding other non-gambling elements to their games.

"Diablo Immortal," developed by Blizzard Entertainment in collaboration with NetEase, serves as an example where a game evades certain regulations by tweaking its loot box mechanisms. The game employs a system where, instead of purchasing loot boxes directly, players buy "legendary crests," which can be used to enhance the rewards from in-game activities known as "rifts." This method indirectly influences the game's reward system, affecting the type of loot players receive from participating in these rifts. The primary concern with this approach is that it sidesteps the direct sale of loot boxes, which are subject to increasing scrutiny and regulation in several jurisdictions around the world.

3.3 What Need to Be Done?

What should we do in the face of the challenge of coming up with appropriate regulations for these gambling-like mechanics in video games? This is an intricate challenge that cannot be fully addressed in a brief discussion. However, initiating conversation on this topic and raising awareness about its significance is crucial. In this section, a few starting points are presented for further consideration.

Rather than regulating loot boxes as gambling and potentially banning their sale—a move that could negatively impact consumer freedom and commercial interests—an alternative consumer protection measure has been suggested: requiring the disclosure of loot box probabilities. This measure would inform players about their chances of obtaining randomized rewards. This less restrictive approach aims to strike a balance between preserving consumer freedom to purchase loot boxes and providing protection from the potential gambling-related harm they can cause. Such a measure, which could be implemented either through industry self-regulation or by law, offers a more considerable and comprehensive solution than simply extending gambling regulations. Currently, China is the only country that has legally mandated video gaming service providers to publish the probabilities of obtaining randomized rewards from loot boxes (Xiao, 2022; 2023). This underscores the global lack of legal regulations surrounding loot box practices.

Drummond et al. (2019) recommend having a loot box limit-setting. This approach involves gamblers specifying, either voluntarily or as a requirement, the maximum amount they wish to spend before starting to gamble. Once this limit is reached, the system sends a reminder and initiates a cooling-off period during which the player cannot continue gambling. Limit-setting has proven broadly effective in reducing excessive spending and is generally well-received by gamblers. A similar spending cap might be effective in preventing overspending on loot boxes, gacha games, and various kinds of social casinos.

The introduction of age restrictions for games featuring gambling-like mechanisms could protect younger audiences from early exposure to these potentially addictive elements. This could be complemented by clear labeling on such games, providing explicit warnings about the inclusion of gambling elements and their associated risks. Moreover, setting time limits on gameplay can help mitigate the prolonged engagement that often exacerbates gambling tendencies in players, especially minors. Ensuring that items obtained through these mechanisms cannot be converted into real money might also help these games steer clear of legal definitions of gambling.

Guidelines for ethical game design are in critical need. Such guidelines should discourage practices that exploit psychological and neurological vulnerabilities. Regular audits and compliance checks could help enforce adherence to these guidelines. In addition, educational initiatives aimed at players could elucidate the risks involved in playing such games, fostering a more informed gamer community. Lastly, supporting research into the psychological, neurological, and social impacts of loot boxes and gacha systems would provide valuable insights, guiding future regulations and protective measures. While this discussion merely scratches the surface of a multifaceted issue, it aims to start a dialogue and highlight the need for thoughtful consideration and action in this area.

4 Ethical and Neuroethical Concerns About Social Stigma of Gaming Communities

There are many pervasive and negative stereotypes of gamers and gaming communities that have contributed to a broader societal stigmatization of gaming. Such stereotypes include but are not limited to gamers being socially incompetent, more violent and aggressive, unproductive, juvenile, and physically unhealthy. While the values and harm of video games are matters of ongoing complex debates, this much is clear: many of these stereotypes stem from outdated misconceptions and misrepresentations of gaming activities. As Griffiths (2005) notes, there is little evidence that moderate play has any serious adverse effects on children and adolescents. Thus, video gaming is safe for most players. Such inaccurate and exaggerated views of gamers not only reduce the perceived value of gaming but also marginalize its diverse communities, overlooking the social, cognitive, and educational benefits that gaming can provide. For example, gamers are often portrayed as socially awkward and incompetent. In contrast, gaming has evolved into a highly social pursuit, with a large player base actively engaging in online multiplayer games and esports, watching and interacting with streamers, as well as building vibrant gaming communities. Several studies (Gentile et al., 2009; Kowert & Oldmeadow, 2013, 2015; Sergeyeva et al., 2018; Zhang & Frederick, 2018) challenged the generalized stereotype of social ineptitude among online video game players. In fact, many have argued that the online gaming space can offer a social space that encourages closeness, belonging, and security.

4.1 Roots of Stigma: Skewed Media Portrait

Despite numerous research studies showing potential benefits of gaming, many stereotypes persist in popular culture and continue to cause harm to the gaming communities. Part of the reason for the deeply entrenched stigma comes from how media and popular culture portray the gaming communities. In films, TV shows, and ads, gamers are often called "nerds," "geeks," and depicted as reclusive, weird, and overly obsessed. These portrayals fail to reflect the diversity of experiences and identities within the gaming community. Additionally, media coverage often focuses on extreme cases of gaming disorder, violence, or addiction, creating a distorted view of gaming as a predominantly harmful activity. These exaggerated narratives conceal the reality that gaming can be a healthy, enriching, and transformative activity for many. Generational divides further exacerbate these misunderstandings, as older generations often struggle to appreciate the cultural, artistic, and social dimensions of gaming. The swift advancement of gaming technologies and culture has also increased this disconnect, leaving those outside the gaming communities with a limited understanding of their positive contributions. Despite the growing acceptance of gaming today, the media often sensationalizes negative aspects, overshadowing its positive contributions and reinforcing the stigma surrounding gamers. The challenge is to strike a balance between addressing concerns of gaming and acknowledging gaming's potential for creativity, education, and social connection.

4.2 Harm of Negative Stigma to Gaming Communities

The negative social stigma attached to gaming can profoundly impact both the individual gamers and the broader gaming communities. Many gamers might feel embarrassed or ashamed for openly discussing their hobby because they do not want to be associated with negative stereotypes. This prevents gamers from fully accepting their identity and finding community support, thus depriving them of the support and friendship that gaming communities could offer.

Furthermore, hiding one's true interests and identity can create cognitive dissonance and psychological distress, adversely affecting gamers' mental health and self-esteem. These stereotypes can discourage individuals from seeking mental health support, as they might fear additional judgment or misunderstanding from those who do not engage in gaming. Such chronic, prolonged social isolation and negative emotional states can have detrimental effects on the health of the brain, potentially resulting in cognitive decline and problems with mood regulation. Public stigma is known to deter help-seeking behaviors and negatively impact self-perception among disordered gamblers. Peter et al.'s study (2019) examines the public stigma associated with traditional casino gambling, eSports gambling, and internet gaming compared to individuals experiencing financial crises without behavioral addictions. Using a vignette-based experimental design, the study found that all three behavioral addictions were more stigmatized than the control group. These findings emphasize the need to reduce public stigma associated with gambling and gaming addictions and offer valuable insights for developing interventions aimed at addressing and mitigating stigma.

Besides potentially affecting social identity, mental wellbeing, and neurological health, the view of gaming as a "waste of time" can lead to real disadvantages in professional and academic settings. Prejudices against gamers may negatively impact decisions related to hiring, promotions, or academic evaluations of individuals known to be gamers. Finally, the reinforcement of stereotypes due to stigma perpetuates a cycle where these negative views are further entrenched, influencing public perception and potentially affecting the development of games and gamer culture. This cycle discourages diversity in gaming content and limits the growth of a more inclusive gaming community.

4.3 The Dark Side of Neuro-Malpractice: Electroshock Therapy and the Healing Power of Interactive Advocacy

In one of the most disturbing incidents highlighting the severe social stigma associated with gaming, Dr. Yang Yongxin, a Chinese psychiatrist, gained international notoriety for his controversial use of electroshock therapy. Proclaimed as a treatment for "internet addiction," Yang subjected approximately 3000 adolescents to extreme and painful procedures involving high levels of electric shocks and verbal and physical abuse. These harsh treatments inflicted significant psychological and physical trauma on the patients (Bax, 2018; Branigan, 2009a, b). From a neuroethical standpoint, Yang's methods raise profound concerns. The use of such invasive and punitive treatment without robust scientific justification violates basic bioethical principles. The willingness of parents to send their only child to Yang's facility, often against their child's will, reveals the deeply rooted negative perceptions of gaming within Chinese society. This extreme reaction highlights an urgent need for a more nuanced and objective perspective on gaming.

This tragic episode inspired the creation of "Breakout 13," an interactive video game designed to share the experiences of the young gamers at Yang's facility. Developed by ALT Lab, the game offers players an immersive journey through the lives of these adolescents, portraying their struggles and resilience (Bailey, 2023). "Breakout 13" not only aims to raise awareness about the dire consequences of unethical medical practices but also challenges the prevailing stigmas around gaming by showcasing its potential as a powerful medium for storytelling and social advocacy. Breakout 13 has been widely praised by players worldwide, particularly in China, for its profound cultural impact and its ability to provide a truthful yet therapeutic perspective for those who have experienced harm and trauma from stigmatization. Using video games as a powerful medium to confront uncomfortable truths highlights the resilience and courage of a gaming community challenging the unfair stigma it often faces.

References

Bailey, D. (2023). *This live-action game about escaping a Chinese internet addiction treatment prison is breaking out on steam.* Gamesradar+. https://www.gamesradar.com/this-live-action-game-about-escaping-a-chinese-internet-addiction-treatment-prison-is-breaking-out-on-steam/

Bavelier, D., & Green, C. S. (2003). Action video game modifies visual selective attention. *Nature (London), 423*(6939), 534–537. https://doi.org/10.1038/nature01647

Bavelier, D., Green, C., Pouget, A., & Schrater, P. (2012). Brain plasticity through the life span: Learning to learn and action video games. *Annual Review of Neuroscience, 35*(1), 391–416. https://doi.org/10.1146/annurev-neuro-060909-152832

Bax, T. (2018). War with the internet demon: Shock-and-awe of China's youth. In B. Bakken (Ed.), *Crime and the Chinese dream* (1st ed., pp. 124–154). Hong Kong University Press. https://doi.org/10.1515/9789888455119-008

BBC. (2020). *Clearview AI: Face-collecting company database hacked.* https://www.bbc.com/news/technology-51658111

Boot, W. R., Kramer, A. F., Simons, D. J., Fabiani, M., & Gratton, G. (2008). The effects of video game playing on attention, memory, and executive control. *Acta Psychologica, 129*(3), 387–398. https://doi.org/10.1016/j.actpsy.2008.09.005

Brady, A., & Prentice, G. (2021). Are loot boxes addictive? Analyzing participant's physiological arousal while opening a loot box. *Games and Culture, 16*(4), 419–433. https://doi.org/10.1177/1555412019895359

Branigan, T. (2009a). Guardian weekly: International news: China online: Internet addiction grips nation: "Flawed" diagnosis of dependency fuels spread of bogus therapy. *The Guardian (London).* https://go.exlibris.link/yCkjGcQF

Branigan, T. (2009b). International: China bans electric shock treatment used to "cure" young internet addicts: Ministry of health steps in after safety fears raised: Definition of addiction is too vague, say experts. *The Guardian (London).* https://go.exlibris.link/sYHDjlY2

Cartwright, P., & Hyde, R. (2022). Virtual coercion and the vulnerable consumer: 'Loot boxes' as aggressive commercial practices. *Legal Studies (Society of Legal Scholars), 42*(4), 555–575. https://doi.org/10.1017/lst.2022.7

Coenen, F., Scheepers, F. E., Palmen, S. J. M., de Jonge, M. V., & Oranje, B. (2020). Serious games as potential therapies: A validation study of a Neurofeedback game. *Clinical EEG and Neuroscience, 51*(2), 87–93. https://doi.org/10.1177/1550059419869471

Cuesta Gómez, J. L., Sánchez Fuentes, S., Santos Vázquez, S., Sancho Requena, P., & Orozco Gómez, M. L. (2021). The use of videogames as digital leisure among people with autism spectrum disorder. *Siglo Cero, 52*(3), 101–117. https://doi.org/10.14201/scero2021523101117

Drummond, A., Sauer, J. D., & Hall, L. C. (2019). Loot box limit-setting: A potential policy to protect video game users with gambling problems? *Addiction (Abingdon, England), 114*(5), 935–936. https://doi.org/10.1111/add.14583

Durkin, K. (2010). Videogames and young people with developmental disorders. *Review of General Psychology, 14*(2), 122–140. https://doi.org/10.1037/a0019438

Egami, H., Wakabayashi, T., Yamamoto, T., Rahman, M. S., & Egami, C. (2023). Video game play positively affects well-being: Evidence from a natural experiment in Japan. In *Proceedings of annual conference, Digital Game Research Association JAPAN* (pp. 35–50). https://doi.org/10.57518/digrajproc.13.0_35

Egami, H., Rahman, M. S., Yamamoto, T., Egami, C., & Wakabayashi, T. (2024). Causal effect of video gaming on mental well-being in Japan 2020–2022. *Nature Human Behaviour, 8*(10), 1943–1956. https://doi.org/10.1038/s41562-024-01948-y

Engelhardt, C. R., Mazurek, M. O., & Hilgard, J. (2017). Pathological game use in adults with and without autism spectrum disorder. *PeerJ, 5*, e3393. https://doi.org/10.7717/peerj.3393

Gainsbury, S. M. (2015). Online gambling addiction: The relationship between internet gambling and disordered gambling. *Current Addiction Reports, 2*(2), 185–193. https://doi.org/10.1007/s40429-015-0057-8

Gainsbury, S. M., Russell, A. M. T., King, D. L., Delfabbro, P., & Hing, N. (2016). Migration from social casino games to gambling: Motivations and characteristics of gamers who gamble. *Computers in Human Behavior, 63*, 59–67. https://doi.org/10.1016/j.chb.2016.05.021

Gentile, D. A., Anderson, C. A., Yukawa, S., Ihori, N., Saleem, M., Ming, L. K., Shibuya, A., Liau, A. K., Khoo, A., Bushman, B. J., Rowell Huesmann, L., & Sakamoto, A. (2009). The effects of prosocial video games on prosocial behaviors: International evidence from correlational, longitudinal, and experimental studies. *Personality & Social Psychology Bulletin, 35*(6), 752–763. https://doi.org/10.1177/0146167209333045

Gibson, E., Griffiths, M. D., Calado, F., & Harris, A. (2022). The relationship between videogame micro-transactions and problem gaming and gambling: A systematic review. *Computers in Human Behavior, 131*, 107219. https://doi.org/10.1016/j.chb.2022.107219

González-Cabrera, J., Basterra-González, A., Ortega-Barón, J., Caba-Machado, V., Díaz-López, A., Pontes, H. M., & Machimbarrena, J. M. (2023). Loot box purchases and their relationship with internet gaming disorder and online gambling disorder in adolescents: A prospective study. *Computers in Human Behavior, 143*, 107685. https://doi.org/10.1016/j.chb.2023.107685

Griffiths, M. (2005). Video games and health: Video gaming is safe for Most players and can be useful in health care. *BMJ (Online), 331*(7509), 122–123. https://doi.org/10.1136/bmj.331.7509.122

Gruzelier, J. H. (2014). EEG-neurofeedback for optimising performance. I: A review of cognitive and affective outcome in healthy participants. *Neuroscience and Biobehavioral Reviews, 44*, 124–141. https://doi.org/10.1016/j.neubiorev.2013.09.015

Gwilliam, M. (2022). *Overwatch 2 dev reveals how Symmetra's autism is showcased on the battlefield.* https://www.dexerto.com/overwatch/overwatch-2-dev-reveals-how-symmetras-autism-is-showcased-on-the-battlefield-1947087/

Harvey, D. J. (2021). Should loot boxes be considered gambling or can self-regulation and corporate social responsibility solve the loot box issue? A review of current UK law and international legislation. *Interactive Entertainment Law Review, 4*(1), 48–62. https://doi.org/10.4337/ielr.2021.01.03

Hing, N., Rockloff, M. J., Russell, A. M. T., Browne, M., Newall, P. W. S., Greer, N., King, D. L., & Thorne, H. (2022). Loot box purchasing is linked to problem gambling in adolescents when controlling for monetary gambling participation. *Journal of Behavioral Addictions, 11*(2), 396–405. https://doi.org/10.1556/2006.2022.00015

Hong, J., Su, Y., Wang, J., Xu, X., Qu, W., Fan, H., Tan, Y., Wang, Z., Zhao, Y., & Tan, S. (2023). Association between video gaming time and cognitive functions: A cross-sectional study of Chinese children and adolescents. *Asian Journal of Psychiatry, 84*, 103584. https://doi.org/10.1016/j.ajp.2023.103584

Honer, P. (2021). Limiting the loot box: Overview and difficulties of a common EU response. *Interactive Entertainment Law Review, 4*(1), 63–83. https://doi.org/10.4337/ielr.2021.01.04

Jiménez-Muñoz, L., Peñuelas-Calvo, I., Calvo-Rivera, P., Díaz-Oliván, I., Moreno, M., Baca-García, E., & Porras-Segovia, A. (2022). Video games for the treatment of autism Spectrum disorder: A systematic review. *Journal of Autism and Developmental Disorders, 52*(1), 169–188. https://doi.org/10.1007/s10803-021-04934-9

Jirayucharoensak, S., Israsena, P., Pan-Ngum, S., Hemrungrojn, S., & Maes, M. (2019). A game-based neurofeedback training system to enhance cognitive performance in healthy elderly subjects and in patients with amnestic mild cognitive impairment. *Clinical Interventions in Aging, 14*(Journal Article), 347–360. https://doi.org/10.2147/CIA.S189047

Junttila, K., Smolander, A.-R., Karhila, R., Giannakopoulou, A., Uther, M., Kurimo, M., & Ylinen, S. (2022). Gaming enhances learning-induced plastic changes in the brain. *Brain and Language, 230*, 105124. https://doi.org/10.1016/j.bandl.2022.105124

Ji, H., Wu, S., Won, J., Weng, S., Lee, S., Seo, S., & Park, J.-J. (2023). The effects of exergaming on attention in children with attention deficit/hyperactivity disorder: randomized controlled trial. *JMIR Serious Games, 11*, e40438. https://doi.org/10.2196/40438

Khong, A., Jiangnan, L., Thomas, K. P., & Vinod, A. P. (2014). BCI based multi-player 3-D game control using EEG for enhancing attention and memory. In *2014 IEEE international conference on systems, man, and cybernetics (SMC)* (pp. 1847–1852). https://doi.org/10.1109/SMC.2014.6974189

Kowert, R., & Oldmeadow, J. A. (2013). (A)social reputation: Exploring the relationship between online video game involvement and social competence. *Computers in Human Behavior, 29*(4), 1872–1878. https://doi.org/10.1016/j.chb.2013.03.003

Kowert, R., & Oldmeadow, J. A. (2015). Playing for social comfort: Online video game play as a social accommodator for the insecurely attached. *Computers in Human Behavior, 53*, 556–566. https://doi.org/10.1016/j.chb.2014.05.004

Kristiansen, S., & Severin, M. C. (2020). Loot box engagement and problem gambling among adolescent gamers: Findings from a national survey. *Addictive Behaviors, 103*, 106254–106254. https://doi.org/10.1016/j.addbeh.2019.106254

Lakić, N., Bernik, A., & Čep, A. (2023). Addiction and spending in Gacha games. *Information (Basel), 14*(7), 399. https://doi.org/10.3390/info14070399

Leahy, D. (2022). Rocking the boat: Loot boxes in online digital games, the regulatory challenge, and the EU'S unfair commercial practices directive. *Journal of Consumer Policy, 45*(3), 561–592. https://doi.org/10.1007/s10603-022-09522-7

Lyu, Y., An, P., Xiao, Y., Zhang, Z., Zhang, H., Katsuragawa, K., & Zhao, J. (2023). Eggly: Designing mobile augmented reality Neurofeedback training games for children with autism Spectrum disorder. *Proceedings of ACM on Interactive, Mobile, Wearable and Ubiquitous Technologies, 7*(2), 1–29. https://doi.org/10.1145/3596251

Marston, H. R., & Kowert, R. (2023). What role can videogames play in the COVID-19 pandemic? *Emerald Open Research, 1*(2). https://doi.org/10.1108/EOR-02-2023-0011

Mazurek, M. O., & Engelhardt, C. R. (2013). Video game use in boys with autism spectrum disorder, ADHD, or typical development. *Pediatrics (Evanston), 132*(2), 260–266. https://doi.org/10.1542/peds.2012-3956

Mazurek, M. O., & Wenstrup, C. (2013). Television, video game and social media use among children with ASD and typically developing siblings. *Journal of Autism and Developmental Disorders, 43*(6), 1258–1271. https://doi.org/10.1007/s10803-012-1659-9

Merzenich, M. M., Van Vleet, T. M., & Nahum, M. (2014). Brain plasticity-based therapeutics. *Frontiers in human neuroscience, 8*, 385. https://doi.org/10.3389/fnhum.2014.00385

Mazurek, M. O., Engelhardt, C. R., & Clark, K. E. (2015). Video games from the perspective of adults with autism spectrum disorder. *Computers in Human Behavior, 51*, 122–130. https://doi.org/10.1016/j.chb.2015.04.062

Murch, W. S., & Clark, L. (2016). Games in the brain: Neural substrates of gambling addiction. *The Neuroscientist, 22*(5), 534–545. https://doi.org/10.1177/1073858415591474

Nahum, M., & Bavelier, D. (2020). Video games as rich environments to foster brain plasticity. In N. F. Ramsey & J. d. R. Millán (Eds.), *Handbook of clinical neurology* (Vol. 168, pp. 117–136). Elsevier. https://doi.org/10.1016/B978-0-444-63934-9.00010-X

Peter, S. C., Li, Q., Pfund, R. A., Whelan, J. P., & Meyers, A. W. (2019). Public stigma across addictive behaviors: Casino gambling, eSports gambling, and internet gaming. *Journal of Gambling Studies, 35*(1), 247–259. https://doi.org/10.1007/s10899-018-9775-x

Potenza, M. N., Balodis, I. M., Derevensky, J., Grant, J. E., Petry, N. M., Verdejo-Garcia, A., & Yip, S. W. (2019). Gambling disorder. *Nature Reviews. Disease Primers, 5*(1), 51–51. https://doi.org/10.1038/s41572-019-0099-7

Ringland, K. E. (2019). "Autsome": Fostering an autistic identity in an online minecraft community for youth with autism. In N. G. Taylor, N. G. Taylor, M. H. Martin, C. Christian-Lamb, C. Christian-Lamb, M. H. Martin, B. Nardi, & B. Nardi (Eds.), *Information in contemporary society* (Vol. 11420, pp. 132–143). Springer International Publishing AG. https://doi.org/10.1007/978-3-030-15742-5_12

Rodrigo-Yanguas, M., González-Tardón, C., Bella-Fernández, M., & Blasco-Fontecilla, H. (2022). Serious video games: angels or demons in patients with attention-deficit hyperactivity disorder? A quasi-systematic review. *Frontiers in Psychiatry, 13*, 798480. https://doi.org/10.3389/fpsyt.2022.798480

Sanmartin, F. J., Cuadrado, F., Larriva, V. D., Galvez-Lara, M., Moriana, J. A., & Velasco, J. (2023). Loot boxes use as a new form of gambling within video games/El consumo de loot boxes Como una nueva forma de azar en los videojuegos. *Adicciones (Palma de Mallorca), 34*(4), 407.

Schoneveld, E. A., Malmberg, M., Lichtwarck-Aschoff, A., Verheijen, G. P., Engels, R. C. M. E., & Granic, I. (2016). A neurofeedback video game (MindLight) to prevent anxiety in children: A randomized controlled trial. *Computers in Human Behavior, 63*, 321–333. https://doi.org/10.1016/j.chb.2016.05.005

Schwiddessen, S., & Karius, P. (2018). Watch your loot boxes! – Recent developments and legal assessment in selected key jurisdictions from a gambling law perspective. *Interactive Entertainment Law Review, 1*(1), 17–43. https://doi.org/10.4337/ielr.2018.01.02

Sergeyeva, O., Tsareva, A., Zinoveva, N., & Kononova, O. (2018). Social skills amongst MMORPG-gamers: Empirical study. *SHS Web of Conferences, 50*, 1008. https://doi.org/10.1051/shsconf/20185001008

Shaffer, H. J., & Martin, R. (2011). Disordered gambling: Etiology, trajectory, and clinical considerations. *Annual Review of Clinical Psychology, 7*(1), 483–510. https://doi.org/10.1146/annurev-clinpsy-040510-143928

Spicer, S. G., Nicklin, L. L., Uther, M., Lloyd, J., Lloyd, H., & Close, J. (2022). Loot boxes, problem gambling and problem video gaming: A systematic review and meta-synthesis. *New Media & Society, 24*(4), 1001–1022. https://doi.org/10.1177/14614448211027175

Suhail, T. A., & Vinod, A. P. (2024). Enhancing overt and covert attention using a real-time neurofeedback game with consumer-grade EEG. *IEEE Transactions on Cognitive and Developmental Systems, 16*(1), 223–231. https://doi.org/10.1109/TCDS.2023.3260441

Sundberg, M. (2018). Online gaming, loneliness and friendships among adolescents and adults with ASD. *Computers in Human Behavior, 79*, 105–110. https://doi.org/10.1016/j.chb.2017.10.020

Tandon, O., & Kumar, A. (2018). *How coin master disrupted social casino and pocketed $100M.* Deconstructor of Fun. https://www.deconstructoroffun.com/blog/2019/3/4/is-coin-master-the-new-face-of-social-casino

Thomas, K. P., Vinod, A. P., & Guan, C. (2013). Design of an online EEG based neurofeedback game for enhancing attention and memory. In *2013 35th Annual international conference of the IEEE engineering in medicine and biology society (EMBC), 2013* (pp. 433–436). https://doi.org/10.1109/EMBC.2013.6609529

Teruel, M. A., Navarro, E., Romero, D., García, M., Fernández-Caballero, A., & González, P. (2017). An Innovative Tool to Create Neurofeedback Games for ADHD Treatment. In J. Toledo Moreo, H. Adeli, J. R. Álvarez-Sánchez, J. M. Ferrández Vicente, & F. de la Paz López (Eds.), *Natural and Artificial Computation for Biomedicine and Neuroscience* (Vol. 10337, pp. 183–192). Springer International Publishing AG. https://doi.org/10.1007/978-3-319-59740-9_18

von Meduna, M., Steinmetz, F., Ante, L., Reynolds, J., & Fiedler, I. (2020). Loot boxes are gambling-like elements in video games with harmful potential: Results from a large-scale population survey. *Technology in Society, 63*, 101395. https://doi.org/10.1016/j.techsoc.2020.101395

Wang, Q., Sourina, O., & Nguyen, M. K. (2011). Fractal dimension based neurofeedback in serious games. *The Visual Computer, 27*(4), 299–309. https://doi.org/10.1007/s00371-011-0551-5

Walsh, O., Linehan, C., & Ryan, C. (2025). Is there evidence that playing games promotes social skills training for autistic children and youth? *Autism : The International Journal of Research and Practice, 29*(2), 329–343. https://doi.org/10.1177/13623613241277309

Wang, J., Wang, W., & Hou, Z.-G. (2020). Toward improving engagement in neural rehabilitation: Attention enhancement based on brain-computer Interface and audiovisual feedback. *IEEE Transactions on Cognitive and Developmental Systems, 12*(4), 787–796. https://doi.org/10.1109/TCDS.2019.2959055

Weinstein, A., & Weizman, A. (2012). Emerging association between addictive gaming and attention-deficit/hyperactivity disorder. *Current Psychiatry Reports, 14*(5), 590–597. https://doi.org/10.1007/s11920-012-0311-x

Wijnhoven, L. A. M. W., Creemers, D. H. M., Vermulst, A. A., Lindauer, R. J. L., Otten, R., Engels, R. C. M. E., & Granic, I. (2020). The effects of the video game 'Mindlight' on anxiety of children with an autism spectrum disorder: A randomized controlled trial. *Journal of Behavior Therapy and Experimental Psychiatry, 68*, 101548–101548. https://doi.org/10.1016/j.jbtep.2020.101548

Xiao, L. Y. (2022). Drafting video game loot box regulation for dummies: A Chinese lesson. *Information & Communications Technology Law, 31*(3), 343–381. https://doi.org/10.1080/13600834.2022.2088062

Xiao, L. Y. (2023). Breaking ban: Belgium's ineffective gambling law regulation of video game loot boxes. *Collabra Psychology, 9*(1). https://doi.org/10.1525/collabra.57641

Xiao, L. Y., Fraser, T. C., & Newall, P. W. S. (2023). Opening Pandora's loot box: Weak links between gambling and loot box expenditure in China, and player opinions on probability disclosures and pity-timers. *Journal of Gambling Studies, 39*(2), 645–668. https://doi.org/10.1007/s10899-022-10148-0

Zendle, D., Meyer, R., Cairns, P., Waters, S., & Ballou, N. (2020). The prevalence of loot boxes in mobile and desktop games. *Addiction (Abingdon, England), 115*(9), 1768–1772. https://doi.org/10.1111/add.14973

Zhang, T., & Frederick, C. M. (2018). Busting the myth of the non-social gamer: Comparing friendship quality between gamers and non-gamers. *Proceedings of the Human Factors and Ergonomics Society Annual Meeting, 62*(1), 746–746. https://doi.org/10.1177/1541931218621169

Zhao, Y., Feng, H., Wu, X., Du, Y., Yang, X., Hu, M., Ning, H., Liao, L., Chen, H., & Zhao, Y. (2020). Effectiveness of exergaming in improving cognitive and physical function in people with mild cognitive impairment or dementia: systematic review. *JMIR serious games, 8*(2), e16841. https://doi.org/10.2196/16841

Part V
Public Health Perspectives on Gaming

Chapter 16
Epidemiology of Problem Gaming: Prevalence, Patterns, and Trends

Matthew W. R. Stevens, Paul H. Delfabbro, and Daniel L. King

Abstract Video games have transformed from a niche entertainment form to a globally popular pastime enjoyed by over one-third of the world's population. The health and social consequences of gaming can vary greatly and depend on a range of individual and social factors, as well as structural characteristics of the game itself. This chapter explores industry developments that have led to wider adoption of gaming and critically examines the psychological, social, and health implications of digital game play. This chapter will discuss the motivations, patterns, and potential risks associated with gaming, highlighting the dual nature of gaming as both adaptive and potentially problematic under certain conditions. This chapter will summarize research on the prevalence of problem gaming based on population studies and meta-analytic reviews. The chapter aims to provide researchers, clinicians, and policymakers with a balanced overview of modern gaming from a public health perspective.

Keywords Gaming · Problem gaming · Public health · Psychological impacts · Social impacts

M. W. R. Stevens (✉)
School of Biomedicine, The University of Adelaide, Adelaide, SA, Australia

College of Education, Psychology, & Social Work, Flinders University, Adelaide, SA, Australia
e-mail: matthew.stevens@adelaide.edu.au

P. H. Delfabbro
School of Psychology, The University of Adelaide, Adelaide, SA, Australia

D. L. King
College of Education, Psychology, & Social Work, Flinders University, Adelaide, SA, Australia

L. Murrins Marques et al. (eds.), *Social and Affective Neuroscience of Gaming*,
https://doi.org/10.1007/978-3-032-00325-6_16

Introduction

Commercial video games began with arcade games in 1980s, which were immensely popular among younger male demographics (Braun & Giroux, 1989). Many advancements to the basic gaming technology, including improvements in memory, audio and graphical quality, and artificial intelligence, have contributed to the growth of gaming participation since then. The increased accessibility of digital gaming technologies, including consoles and later mobile gaming, has also expanded the population of individuals who play video games. Market estimates suggest there are over 3.4 billion people worldwide who engage in gaming; a figure which has increased from 3.1 billion in 2022 and is forecasted to reach 3.8 billion by 2027 (Newzoo, 2024). This figure represents over one-third of the global population. In highly developed regions, where access to digital technologies is widespread, rates of gaming participation are in excess of 50% of the population, indicating the growing integration of gaming with other technologies and their presence in everyday life. In the United States, estimates suggest that 61% of Americans play video games for at least 1 h per week, which increased from 59% in 2014 (Entertainment Software Association [ESA], 2014, 2024). Household device penetration in the USA has also increased, with the proportion of households owning a dedicated gaming PC, smartphone, or console increasing from 51% in 2014 to 75% in 2024 (ESA, 2014, 2024). Mobile gaming has emerged as particularly popular, with 67% of players using mobile devices, followed by PC (53%), console (36%), and other forms (ESA, 2024).

Changes in access and availability of gaming, particularly on mobile devices, have coincided with changes to gaming demographics worldwide. Gaming is becoming increasingly popular among older age groups, with recent estimates suggesting that the average age of gamers in the USA has increased in the last 20 years, from 29 years in 2004 to 36 years in 2024; while almost one-third of players are now aged over 50 years (ESA, 2004, 2024). Gaming has also become increasingly popular with female players, which is reflected in a closing gender gap between participation rates. For example, in 2004, market estimates suggested the proportion of female gamers was less than 40% (ESA, 2004); whereas today, nearly 46% are female (ESA, 2024). Much of this has been driven by the diversification of gaming platforms and genres, which now include more casual and mobile games that appeal to a broader range of preferences, and the breaking down of traditional masculine gaming stereotypes that previously made gaming feel less accessible to women (Lopez-Fernandez et al., 2019).

Gaming has evolved from solo entertainment to become a social activity. Market estimates suggest that 88% of players are engaging in online play, and 74% are playing with others, both online and in-person (ESA, 2024). Single-player games have incorporated social elements via streaming and sharing content with others online, and popular games tend to have large online communities or fanbases that support and discuss the game in online forums and other social media. In tandem with technologies that have enabled online play, gaming is an activity that is enjoyed in the company of others, and is even viewed by some as a bonding activity between parents and children. For example, the majority of parents who play video games report playing with their children (70%), and 40% do so at least weekly (ESA, 2024).

Game Design and Monetization

The expansion of gaming access and availability through digital platforms has produced a variety of innovative business models. As a global market, the gaming industry is expected to reach $187.7 billion in 2024, with projected growth to $312 billion estimated by 2027 (Newzoo, 2024). To attract diverse consumers, gaming companies now offer various ways to purchase games. These include physical and digital copies, in-game purchases, and subscriptions to services (e.g., PlayStation Plus and Xbox Game Pass), while mobile games also generate revenue through paid downloads and in-app purchases. A typical console game might retail for USD$40 and $70 (Statista, 2022), whereas a digital copy (e.g., on Steam) might be available for less and at a discounted price.

Some gaming business models have attracted controversy for their monetized elements that involve continuous spending mechanisms, particularly for games that appear primarily targeted at underage users. For example, pay-to-play (P2P) and free-to-play (F2P) models have meant that developers can generate ongoing revenues through subscriptions and in-game transactions. While F2P games may have lowered the up-front cost and thus expanded the appeal of many games, they rely on revenue generation through microtransactions, which have introduced a variety of ethical design challenges. So-called "predatory" design features (King & Delfabbro, 2018), including loot-boxes, exploit principles of reinforcement learning and behavioral economics (e.g., sunk cost bias) to generate continuous revenues from players. Such features have been likened to gambling and have raised concerns about player manipulation and the need for responsible game design (Drummond & Sauer, 2018). As gaming continues to grow in popularity, there is a need for more research into prevention and early intervention strategies to mitigate potential harms associated with immersive and potentially addictive game elements (King & Delfabbro, 2019a).

Healthy Gaming

Public health conceptualizes health not merely as the absence of disease or morbidity, but rather as a state of complete well-being encompassing physical, social, and emotional domains. Research shows that playing video games can positively contribute to an individual's overall health and well-being in several ways. Cognitively, gaming has been shown to enhance a range of complex brain functions, including attention mechanisms, working memory, decision-making, and visuospatial representation (Bediou et al., 2018; Green & Bavelier, 2012; Palaus et al., 2017). These neurological improvements suggest that certain games may have the potential to provide dynamic cognitive training environments (Smith & Basak, 2023) and may even help to attenuate cognitive decline in older adults (Basak et al., 2008).

Socially, many modern gaming platforms are designed around connection and community. Online multiplayer games often require players to build relationships with other players and participate in online communities. For example, success in many group-based role-playing games requires players to form partnerships with other players in order to coordinate roles and responsibilities to complete complex tasks. This social requirement enhances many gamers' experiences and facilitates social connectedness and real-world friendships (Tushya et al., 2023). Gaming under certain conditions also has the potential to facilitate healthier parent–child relationships and interactions (Buswell et al., 2012; Coyne et al., 2011; Musick et al., 2021), which may protect against problematic gaming in adolescence (King & Delfabbro, 2017).

Psychologically, video games can provide significant mood enhancement. Gaming has been shown to be effective as a stress-reduction and anxiety-management strategy (Villani et al., 2018), including as a distraction from anxiety and pain in medical settings. Other reviews have found gaming has mental health benefits (Pine et al., 2020); and in some cases, emotion regulation was associated with playing strategy games (Pallavicini et al., 2018). Gaming may also promote mindfulness (Cruea, 2020) through the immersive nature of the activity and its ability to induce flow (see Csikszentmihalyi, 1990). The benefits of mindful practice itself are well-documented, including improvements in mood, coping with stress, attention, optimism, and reducing anxiety and fatigue (Sumantry & Stewart, 2021).

However, although gaming has many clear health benefits, some of the same cognitive and emotional mechanisms that make gaming beneficial for some, under certain circumstances, may also contribute to maladaptive behaviors for others. For example, some players who use gaming as a coping mechanism for negative mood states may struggle to develop other strategies, leading to potential issues in situations when gaming is unavailable. It is therefore critical to examine not only the aspects of games themselves that impact on health, but also characteristics of the individual, and the interaction between the two, to better understand how the health effects of gaming are enabled (King & Delfabbro, 2019b).

Motivation and Engagement

The degree to which individuals engage in gaming occurs along a continuum, ranging from casual or recreational at one extreme to unhealthy and disordered at the other. For the majority who play games, the activity begins casually and rarely escalates or persists regularly beyond a recreational level of engagement with occasional periods of extended play (e.g., during holidays). However, some players transition from casual play into more intensive and risky levels of engagement. This progression may not necessarily be best reflected by the change in the amount of time spent playing (Baggio et al., 2016), although it is typical for risks to be associated with greater frequency of involvement. Instead, risky gaming is often driven by other psychological (Brand et al., 2019), and motivational drives (Ballabio et al., 2017).

The extent to which gaming leads to harm is likely to be influenced by the reasons players have for gaming, and these motivations have been found to vary widely. For example, a report by the ESA (2023) of nearly 13,000 participants aged 16 or older from 12 countries (mostly Western, European, and East Asian) reported that the primary motivations for players were for enjoyment, to pass the time, and to alleviate stress (which are generally healthy). By contrast, using games as an escape from the real world or to avoid negative emotional states can transform gaming from a coping mechanism into a potential dependency (Melodia et al., 2022), which indicates the presence of less healthy motivations. These observations have led to the development of motivational typologies, with classifications including: psychological (e.g., achievement or mastery), social (e.g., community connection), and physiological (e.g., to alleviate stress or anxiety). Within these groupings, it can be seen that, while most motivations tend to reflect adaptive responses (see Demetrovics et al., 2011), others may also signal problematic usage, particularly in individuals experiencing psychological distress (Király et al., 2015; Slack et al., 2022).

Considering the continuum of gaming engagement, gamers can be categorized into four basic groups:

1. Recreational gamers: casual gamers who engage intermittently and infrequently.
2. Intensive gamers: highly engaged and frequent or habitual players, and those whose gaming may be a source of income (e.g., e-Sports professionals and streamers).
3. At-risk gamers: gamers whose involvement in gaming is beginning to show signs of problems but not yet at the level or for a sustained period required to meet health diagnoses. This includes gamers who regularly engage in "binge" gaming behaviors.
4. Disordered gamers: individuals experiencing impaired control over their gaming and who report (or who have others who report on their behalf) that they experience significant life disruptions due to gaming activities.

While these categories provide a framework for understanding gaming engagement, they also highlight how individuals might transition from recreational to more intensive or potentially disordered gaming patterns. However, it is important to recognize that the frequency of play does not necessarily indicate harm or disorder. This player typology can be used to guide public health approaches to meet the needs of each group by recognizing that each group has different relationships with gaming activities and would benefit from different approaches to reduce risk or harm.

Problematic Gaming

Definitions and Epidemiology

Research has increasingly focused on understanding the potential negative effects of excessive gaming behavior. Therefore, it is important to examine and define what distinguishes healthy engagement from problematic use. In 2013, the American Psychiatric Association acknowledged problematic internet use in the fifth edition of the Diagnostic and Statistical Manual of Mental Disorders (DSM-5), with the inclusion of internet gaming disorder (IGD) in Part III (American Psychiatric Association, 2013). Although IGD is not currently recognized in the DSM-5-TR, in 2018, the World Health Organization (WHO) included Gaming Disorder (GD) in the 11th revision of the International Classification of Diseases (ICD-11; World Health Organization, 2018). Gaming disorder is defined as a persistent and recurrent pattern of gaming behavior over a 12-month period, which is characterized by:

- A loss of control over gaming activities,
- Increased prioritization of gaming at the expense of other life areas; and
- Continued gaming despite the knowledge of negative consequences.

Importantly, to meet the diagnosis, an individual must have experienced significant impairment to their regular functioning. The ICD-11 also refers to additional clinical features, including withdrawal and tolerance, which are not essential for diagnosis but may be relevant to the conceptualization of some individuals.

Trends in Prevalence

Research into population rates of problematic gaming has had several challenges. The first and most widespread issue was the inconsistent conceptualization of problem gaming. The overlapping nature of problematic internet and gaming activities often led to them being measured together, thereby making it difficult to determine whether gaming was the primary issue. Given that IGD was only first described in 2013, earlier studies conducted in the late 1990s and early 2000s often referred to broader concepts of internet addiction. A related issue was a lack of consistent measurement approaches to problem gaming, and sometimes inconsistent use of the same tools for measuring problem gaming.

Many reviews of the prevalence of problem gaming have reported highly varying prevalence rates. For example, a systematic review by Feng et al. (2017) included 27 studies, predominantly from European and Asian contexts, that revealed prevalence estimates ranging from 0.7% to as high as 15.6% among young people. Similarly, in 2017, Mihara and Higuchi systematically reviewed 50 studies and estimated the prevalence rate to range between 0.7% and 27.5%. In each report, the authors noted that variability in the choice of measurement created uncertainty regarding the reliability of prevalence estimates. In 2018, Fam conducted a meta-analysis of prevalence rates among 16 studies with adolescent samples and found an average estimate of 4.6%, with higher rates among studies with male samples (6.8%) versus female (1.3%) samples.

More recent systematic reviews have reported that the prevalence of gaming disorder tends to fall between 2% and 3%. In 2021, Stevens et al. conducted a meta-analysis of 53 studies in the decade between 2009 and 2019 and reported a pooled prevalence rate of around 3%.[1] When considering the most stringent sampling criteria, estimates were slightly below 2%. Kim et al. (2022) analyzed 61 studies (comprising 71 unique populations) and reported a pooled prevalence estimate of 3.3% (which was reduced to 2.4% when only the most stringent samples were included). Limone et al. (2023) analyzed 27 studies, focusing on a broad range of psychological, social, and individual difference characteristics associated with problem gaming, and reported a prevalence estimate of 5%. Consistent with previous reviews, the authors of all three meta-analyses identified higher prevalence rates among male samples compared with female samples (sometimes two to three times higher). Moreover, these reviews identified that rates of prevalence were higher among younger cohorts, studies with smaller sample sizes, and studies of poorer quality.

Another important finding from these reviews was that higher rates of prevalence were found among Asian studies, compared with other regions. Liao et al. (2022) conducted a systematic review of 26 individual studies from East Asian regions and identified an average prevalence rate of 12% (reduced to 6% when following more stringent criteria). Similarly, the prevalence rate for males was twice as high as for females, and notably, the prevalence rates tended to approach zero after age 30.

Reviews of longitudinal studies have investigated the changes in prevalence rates over time and have highlighted some important findings. For instance, while male prevalence rates have tended to remain stable, female gaming disorder rates have gradually increased from an average of 1.75% in 2009 to 4.47% in 2017 (Stevens et al., 2021), highlighting the changing participation and potential vulnerability among female populations that may have been previously less recognized or not identified due to sampling bias.

Methodological challenges have also been a commonly highlighted issue in prevalence research. The field has been characterized by significant variability in screening tools, with more than 35 different methods identified in one comprehensive review (King et al., 2020). Many of these tools were adapted from criteria used in other behavioral addiction contexts, such as gambling disorders, which introduces potential measurement inconsistencies. It should also be noted that most, if not all, studies included in each of the meta-analyses described here were based on the DSM-5 criteria, rather than the ICD-11 criteria. As a result, a question remains as to whether stricter diagnostic criteria outlined in the ICD-11 would yield different estimates.

[1] A recent corrigendum by Stevens et al. (2023) noted that the prevalence rate reduced from 3.05% to 2.96% with the exclusion of two studies which should have been excluded based on exclusion criteria.

Vulnerable Populations

From an epidemiological perspective, recognizing the populations who may be at greatest risk of a transition from healthy to disordered gaming helps inform future prevention and early intervention approaches. Research has highlighted several key demographic risk factors that have implications for public health. First, males have a higher likelihood of developing a gaming disorder than females. The reasons for this are not entirely clear, but it may be due to how males typically engage with games and features of games themselves to which males are drawn. For example, research suggests that males tend to gravitate toward competitive and action-oriented games, such as Massive Multiplayer Online Role-Playing Games (MMORPGs), which are linked with higher rates of problematic gaming than other types of games (King et al., 2019; Rehbein et al., 2021). Others have suggested that higher rates of unresolved mental health issues, less social support, and stress vulnerability among males create pathways for gaming disorder to develop (Chen et al., 2018).

Second, young people have been shown to be more susceptible to gaming disorder than older people. This finding is not unsurprising given that young people spend more time interacting with digital technologies, including gaming and social media, than older people (Auxier & Anderson, 2021; ESA, 2024), and this group tends to have more disposable time and fewer responsibilities. Moreover, biological changes in the human brain occurring through childhood and adolescence (particularly the pre-frontal cortex, which is implicated in decision-making and impulse control) have not typically reached full maturation until around age 25 years (Arain et al., 2013). As a consequence, individuals are most susceptible to acquiring addictive disorders, including gaming, between the ages of 19 and 25 (Paus et al., 2008; Solmi et al., 2022).

Third, people living in certain regions (e.g., Asia) are more susceptible than other parts of the world. There are several possible explanations for this. First, gaming, streaming, and e-Sports culture are more popular and developed in these regions (Jin & Kim, 2024; Statista, 2023), which may make it more likely that younger players seek out this type of activity and engagement. Second, family dynamics, including single-child families, where gaming typically becomes a social outlet in lieu of spending time interacting with siblings, may facilitate more problematic engagement (Lim, 2012). Third, an emphasis on obedience and authoritarian parenting styles, which are more common in certain Asian regions, may lead to problem gaming more readily (Yang et al., 2020). Finally, in some cases, a strong focus on achievement and academic success in these cultures may generate additional pressures and anxieties, which gaming can be used to escape from or alleviate temporarily (Giardina et al., 2024; Lim, 2012).

Conclusions

This chapter has attempted to highlight some of the developments and trends that have facilitated greater engagement in gaming in the general population, including the notable increase in participation among older and female players. As digital technologies have proliferated and increased access and availability, gaming is increasingly becoming part of everyday life. The number of households owning and using dedicated gaming devices continues to increase, with the demographics of those who enjoy video games becoming ever more balanced in terms of age and gender. These increases are likely due to the growing popularity of online services, social media, streaming, and e-Sports worldwide.

This chapter highlighted some of the health and well-being benefits of gaming. Although there are clear cognitive, social, and psychological effects, the extent to which these are realized depends on the characteristics of the individual. While the majority of individuals engage with games in adaptive and beneficial ways, a small subset may develop problematic patterns of use. For example, while an online multiplayer game could serve as a valuable social connection tool for an introverted individual seeking structured social interaction, it could also have a negative impact if played excessively by a person who uses it as a coping mechanism for social anxiety. Understanding the characteristics and vulnerabilities of different gaming individuals is helpful from an epidemiological and public health perspective, as it can inform prevention and early intervention approaches within regulatory and clinical frameworks. Ultimately, understanding gaming from a public health perspective involves a balanced view that recognizes it as a leisure and cultural activity, with potential for benefits and harms.

References

American Psychiatric Association. (2013). *Diagnostic and statistical manual of mental disorders: DSM-5* (Vol. 5, No. 5). American Psychiatric Association.

Arain, M., Haque, M., Johal, L., Mathur, P., Nel, W., Rais, A., et al. (2013). Maturation of the adolescent brain. *Neuropsychiatric Disease and Treatment, 9*, 449–461.

Auxier, B., & Anderson, M. (2021). Social media use in 2021. *Pew Research Center, 1*(1), 1–4.

Baggio, S., Dupuis, M., Studer, J., Spilka, S., Daeppen, J. B., Simon, O., et al. (2016). Reframing video gaming and internet use addiction: Empirical cross-national comparison of heavy use over time and addiction scales among young users. *Addiction, 111*(3), 513–522.

Ballabio, M., Griffiths, M. D., Urbán, R., Quartiroli, A., Demetrovics, Z., & Király, O. (2017). Do gaming motives mediate between psychiatric symptoms and problematic gaming? An empirical survey study. *Addiction Research & Theory, 25*(5), 397–408.

Basak, C., Boot, W. R., Voss, M. W., & Kramer, A. F. (2008). Can training in a real-time strategy video game attenuate cognitive decline in older adults? *Psychology and Aging, 23*(4), 765.

Bediou, B., Adams, D. M., Mayer, R. E., Tipton, E., Green, C. S., & Bavelier, D. (2018). Meta-analysis of action video game impact on perceptual, attentional, and cognitive skills. *Psychological Bulletin, 144*(1), 77.

Brand, M., Wegmann, E., Stark, R., Müller, A., Wölfling, K., Robbins, T. W., & Potenza, M. N. (2019). The Interaction of Person-Affect-Cognition-Execution (I-PACE) model for addictive behaviors: Update, generalization to addictive behaviors beyond internet-use disorders, and specification of the process character of addictive behaviors. *Neuroscience & Biobehavioral Reviews, 104*, 1–10.

Braun, C. M., & Giroux, J. (1989). Arcade video games: Proxemic, cognitive and content analyses. *Journal of Leisure Research, 21*(2), 92–105.

Buswell, L., Zabriskie, R. B., Lundberg, N., & Hawkins, A. J. (2012). The relationship between father involvement in family leisure and family functioning: The importance of daily family leisure. *Leisure Sciences, 34*(2), 172–190.

Chen, K. H., Oliffe, J. L., & Kelly, M. T. (2018). Internet gaming disorder: An emergent health issue for men. *American Journal of Men's Health, 12*(4), 1151–1159.

Coyne, S. M., Padilla-Walker, L. M., Stockdale, L., & Day, R. D. (2011). Game on… girls: Associations between co-playing video games and adolescent behavioral and family outcomes. *Journal of Adolescent Health, 49*(2), 160–165.

Cruea, M. D. (2020). Gaming the mind and minding the game: Mindfulness and flow in video games. In *Video games and well-being: Press start* (pp. 97–107). Palgrave Pivot.

Csikszentmihalyi, M. (1990). *Flow: The psychology of optimal experience*. Haper and Row.

Demetrovics, Z., Urbán, R., Nagygyörgy, K., Farkas, J., Zilahy, D., Mervó, B., et al. (2011). Why do you play? The development of the motives for online gaming questionnaire (MOGQ). *Behavior Research Methods, 43*, 814–825.

Drummond, A., & Sauer, J. D. (2018). Video game loot boxes are psychologically akin to gambling. *Nature Human Behaviour, 2*(8), 530–532.

Entertainment Software Association. (2004). *Essential facts about the computer and video game industry, 2004*. Washington, DC. Retrieved from: https://archive.org/details/esa-essential-facts/ESA2004_Essential_facts

Entertainment Software Association. (2014). *Essential facts about the computer and video game industry, 2014*. Washington, DC. Retrieved from: https://docslib.org/doc/1762666/a-2014-study-by-the-entertainment-software-association-esa

Entertainment Software Association. (2023). *Power of play, global report 2023*. Washington, DC. Retrieved from: https://www.theesa.com/resources/power-of-play-global-report-2023/

Entertainment Software Association. (2024). *Essential facts about the U.S. video game industry, 2024*. Washington, DC. Retrieved from: https://www.theesa.com/resources/essential-facts-about-the-us-video-game-industry/2024-data/

Fam, J. Y. (2018). Prevalence of internet gaming disorder in adolescents: A meta-analysis across three decades. *Scandinavian Journal of Psychology, 59*(5), 524–531.

Feng, W., Ramo, D. E., Chan, S. R., & Bourgeois, J. A. (2017). Internet gaming disorder: Trends in prevalence 1998–2016. *Addictive Behaviors, 75*, 17–24.

Giardina, A., Fournier, L., Starcevic, V., King, D. L., Di Blasi, M., Schimmenti, A., & Billieux, J. (2024). From active escapism to virtual withdrawal: Validation of the Compensatory-Dissociative Online Gaming scales (C-DOGs). *Journal of Behavioral Addictions, 13*, 1028.

Green, C. S., & Bavelier, D. (2012). Learning, attentional control, and action video games. *Current Biology, 22*(6), R197–R206.

Jin, Y., & Kim, S. J. (2024). Esports culture in Asia. In *Routledge handbook of esports* (pp. 591–604). Routledge.

Kim, H. S., Son, G., Roh, E. B., Ahn, W. Y., Kim, J., Shin, S. H., et al. (2022). Prevalence of gaming disorder: A meta-analysis. *Addictive Behaviors, 126*, 107183.

King, D. L., & Delfabbro, P. H. (2017). Features of parent-child relationships in adolescents with internet gaming disorder. *International Journal of Mental Health and Addiction, 15*, 1270–1283.

King, D. L., & Delfabbro, P. H. (2018). Predatory monetization features in video games (e.g., "loot boxes") and internet gaming disorder. *Addiction, 113*, 1967–1969.

King, D. L., & Delfabbro, P. H. (2019a). Loot box limit-setting is not sufficient on its own to prevent players from overspending: A reply to Drummond, Sauer & Hall. *Addiction, 114*, 1324–1325.
King, D. L., & Delfabbro, P. H. (2019b). *Internet gaming disorder: Theory, assessment, treatment, and prevention*. Elsevier Academic Press.
King, D. L., Delfabbro, P. H., Perales, J. C., Deleuze, J., Király, O., Krossbakken, E., & Billieux, J. (2019). Maladaptive player-game relationships in problematic gaming and gaming disorder: A systematic review. *Clinical Psychology Review, 73*, 101777.
King, D. L., Chamberlain, S. R., Carragher, N., Billieux, J., Stein, D., Mueller, K., et al. (2020). Screening and assessment tools for gaming disorder: A comprehensive systematic review. *Clinical Psychology Review, 77*, 101831.
Király, O., Urbán, R., Griffiths, M. D., Ágoston, C., Nagygyörgy, K., Kökönyei, G., & Demetrovics, Z. (2015). The mediating effect of gaming motivation between psychiatric symptoms and problematic online gaming: An online survey. *Journal of Medical Internet Research, 17*(4), e88.
Liao, Z., Chen, X., Huang, Q., & Shen, H. (2022). Prevalence of gaming disorder in East Asia: A comprehensive meta-analysis. *Journal of Behavioral Addictions, 11*(3), 727–738.
Lim, S. S. (2012). Regulatory initiatives for managing online risks and opportunities for youths – The East Asian experience. In M. Walrave (Ed.), *e-Youth: Balancing between opportunities and risks?* (pp. 271–290). Peter Lang.
Limone, P., Ragni, B., & Toto, G. A. (2023). The epidemiology and effects of video game addiction: A systematic review and meta-analysis. *Acta Psychologica, 241*, 104047.
Lopez-Fernandez, O., Williams, A. J., Griffiths, M. D., & Kuss, D. J. (2019). Female gaming, gaming addiction, and the role of women within gaming culture: A narrative literature review. *Frontiers in Psychiatry, 10*, 454.
Melodia, F., Canale, N., & Griffiths, M. D. (2022). The role of avoidance coping and escape motives in problematic online gaming: A systematic literature review. *International Journal of Mental Health and Addiction, 20*, 996–1022.
Mihara, S., & Higuchi, S. (2017). Cross-sectional and longitudinal epidemiological studies of internet gaming disorder: A systematic review of the literature. *Psychiatry and Clinical Neurosciences, 71*(7), 425–444.
Musick, G., Freeman, G., & McNeese, N. J. (2021). Gaming as family time: Digital game co-play in modern parent-child relationships. *Proceedings of the ACM on Human-Computer Interaction, 5*, 1–25.
Newzoo. (2024). *Global market games report*. Retrieved from: https://newzoo.com/resources/trend-reports/newzoos-global-games-market-report-2024-free-version
Palaus, M., Marron, E. M., Viejo-Sobera, R., & Redolar-Ripoll, D. (2017). Neural basis of video gaming: A systematic review. *Frontiers in Human Neuroscience, 11*, 231323.
Pallavicini, F., Ferrari, A., & Mantovani, F. (2018). Video games for well-being: A systematic review on the application of computer games for cognitive and emotional training in the adult population. *Frontiers in Psychology, 9*, 407892.
Paus, T., Keshavan, M., & Giedd, J. N. (2008). Why do many psychiatric disorders emerge during adolescence? *Nature Reviews Neuroscience, 9*(12), 947–957.
Pine, R., Fleming, T., McCallum, S., & Sutcliffe, K. (2020). The effects of casual videogames on anxiety, depression, stress, and low mood: A systematic review. *Games for Health Journal, 9*(4), 255–264.
Rehbein, F., King, D. L., Staudt, A., Hayer, T., & Rumpf, H. J. (2021). Contribution of game genre and structural game characteristics to the risk of problem gaming and gaming disorder: A systematic review. *Current Addiction Reports, 8*(2), 263–281.
Slack, J. D., Delfabbro, P., & King, D. L. (2022). Toward a delineation of the differences between high engagement and problem gaming. *Addictive Behaviors Reports, 16*, 100462.
Smith, E. T., & Basak, C. (2023). A game-factors approach to cognitive benefits from video-game training: A meta-analysis. *PLoS One, 18*(8), e0285925.

Solmi, M., Radua, J., Olivola, M., Croce, E., Soardo, L., Salazar de Pablo, G., et al. (2022). Age at onset of mental disorders worldwide: Large-scale meta-analysis of 192 epidemiological studies. *Molecular Psychiatry, 27*(1), 281–295.

Statista. (2022). *Average sales price (ASP) of video game software at year of release as of August 2022, by platform* [infographic]. Statista. Retrieved from: https://www.statista.com/statistics/1388073/average-price-of-video-games-by-platform/

Statista. (2023). *Share of eSports frequent viewers and enthusiasts worldwide in 2019, by region* [infographic]. Statista. Retrieved from: https://www.statista.com/statistics/673740/share-of-esports-viewers-by-region-worldwide/

Stevens, M. W., Dorstyn, D., Delfabbro, P. H., & King, D. L. (2021). Global prevalence of gaming disorder: A systematic review and meta-analysis. *Australian and New Zealand Journal of Psychiatry, 55*(6), 553–568.

Stevens, M. W., Dorstyn, D., Delfabbro, P. H., & King, D. L. (2023). Corrigendum to: Stevens, M. W. R., Dorstyn, D., Delfabbro, P. H., & King, D. L. (2021). Global prevalence of gaming disorder: A systematic review and meta-analysis. *Australian and New Zealand Journal of Psychiatry, 55*, 553–568. *Australian and New Zealand Journal of Psychiatry, 57*(6), 928–928.

Sumantry, D., & Stewart, K. E. (2021). Meditation, mindfulness, and attention: A meta-analysis. *Mindfulness, 12*, 1332–1349.

Tushya, Chhabra, D., & Abraham, B. (2023). Social networking or social isolation? A systematic review on socio-relational outcomes for members of online gaming communities. *Games and Culture, 20*, 355–383.

Villani, D., Carissoli, C., Triberti, S., Marchetti, A., Gilli, G., & Riva, G. (2018). Videogames for emotion regulation: A systematic review. *Games for Health Journal, 7*(2), 85–99.

World Health Organization. (2018). *Gaming disorder*. Retrieved from: http://www.who.int/features/qa/gaming-disorder/en/

Yang, X., Jiang, X., Mo, P. K. H., Cai, Y., Ma, L., & Lau, J. T. F. (2020). Prevalence and interpersonal correlates of internet gaming disorders among Chinese adolescents. *International Journal of Environmental Research and Public Health, 17*(2), 579.

Chapter 17
Virtual Reality Games as a Tool for Health Promotion and Education in Disability Care

Giovanna de Paula Vidigal, Eduardo Dati Dias, Juliana Perez Weingartner, and Carlos Bandeira de Mello Monteiro

Abstract Virtual reality (VR) games offer innovative possibilities for promoting health, education, and inclusion in disability care. This chapter explores how VR gaming leverages neuroplasticity to drive cognitive, motor, and emotional improvements, highlighting its therapeutic potential for individuals with disabilities. It examines the role of engagement, motivation, and adaptive gameplay in making rehabilitation more accessible and enjoyable, while emphasizing the growing importance of digital biomarkers in monitoring therapeutic outcomes. Furthermore, it discusses how VR gamification can enhance mental health by reducing stress, fostering emotional resilience, and mitigating social isolation. By integrating neuroscience, motivational strategies, and inclusive design, VR games emerge as powerful tools for advancing health promotion and education for people with disabilities. Future directions call for deeper personalization, advanced technologies, long-term engagement strategies, and inclusive social features to maximize the impact of adapted games.

Keywords Virtual reality · Disability rehabilitation · Neuroplasticity · Digital health · Gamification

G. de Paula Vidigal (✉)
Faculty of Medicine, University of São Paulo – FMUSP, São Paulo, Brazil
e-mail: giovanna.vidigal@usp.br

E. D. Dias · J. P. Weingartner
School of Arts, Sciences and Humanities, University of São Paulo, São Paulo, Brazil

C. B. de Mello Monteiro
Faculty of Medicine, University of São Paulo – FMUSP, São Paulo, Brazil

School of Arts, Sciences and Humanities, University of São Paulo, São Paulo, Brazil

L. Murrins Marques et al. (eds.), *Social and Affective Neuroscience of Gaming*,
https://doi.org/10.1007/978-3-032-00325-6_17

Introduction

The transformative potential of gaming for individuals with disabilities lies in its ability to merge entertainment with meaningful therapeutic, educational, and social outcomes. Virtual reality (VR) games offer unique opportunities to address challenges in rehabilitation, mental health, and inclusion by creating adaptive and engaging environments tailored to individual needs. For people with disabilities, digital games can serve as tools for enhancing neuroplasticity, fostering cognitive and motor improvements, and promoting emotional resilience through structured and purposeful activities. These technologies not only support physical and cognitive rehabilitation but also provide powerful mechanisms for motivating, engagement, and sustained participation (González-González et al., 2019).

This chapter explores how VR games have evolved into a dynamic platform for improving the quality of life in disability care. It begins by delving into the neuroscience behind virtual games, emphasizing its ability to stimulate the brain's adaptive capacity and foster functional recovery. The discussion then transitions to the role of engagement and motivation in rehabilitation, showing how gamified interventions transform repetitive tasks into rewarding and empowering experiences. Further, the chapter highlights the mental health benefits of VR games, particularly its capacity to reduce stress, alleviate social isolation, and foster emotional balance. Together, these insights demonstrate how gaming can be harnessed as a powerful tool to promote health, education, and inclusion for individuals with disabilities.

Virtual Reality Gaming Neuroscience for People with Disabilities

Digital games have demonstrated their ability to stimulate neuroplasticity, fostering the brain's capacity to reorganize and create new neural pathways. This process is particularly valuable for individuals with disabilities, where targeted activities in VR games can activate specific brain regions associated with motor, cognitive, and emotional functions. Systematic reviews have highlighted the potential of VR in promoting neuroplasticity and recovery. According to Feitosa et al. (2022), VR interventions for motor rehabilitation, as revealed through fMRI studies, induced significant neural reorganization in both the ipsilesional and contralesional hemispheres, with key activation changes observed in the primary motor cortex, sensorimotor cortex, and supplementary motor area. Complementing these findings, Drigas and Sideraki (2024) demonstrated that immersive and interactive VR experiences can drive neuroplastic changes across diverse populations, leading to measurable improvements in motor skills, cognitive function, and emotional regulation.

By providing structured and adaptive tasks, virtual games offer consistent and purposeful stimulation necessary for reinforcing neural circuits, making them essential tools for rehabilitation. These tailored interventions not only facilitate functional recovery but also provide deeper insights into how the brain adapts under specific therapeutic conditions (Mang et al., 2013; Prajjwal et al., 2014; Fu & Ji, 2023; Kamari et al., 2024). Through targeted activities, virtual reality games effectively support motor and cognitive rehabilitation. For instance, repetitive motor tasks strengthen neural pathways associated with balance and coordination, while problem-solving challenges enhance cognitive skills such as memory and decision-making. Over time, these structured experiences lead to measurable changes in brain activity and structure, including improved neural connectivity and increased activation in motor and executive areas, underscoring the transformative potential of gaming in driving neuroplasticity and recovery (Warraich & Kleim, 2010; Aguilera-Rubio et al., 2024; Kamari et al., 2024).

Digital Biomarkers

In this context, the use of biomarkers helps to capture neural responses and motor and social behaviors during VR game interactions. Biomarkers are essential for evaluating these behaviors, offering real-time insights through metrics such as movement patterns, reaction times, and physiological responses. For example, VR biomarkers have proven effective in identifying subtle deficits in instrumental activities of daily living, such as challenges with mobility or task completion, which are critical for early intervention and therapeutic adjustments (Park et al., 2024). By capturing body movements through 3D sensor cameras and translating them into game analytics, these metrics serve as digital biomarkers linked to physical and cognitive performance. This process ensures that game-based therapies remain challenging yet appropriately tailored to individual needs, maximizing their effectiveness (Papa et al., 2023; Almeida et al., 2024).

Recent research highlights the expanding potential of digital biomarkers in assessing health states. Gielis et al. (2021), for instance, used digital biomarkers derived from virtual card game data to detect mild cognitive impairment (MCI) in older adults. Their machine-learning approach successfully distinguished individuals with MCI from their healthy peers. Similarly, Petsani et al. (2022) explored how digital biomarkers derived from exergame interactions can provide meaningful insights into participants' physical and cognitive states.

Together, neuroplasticity and digital biomarkers form a dynamic feedback loop within the gaming experience. Neuroplasticity drives the adaptation and strengthening of neural pathways, while biomarkers provide the evidence and metrics to measure these changes. This relationship enables researchers and clinicians to monitor the progression of neurological conditions with unprecedented accuracy, refining therapeutic strategies in real time. Virtual games, in this context, become more than mere tools for rehabilitation—they evolve into critical components of neuroscience-based therapies that induce meaningful brain changes while generating quantifiable data to validate and improve their efficacy.

Engagement and Motivation for Rehabilitation

Engagement and motivation are pivotal factors in the success of health promotion and education programs, particularly for disabled individuals who may face unique physical and cognitive barriers.

Traditional health interventions often struggle to capture sustained attention, leading to poor adherence and suboptimal outcomes (Zuki et al., 2024). VR games and serious games have emerged as powerful tools to address these challenges, transforming tedious or repetitive tasks into interactive and rewarding experiences. By leveraging VR game mechanics such as rewards, progress tracking, and social features, these platforms offer a novel way to maintain interest and drive behavioral change (Mouatt et al., 2020; Zuki et al., 2024).

Enjoyment is a crucial factor in the success of rehabilitation, especially for individuals with Autism Spectrum Disorder (Almeida et al., 2024) and Cerebral palsy (Silva et al., 2021) who often face challenges such as low motivation and interest in physical activity fostering a fun and flexible environment can improve participation and outcomes. Virtual games have demonstrated the ability to enhance motivation and emotional well-being by making rehabilitation tasks fun and interactive. Virtual tasks have shown potential to reduce tension, increase vigor, and improve mood, emphasizing the importance of enjoyment in fostering participation and achieving better therapeutic outcomes (Almeida et al., 2024; Silva et al., 2021). In a longitudinal study, Moraes et al. (2022) analysed the motor performance, physical activity, and enjoyment of people with Autism Spectrum Disorder while practicing a virtual reality game. The researchers observed that enjoyment can predict greater physical activity involvement, leading to better health across all ages and abilities.

Virtual Reality Game Elements Driving Engagement for Disabled People

The effectiveness of VR games in sustaining attention and motivation lies in its ability to combine various elements of game design. These elements are not only entertaining but also aligned with therapeutic and educational goals. Key components include:

Rewards and Incentives

Rewards and incentives are critical components of gamification, acting as motivators to maintain engagement and drive consistent participation. For individuals with disabilities, rewards can transform repetitive or challenging tasks into enjoyable experiences, fostering progress, self-esteem, and motivation. By triggering the brain's reward pathways, these mechanisms create a positive feedback loop, encouraging continued involvement in health or education interventions (Feitosa et al., 2022).

Effective rewards include tangible incentives like vouchers, digital achievements such as points or badges, and social recognition through leaderboards or peer acknowledgment. In a systematic review by Sardi et al. (2017), which analyzed studies employing gamification in e-Health applications, it was reported that 93% of the articles included real-time feedback systems and virtual rewards (such as points).

A study by Palidis et al. (2024) explored the use of extrinsic feedback and rewards in virtual reality-based motor rehabilitation for post-stroke patients. Their system utilized visual and auditory rewards alongside gamified scoring mechanisms to encourage the use of impaired limbs, reducing reliance on compensatory strategies. These reward systems supported engagement and neuroplasticity by reinforcing desired movement patterns and improving voluntary action selection. By aligning rewards with therapeutic goals, the intervention ensured both user engagement and clinical effectiveness.

Cascella et al. (2023) demonstrated the value of leaderboards and digital badges in sustaining interest in critical care settings, showcasing the versatility of these strategies across contexts. Furthermore, visual and interactive elements, like animations or celebratory graphics, amplify the emotional impact of rewards, making achievements more satisfying (Mouatt et al., 2020). A systematic review by Wiley et al. (2021) on the evaluation of digital games for assessing attention highlighted that rewards, sound effects, and graphic elements are key factors in enhancing engagement and motivation.

Different game genres leverage rewards uniquely, from unlocking levels in puzzle games to gaining resources in simulations, catering to diverse motivations. Personalization is key, particularly for disabled users. Aligning rewards with user preferences and capabilities enhances their appeal, ensuring that gamified systems are engaging, accessible, and impactful over the long term (Mahmoudi et al., 2024). By incorporating adaptive gameplay and personalized feedback systems, gamified interventions maintain a balance between challenge and accessibility, as noted by Zuki et al. (2024), enabling diverse users to experience meaningful engagement.

Personalization and Adaptive Gameplay

Personalization and adaptive gameplay are essential features of gamified interventions, allowing content to be tailored to the specific needs and abilities of each user. By dynamically adjusting difficulty levels and tasks in real time, these systems ensure that activities remain achievable yet challenging, fostering engagement and preventing frustration. For example, a game designed for physical or cognitive rehabilitation might use motion-sensing technology to track user progress, gradually increasing the complexity of motor or cognitive tasks (Crocetta et al., 2018; Aranha et al., 2023). Personalization also extends to customization of avatars, game environments, or narratives, empowering users to engage with content that reflects their interests and capabilities, thereby enhancing their emotional connection to the intervention.

According to Steiner et al. (2020), avatars, along with other game elements like points and tasks, are among the most commonly used components to support patients' adherence and progress across various rehabilitation settings. Creating a personalized and interactive experience, avatars help tailor therapy to individual needs, fostering a sense of presence and agency that encourages consistent participation and improves outcomes. For disabled individuals, these features transform gamified systems into supportive, empowering tools that encourage consistent participation, enabling health and education interventions to be both impactful and enjoyable (Mekler et al., 2013; Hellín et al., 2023; Carvalho et al., 2024).

The study by Aranha et al. (2023) provides a compelling example of the potential of adaptive systems. They developed a non-immersive virtual reality game for motor rehabilitation, incorporating effective adaptation to personalize the gaming experience for each patient. Designed with a user-friendly interface, the game allows physiotherapists to implement adaptations without modifying the source code, making it both accessible and intuitive. Experimental evaluations demonstrated its promise: developers found the framework practical and effective for integrating affective adaptations, while patients reported improved engagement during rehabilitation.

Adaptive gameplay frameworks are inherently versatile, offering the ability to address a wide range of disabilities by personalizing games to individual needs. These systems dynamically adjust gameplay elements, such as task complexity, interaction methods, and feedback mechanisms, ensuring that they align with the physical, cognitive, and emotional capabilities of each user. For instance, frameworks designed for rehabilitation can adapt tasks for users with motor impairments by using alternative input methods or adjusting cognitive challenges for individuals with varying processing abilities. Studies, such as those by Aranha et al. (2023) and González-González et al. (2019), demonstrate that by focusing on individual characteristics and preferences, adaptive systems can provide meaningful and effective experiences for diverse user groups. This personalization ensures that gamified interventions are inclusive and impactful, regardless of the type or severity of the disability being addressed.

Additionally, Kira et al. (2024) emphasize the value of automated adaptations in rehabilitation games, validated through usability tests with physiotherapists and patients. These systems are particularly effective for targeting specific motor rehabilitation goals. Usability feedback highlighted the importance of intuitive designs and dynamic adjustments, which further reinforce the value of personalized adaptations for optimizing therapeutic outcomes.

Progress Tracking and Feedback

Progress tracking and feedback are core elements of gamified health and education interventions, especially for disabled individuals, as they provide continuous motivation and a clear sense of achievement. By visualizing progress through graphs, badges, dashboards, or progress bars, users can see tangible evidence of their efforts, which reinforces positive behaviors and fosters long-term engagement. Virtual reality games designed for rehabilitation might display a user's improvement in motor skills over time or show the number of completed therapy sessions, creating a sense of accomplishment that encourages continued participation.

For disabled individuals, progress tracking can also help build confidence by highlighting gradual improvements, even in tasks that may initially seem challenging. Furthermore, detailed tracking systems enable healthcare providers or educators to monitor users' progress remotely, identifying areas of success or difficulty and providing timely interventions when needed. A study by Ravva et al. (2024) demonstrated that progress tracking using wearable sensors and interactive dashboards in VR environments significantly improved motor skills by offering detailed performance metrics to guide therapy adjustments.

In addition to progress tracking, real-time feedback is a crucial feature in rehabilitation virtual games, as it helps maintain engagement and motivation. Adaptive systems further enhance this by personalizing feedback to focus on individual milestones rather than comparisons with others. This approach reduces the potential for discouragement and allows users to concentrate on their unique growth, fostering a more positive and effective rehabilitation experience. Some studies have explored tele-rehabilitation for individuals with Autism Spectrum Disorder (Almeida et al., 2024) and cerebral palsy (da Silva et al., 2021) using virtual reality game platforms. A key feature of these games was the provision of immediate performance feedback, which likely contributed to performance improvements and positive outcomes by fostering motivation and encouraging moderate-intensity physical activity.

Feedback mechanisms such as instant notifications, performance scores, or celebratory animations further enhance this experience, making every small victory visible and meaningful (Zuki et al., 2024). According to Crocetta et al. (2018), feedback in virtual reality rehabilitation enhances motor learning and engagement by providing realistic sensory input that mimics real-world activities. This customizable feedback allows therapists to adjust variables such as speed, size, distance, and sensory stimuli, tailoring exercises to individual needs, addressing specific challenges, and ultimately optimizing rehabilitation outcomes.

By integrating real-time feedback with long-term progress visualization, VR game systems foster an empowering environment for individuals with disabilities, helping them stay motivated, celebrate achievements, and work toward their health and educational goals (Mahmoudi et al., 2024). Similarly, Stanica et al. (2020) highlighted a VR rehabilitation platform combining motion tracking sensors with gamified exercises. This platform provided participants with real-time feedback and progress tracking, allowing them to visualize performance improvements, reinforce motivation, and follow a personalized rehabilitation pathway.

Challenges and Competition

In VR games, challenges and competition are pivotal for enhancing engagement and driving motivation. These elements add a dynamic layer of excitement, encouraging users to push their limits and strive for excellence. Features like leaderboards, time-bound tasks, and multiplayer modes create opportunities for personal accomplishment and social interaction, enriching the overall experience. Cognitive training games, for example, may introduce progressively complex puzzles, allowing users to improve upon previous performance or collaborate with others to solve problems. Fitness games often incorporate recurring activity goals, promoting sustained participation and rewarding significant milestones (Brox et al., 2011a, b).

When designing VR games for individuals with disabilities, it is essential to balance these competitive elements with accessibility. By tailoring game difficulty to the user's abilities, gamified systems can foster an environment where achievements are meaningful and attainable. Research by Yildirim and Surer (2021) underscores the effectiveness of such customization, demonstrating how adaptive games for children with learning challenges improved both engagement and learning outcomes.

Collaboration, rather than direct competition, can also play a key role in inclusive VR games. Structured interactions, such as those explored by Aldabas (2022), reveal that incorporating peer-led play in gamified settings can enhance social communication skills among children with autism. These findings suggest that embedding cooperative challenges into games not only builds essential skills but also encourages teamwork and mutual support.

In therapeutic contexts, competitive frameworks tailored to individual capabilities have shown significant benefits. Lee et al. (2020) highlight how stroke patients engaging in competitive exergames experienced notable improvements in motor skills and motivation compared to traditional rehabilitation methods. Such examples illustrate the potential of gamified therapy to combine effective treatment with a sense of achievement.

Ultimately, challenges and competition within gamification must be thoughtfully designed to ensure inclusivity and engagement. By leveraging personalized difficulty levels, fostering collaboration, and offering opportunities for meaningful competition, gamified systems can inspire users to achieve therapeutic, educational, and social milestones, creating a rewarding and empowering experience (Malone et al., 2021).

Virtual Reality Gamification and Mental Health in Disability

Virtual reality games have emerged as a transformative tool for addressing mental health challenges among individuals with disabilities. By integrating gaming elements into therapeutic contexts, VR gamification fosters a dynamic, engaging, and customizable environment that promotes emotional resilience, reduces stress and anxiety, and counters social isolation. For individuals with disabilities, such interventions offer not only a sense of empowerment but also therapeutic benefits tailored to their specific needs and contexts (Cheng & Ebrahimi, 2023; Caponnetto et al., 2021).

The immersive nature of VR allows users to focus on the present moment, alleviating anxiety and enhancing mood through relaxation-oriented games that incorporate calming visuals, music, and guided interactions. Caponnetto et al. (2021) and Weerdmeester et al. (2021) both highlight the efficacy of VR in managing anxiety disorders, particularly through techniques like controlled exposure and biofeedback integration. These features create safe, replicable therapeutic settings where individuals can develop coping mechanisms, making VR a cost-effective alternative to traditional in vivo therapies. However, the application of these interventions must be carefully tailored to account for individual differences and the limitations of virtual environments to maximize their therapeutic potential.

Incorporating biofeedback into VR games enhances therapeutic outcomes further by providing real-time physiological data, such as heart rate and breathing patterns. These tools equip users with self-regulation skills while fostering engagement and motivation (Weerdmeester et al., 2021; Domínguez et al., 2013). More advanced biofeedback measures, including heart rate variability (HRV) and galvanic skin response (GSR), have proven particularly effective in managing anxiety and stress-related conditions. HRV, as highlighted by Premkumar et al. (2024), serves as a key indicator of autonomic nervous system health, with higher variability linked to better emotional regulation and resilience. GSR, which measures skin conductivity as a proxy for arousal, allows users to monitor their stress levels during therapeutic activities, enabling real-time adjustments to emotional triggers in virtual environments (Wiederhold & Wiederhold, 2025). By integrating these physiological markers into VR therapy, users can develop greater awareness of their bodily responses, practice emotional regulation strategies, and gain a heightened sense of control over their mental health journey. Such applications demonstrate the transformative potential of biofeedback in creating personalized, engaging, and effective therapeutic experiences.

VR gamification also fosters social connection, a critical factor in mental health. Adapted multiplayer environments promote meaningful peer interactions, creating a sense of community and belonging. Customizable avatars, intuitive controls, and gamification elements—such as challenges, rewards, and progress tracking—further enhance user engagement and accessibility (Cheng & Ebrahimi, 2023; Domínguez et al., 2013). These features ensure inclusive participation while sustaining long-term commitment to therapeutic activities.

Moreover, the personalization of VR gamified interventions significantly enhances their effectiveness. Tailoring elements such as difficulty levels, feedback mechanisms, and interactive features ensures that interventions meet individual needs. Cheng and Ebrahimi (2023) emphasize that balancing engagement with accessibility is essential to prevent gaming fatigue or diminished real-world functionality. Similarly, Domínguez et al. (2013) highlight the importance of well-designed systems in maintaining relevance and optimizing therapeutic outcomes.

Future Directions for Research and Development of Adapted Games

The potential of adapted games in rehabilitation and education for individuals with disabilities is vast, but several areas require further research and development to maximize their impact. Future studies should prioritize the following directions:

1. *Personalized Adaptation Frameworks*

Developing sophisticated frameworks for personalized game adaptation is essential. These frameworks should leverage real-time data, such as biomarkers and in-game performance metrics, to adjust difficulty, feedback, and engagement strategies dynamically. By tailoring interventions to individual needs, these systems can optimize therapeutic and educational outcomes, ensuring accessibility for diverse user populations.

2. *Integration of Advanced Technologies*

Incorporating advanced technologies, such as artificial intelligence, machine learning, and biofeedback sensors, can enhance the responsiveness and adaptability of games. These technologies can predict user needs, identify areas for improvement, and provide personalized support, making interventions more effective and engaging.

3. *Exploring Long-Term Engagement Strategies*

Sustaining user engagement over time remains a challenge. Research should explore how gamification elements, such as rewards, progress tracking, and challenges, can be designed to maintain interest while avoiding burnout or gaming fatigue. Understanding the psychological and behavioral factors that drive long-term participation will be key to creating enduringly effective interventions.

4. *Collaborative and Social Features*

Social interaction is a critical component of mental health and rehabilitation. Future developments should focus on enhancing collaborative and multiplayer features, enabling users to connect, cooperate, and compete in ways that foster a sense of community and belonging.

5. *Evaluation of Real-World Impact*

While many studies focus on short-term efficacy, there is a need for longitudinal research assessing the long-term impact of adapted games on physical, cognitive, and emotional well-being. Evaluating outcomes across various settings, such as home-based and clinical environments, will provide a comprehensive understanding of their effectiveness.

6. *Accessibility and Inclusivity*

Ensuring that adapted games are accessible to individuals with a wide range of disabilities is paramount. This includes developing intuitive interfaces, customizable controls, and content designed to accommodate specific needs, such as sensory processing disorders or mobility challenges.

7. *Cross-Disciplinary Collaboration*

Cross-disciplinary collaboration is crucial for ensuring that adapted games address the diverse needs of users with disabilities. By combining the expertise of game developers, neuroscientists, and rehabilitation specialists, teams can create interventions that are both engaging and therapeutically effective. Such collaboration integrates scientific insights, clinical practices, and technological innovation, resulting in games that are accessible, impactful, and tailored to individual user needs.

References

Aguilera-Rubio, Á., Alguacil-Diego, I. M., Mallo-López, A., Jardón Huete, A., Oña, E. D., & Cuesta-Gómez, A. (2024). Use of low-cost virtual reality in the treatment of the upper extremity in chronic stroke: A randomized clinical trial. *Journal of Neuroengineering and Rehabilitation, 21*(1), 12. https://doi.org/10.1186/s12984-024-01303-2

Aldabas, R. (2022). Effects of peer network intervention through peer-led play on basic social communication skills of children with autism spectrum disorder in inclusive classrooms. *Journal of Developmental and Physical Disabilities, 34*, 1121–1141. https://doi.org/10.1007/s10882-022-09840-1

Almeida, A. F. S., Silva, T. D. D., Moraes, Í. A. P., Menezes, L. D. C., Dias, E. D., Araújo, L. V., et al. (2024). Virtual reality as a telerehabilitation strategy for people with autism spectrum disorder during the COVID-19 quarantine scenario: Physical activity, motor performance, and enjoyment. *Disability and Rehabilitation: Assistive Technology, 19*(5), 2046–2056. https://doi.org/10.1080/17483107.2023.2249031

Aranha, R. V., Chaim, M. L., Monteiro, C. B., Silva, T. D., Guerreiro, F. A., Silva, W. S., & Nunes, F. L. (2023). EasyAffecta: A framework to develop serious games for virtual rehabilitation with affective adaptation. *Multimedia Tools and Applications, 82*(2), 2303–2328.

Brox, E., Fernandez-Luque, L., & Tøllefsen, T. (2011a). Healthy gaming—Video game design to promote health. *Applied Clinical Informatics, 2*(2), 128–142. https://doi.org/10.4338/ACI-2010-10-R-0060

Brox, E., Luque, F. L., Evertsen, G. J., & Hernández, J. E. G. (2011b). Exergames for elderly: Social exergames to persuade seniors to increase physical activity. In *2011 5th international conference on pervasive computing technologies for healthcare (PervasiveHealth) and workshops* (pp. 546–549). IEEE. https://doi.org/10.4108/icst.pervasivehealth.2011.246049

Cameron S., Mang Kristin L., Campbell Colin J.D., Ross Lara A., Boyd (2013) Promoting Neuroplasticity for Motor Rehabilitation After Stroke: *Considering the Effects of Aerobic Exercise and Genetic Variation on Brain-Derived Neurotrophic Factor Physical Therapy 93(12)*, 1707-1716. https://doi.org/10.2522/ptj.20130053

Caponnetto, P., Triscari, S., Maglia, M., & Quattropani, M. C. (2021). The simulation game—Virtual reality therapy for the treatment of social anxiety disorder: A systematic review.

International Journal of Environmental Research and Public Health, 18(24), 13209. https://doi.org/10.3390/ijerph182413209

Carvalho, C., Teran, L., Mota, M., & Pereira, R. (2024). Technologies to support adaptable game design: A systematic mapping study. *Journal of the Brazilian Computer Society, 30*(1), 90. https://doi.org/10.5753/jbcs.2024.3090

Cascella, M., Cascella, A., Monaco, F., & Shariff, M. N. (2023). Envisioning gamification in anesthesia, pain management, and critical care: Basic principles, integration of artificial intelligence, and simulation strategies. *Journal of Anesthesia, Analgesia and Critical Care, 3*(1), 33. https://doi.org/10.1186/s44158-023-00118-2

Cheng, C., & Ebrahimi, O. V. (2023). A meta-analytic review of gamified interventions in mental health enhancement. *Computers in Human Behavior, 141*, 107621. https://doi.org/10.1016/j.chb.2022.107621

Crocetta, T. B., de Araújo, L. V., Guarnieri, R., Massetti, T. H. I., De Abreu, L. C., & de Mello Monteiro, C. B. (2018). Virtual reality software package for implementing motor learning and rehabilitation experiments. *Virtual Reality, 22*, 199–209.

Domínguez, A., Saenz-de-Navarrete, J., de-Marcos, L., Fernández-Sanz, L., Pagés, C., & Martínez-Herráiz, J. J. (2013). Gamifying learning experiences: Practical implications and outcomes. *Computers & Education, 63*, 380–392. https://doi.org/10.1016/j.compedu.2012.12.020

Drigas, A., & Sideraki, A. (2024). Brain neuroplasticity leveraging virtual reality and brain-computer interface technologies. *Sensors (Basel), 24*(17), 5725. https://doi.org/10.3390/s24175725

Feitosa, J. A., Fernandes, C. A., Casseb, R. F., & Castellano, G. (2022). Effects of virtual reality-based motor rehabilitation: A systematic review of fMRI studies. *Journal of Neural Engineering, 19*(1), 011002. https://doi.org/10.1088/1741-2552/ac456e

Gielis, K., Vanden Abeele, M.-E., Verbert, K., Tournoy, J., De Vos, M., & Vanden Abeele, V. (2021). Detecting mild cognitive impairment via digital biomarkers of cognitive performance found in Klondike Solitaire: A machine-learning study. *Digital Biomarkers, 5*(1), 44–52. https://doi.org/10.1159/000514105

González-González, C. S., Toledo-Delgado, P. A., Muñoz-Cruz, V., & Torres-Carrion, P. V. (2019). Serious games for rehabilitation: Gestural interaction in personalized gamified exercises through a recommender system. *Journal of Biomedical Informatics, 97*, 103266. https://doi.org/10.1016/j.jbi.2019.103266

Hellín, C. J., Calles-Esteban, F., Valledor, A., Gómez, J., Otón-Tortosa, S., & Tayebi, A. (2023). Enhancing student motivation and engagement through a gamified learning environment. *Sustainability, 15*(19), 14119. https://doi.org/10.3390/su151914119

Kamari, M., Siqueira, V., Bakare, J., & Sebastião, E. (2024). Virtual reality technology for physical and cognitive function rehabilitation in people with multiple sclerosis. *Rehabilitation Research and Practice, 2024*, 2020263. https://doi.org/10.1155/2024/2020263

Kira, A., Pontes, R. G., Araújo, L. V., Monteiro, C. B., Uribe-Quevedo, A., & Nunes, F. L. (2024). An approach for automatic adaptation of serious games applied to virtual motor rehabilitation. In *In 2024 IEEE 12th international conference on serious games and applications for health (SeGAH)* (pp. 1–8). IEEE.

Lee, H. S., et al. (2020). Competitive exergaming for stroke rehabilitation: A randomized controlled trial. *Archives of Physical Medicine and Rehabilitation, 101*(5), 749–757.

Mahmoudi, E., Yoo, P. Y., Chandra, A., Cardoso, R., Dos Santos, C. D., Majnemer, A., & Shikako, K. (2024). Gamification in mobile apps for children with disabilities: Scoping review. *JMIR Serious Games, 12*, e49029. https://doi.org/10.2196/49029

Malone, L. A., Davlyatov, G. K., Padalabalanarayanan, S., & Thirumalai, M. (2021). Active video gaming using an adapted gaming mat in youth and adults with physical disabilities: Observational study. *JMIR Serious Games, 9*(3), e30672. https://doi.org/10.2196/30672

Mekler, E. D., Brühlmann, F., Opwis, K., & Tuch, A. N. (2013). Do points, levels and leaderboards harm intrinsic motivation? In *Proceedings of the first international conference on gameful design, research, and applications – Gamification '13* (pp. 66–73). ACM. https://doi.org/10.1145/2583008.2583017

Moraes, I. A. P., Lima, J. A., Silva, N. M., Simcsik, A. O., Silveira, A. C., Menezes, L. D. C., Araújo, L. V., Crocetta, T. B., Voos, M. C., Tonks, J., Silva, T. D., Dawes, H., & Monteiro, C. B. M. (2022). Effect of longitudinal practice in real and virtual environments on motor performance, physical activity, and enjoyment in people with autism spectrum disorder: A prospective randomized crossover controlled trial. *International Journal of Environmental Research and Public Health, 19*(22), 14668. https://doi.org/10.3390/ijerph192214668

Mouatt, B., Smith, A. E., Mellow, M. L., Parfitt, G., Smith, R. T., & Stanton, T. R. (2020). The use of virtual reality to influence motivation, affect, enjoyment, and engagement during exercise: A scoping review. *Frontiers in Virtual Reality, 1*, 564664. https://doi.org/10.3389/frvir.2020.564664

Palidis, D. J., Gardiner, Z., Stephenson, A., Zhang, K., Boruff, J., & Fellows, L. K. (2024). The use of extrinsic performance feedback and reward to enhance upper limb motor behavior and recovery post-stroke: A scoping review. *Neurorehabilitation and Neural Repair, 39*, 157–173. https://doi.org/10.1177/15459683241298262

Papa, D. C. R., Menezes, L. D. C., Moraes, Í. A. P., Silveira, A. C., Padula, N., Silva, S. O. V., Gaspar, R. C., Dias, E. D., Ferreira, C., Araújo, L. V., Astorino, T. A., Dawes, H., Monteiro, C. B. M., & Silva, T. D. D. (2023). Cardiac autonomic modulation in response to postural transition during a virtual reality task in individuals with spinal cord injury: A cross-sectional study. *PLoS One, 18*(4), e0283820. https://doi.org/10.1371/journal.pone.0283820

Park, B., Kim, Y., Park, J., Choi, H., Kim, S. E., Ryu, H., & Seo, K. (2024). Integrating biomarkers from virtual reality and magnetic resonance imaging for the early detection of mild cognitive impairment using a multimodal learning approach: Validation study. *Journal of Medical Internet Research, 26*, e54538. https://doi.org/10.2196/54538

Petsani, D., Konstantinidis, E., Katsouli, A. M., Zilidou, V., Dias, S. B., Hadjileontiadis, L., & Bamidis, P. (2022). Digital biomarkers for well-being through exergame interactions: Exploratory study. *JMIR Serious Games, 10*(3), e34768. https://doi.org/10.2196/34768

Premkumar, P., Heym, N., Myers, J. A. C., Formby, P., Battersby, S., Sumich, A. L., & Brown, D. J. (2024). Augmenting self-guided virtual-reality exposure therapy for social anxiety with biofeedback: A randomised controlled trial. *Frontiers in Psychiatry, 15*, 1467141. https://doi.org/10.3389/fpsyt.2024.1467141

Ravva, P. U., Kullu, P., Abrar, M. F., & Barmaki, R. L. (2024). A machine learning approach for predicting upper limb motion intentions with multimodal data in virtual reality. *arXiv*. https://doi.org/10.48550/arXiv.2405.13023

Sardi, L., Idri, A., & Fernández-Alemán, J. L. (2017). A systematic review of gamification in e-Health. *Journal of Biomedical Informatics, 71*, 31–48. https://doi.org/10.1016/j.jbi.2017.05.011

Silva, T. D. D., Silva, P. L. D., Valenzuela, E. D. J., Dias, E. D., Simcsik, A. O., De Carvalho, M. G., Fontes, A. M. G. G., Alberissi, C. A. D. O., Araújo, L. V. D., Brandão, M. V. D., Dawes, H., & Monteiro, C. B. D. M. (2021). Serious game platform as a possibility for home-based telerehabilitation for individuals with cerebral palsy during COVID-19 quarantine – A cross-sectional pilot study. *Frontiers in Psychology, 12*, 622678. https://doi.org/10.3389/fpsyg.2021.622678

Stanica, I. C., Moldoveanu, F., Portelli, G. P., Dascalu, M. I., Moldoveanu, A., & Ristea, M. G. (2020). Flexible virtual reality system for neurorehabilitation and quality of life improvement. *Sensors, 20*(21), 6045. https://doi.org/10.3390/s20216045

Steiner, B., Elgert, L., Saalfeld, B., & Wolf, K. H. (2020). Gamification in rehabilitation of patients with musculoskeletal diseases of the shoulder: Scoping review. *JMIR Serious Games, 8*(3), e19914. https://doi.org/10.2196/19914

Warraich, Z., & Kleim, J. A. (2010). Neural plasticity: The biological substrate for neurorehabilitation. *PM&R, 2*(12 Suppl 2), S208–S219. https://doi.org/10.1016/j.pmrj.2010.10.016

Weerdmeester, J. W., van Rooij, M. M. J. W., Maciejewski, D. F., Engels, R. C. M. E., & Granic, I. (2021). A randomized controlled trial assessing the efficacy of a virtual reality biofeedback

video game: Anxiety outcomes and appraisal processes. *Technology, Mind, and Behavior, 2*(2), 1–15. https://doi.org/10.1037/tmb0000028
Wiederhold, B. K., & Wiederhold, M. D. (2025). Virtual reality therapy combined with physiological monitoring provides effective treatment, with objective metrics, for post-traumatic stress disorder. *Expert Review of Medical Devices, 22*, 117–119. https://doi.org/10.1080/17434440.2025.2454930
Wiley, K., Robinson, R., & Mandryk, R. (2021). The making and evaluation of digital games used for the assessment of attention: Systematic review. *JMIR Serious Games, 9*(3), e26449. https://doi.org/10.2196/26449
Yildirim, O., & Surer, E. (2021). Developing adaptive serious games for children with specific learning difficulties: A two-phase usability and technology acceptance study. *JMIR Serious Games, 9*(2), e25997. https://games.jmir.org/2021/2/e25997
Zuki, F. S. M., Sulaiman, S., Mérienne, F., Ricca, A., Guillet, C., & Saad, M. N. M. (2024). Assessing the impact of enriched virtual reality on motivation and engagement in stroke rehabilitation. *IEEE Access, 12*, 143585–143598. https://doi.org/10.1109/ACCESS.2024.3461473

Chapter 18
Addressing Gaming-Related Public Health Concerns and Challenges

Nathasha Ribeiro, Christianne Fernandes Valente Takeda, Douglas Amaral, and Sara Barbosa Franco

Abstract The global growth of gaming has reshaped entertainment and social interaction, raising both opportunities and public health concerns. This chapter explores the complex landscape of gaming disorder, from its classification in the ICD-11 and consideration in the DSM-5-TR to the ongoing debates about diagnosis, prevalence, and intervention. While gaming offers cognitive, emotional, and social benefits, excessive use can lead to significant impairments. We examine epidemiological trends, public health challenges, prevention strategies, and emerging treatment approaches, including cognitive-behavioral therapy, pharmacological interventions, and novel neuromodulation techniques. Emphasizing the importance of early identification, education, and cross-sector collaboration, the chapter highlights the need for balanced responses that address gaming-related risks while preserving gaming's positive contributions to modern life.

Keywords Gaming disorder · Public health · Prevention strategies · Treatment approaches · Behavioral addictions

N. Ribeiro
Public Health, University of Guelph, Guelph, Canada

C. F. V. Takeda
Harvard University, Boston, MA, USA

D. Amaral
Mackenzie Presbyterian University, São Paulo, Brazil

S. B. Franco (✉)
Mental Health Department, Santa Casa de São Paulo School of Medical Sciences, São Paulo, Brazil

L. Murrins Marques et al. (eds.), *Social and Affective Neuroscience of Gaming*,
https://doi.org/10.1007/978-3-032-00325-6_18

Introduction

Our lives have become increasingly digital, which has reshaped everything (Dąbrowska et al., 2022), a transformation recognized worldwide mainly after 2007 with the iPhone launched. As part of this shift, gaming has grown significantly, bringing with it both challenges and solutions. It has been reported that in 2022, approximately 3 billion people played online games worldwide. While this statistic does not directly address gaming disorder, it underscores the massive scale of gaming participation globally.

Despite what we might think, there is little consistent evidence regarding gaming when we think about the prevalence and incidence of gaming-related disorders, especially when we need to classify them as mental health illnesses in their own right. This has become a more relevant discussion in the scientific community recently, when gaming disorder began to be classified in the 11th edition of the International Classification of Diseases (ICD-11).

While it is widely believed that excessive gaming is detrimental due to impoverishment of real-life experiences (Marques et al., 2023; Melodia et al., 2022), the sources reveal that there is ongoing debate surrounding the classification of gaming-related disorders as distinct mental health conditions, such as lack of consensus on defining problematic use, overlapping of symptoms with other mental conditions and the last, and maybe the most relevant of public health, possibility of positive impact of gaming mainly due to several sources also acknowledge the potential benefits of video games, including improved cognitive skills and social connections. This adds another layer of complexity to the debate, as it highlights the need to distinguish between healthy and problematic gaming habits.

However, we can not ignore one of the most prominent meta-analyses on the subject extracted data from 53 studies conducted in 17 different countries and observed an adjusted prevalence of gaming disorders of approximately 2% in the general population (Stevens et al., 2021) which could be comparable to the prevalence of obsessive-compulsive disorders and chemical addictions. The prevalence of adolescent gaming disorders, by country, shows that China has the highest rates, around 11.7% (Satapathy et al., 2024), followed by Spain. Nevertheless, the heterogeneity of the sample, as well as the criteria used for assessing gaming disorder, is too variable between countries to ensure these data are reliable.

In Public Health, we understand that the heterogeneity of the epidemiology of gaming-related disorders can create some noise, including transforming some data into diagnoses that, depending on the subpopulation, do not constitute disorders. Later in this chapter, we will look at the most current discussions on interventions and treatments and then understand why the adequate diagnosis of subpopulations is the main step to be taken in prevention policies.

Gaming Disorder Definition and Classification

The recognition of gaming disorder as a mental health condition has gained increasing attention in recent years. The latest edition of the Diagnostic and Statistical Manual of Mental Disorders (DSM-5-TR) includes Internet Gaming Disorder (IGD) in its "Conditions for Further Study" section, alongside other emerging concerns such as caffeine-related disorders. This inclusion reflects the growing acknowledgment of IGD as a significant mental health issue, although further research is needed to solidify diagnostic criteria and treatment approaches. In contrast, the DSM-5-TR categorizes most addictive disorders under substance-related conditions, including alcohol, tobacco, stimulants, cannabis, and opioids, with Gambling Disorder being the only officially recognized behavioral addiction (APA, 2022).

The World Health Organization (WHO) has taken a more definitive stance by incorporating gaming disorder into the 11th Revision of the International Classification of Diseases (ICD-11) (WHO, 2022). According to the ICD-11, gaming disorder is characterized by impaired control over gaming, prioritization of gaming over other activities to the extent that it takes precedence over daily responsibilities and interests, and the continuation or escalation of gaming despite negative consequences (ICD-11). This formal classification by the WHO underscores the global recognition of gaming disorder as a public health concern.

Gaming disorder shares many characteristics with other addiction problems. It manifests through a persistent preoccupation with gaming, where individuals prioritize gaming above other activities and responsibilities. Withdrawal symptoms such as sadness, anxiety, or irritability may emerge when gaming is restricted or unavailable. Tolerance is another symptom, requiring individuals to increase the amounts of time gaming to achieve the same level of satisfaction. Efforts to reduce or quit gaming are often unsuccessful, leading to the neglect of previously enjoyed activities and social relationships.

These behaviors not only disrupt daily life but can also have long-term physical, emotional, and social consequences. By including gaming disorder in diagnostic and classification manuals, organizations like the APA and WHO emphasize the importance of early identification, awareness, and the development of effective prevention and treatment strategies. This growing recognition aims to foster further research and support for individuals and families affected by this emerging mental health challenge.

Public Health Challenges

Gaming disorder presents significant public health challenges, affecting not only individuals but also communities. Despite being aware of its negative consequences, individuals with gaming disorders often continue to engage in excessive gaming.

They may conceal their gaming habits from family members or others and use gaming as a coping mechanism for negative emotions such as guilt, anxiety, or hopelessness. In severe cases, this behavior can menace personal relationships, job security, and essential life responsibilities, leading to long-term emotional and social repercussions.

A major obstacle in addressing gaming disorder is the widespread lack of awareness among the general public and healthcare professionals once as it becomes noticeable that it is a problem only when it is already a big problem, and one of the reasons is the difficulty in identifying individuals at risk and provide timely intervention. Compounding this issue is the challenge of distinguishing between intensive gaming as a recreational activity and pathological gaming, which disrupts daily life and mental well-being. Without clear diagnostic guidelines and increased training for healthcare providers, many cases of gaming disorder may go undetected or mismanaged, while, in general, young people are losing their lives and capacity to contribute to our society.

The limited availability of specialized treatment programs further exacerbates the problem. Most countries lack dedicated resources or facilities to address gaming disorders effectively, leaving individuals and families without access to the support they need. This shortfall highlights the urgent need for governments and healthcare systems to prioritize the development of evidence-based prevention, diagnostic, and treatment strategies.

Additionally, the evolving nature of gaming itself complicates efforts to tackle gaming disorder. The widespread availability of gaming platforms, the integration of addictive game mechanics, and the rise of immersive virtual environments all contribute to the growing prevalence of this condition. To combat these challenges, public health initiatives must focus on promoting awareness, enhancing professional training, and implementing regulatory measures to reduce gaming-related harm while preserving the positive aspects of gaming.

Global and National Responses

Gaming, when engaged in moderation, offers numerous benefits, including enhancing cognitive abilities (Reynaldo et al., 2021), fostering social connections (Sachan et al., 2025), and serving as an innovative therapeutic tool for managing mental health conditions such as anxiety and depression (Townsend et al., 2022). However, these positive aspects are increasingly overshadowed by the growing risks associated with excessive gaming.

Gaming disorder, characterized by addiction, social withdrawal, and negative effects on both physical and mental health, has prompted governments and global organizations to develop targeted strategies for intervention and prevention. National and international initiatives have emphasized the importance of balanced gaming practices. Educational campaigns aim to raise awareness among gamers, families, and communities about the risks of excessive gaming and the signs of gaming disorder. For example, some countries, such as South Korea and China, have implemented regulatory measures like limiting gaming hours for minors and introducing age-based restrictions to mitigate gaming-related harm (Kattula et al., 2021). These policies highlight the need for proactive approaches to safeguard vulnerable populations while preserving the benefits of gaming as a recreational and educational tool.

At the global level, the inclusion of gaming disorder in the World Health Organization's International Classification of Diseases (ICD-11) has been a significant step forward. This recognition has encouraged countries to integrate gaming disorder into their national health agendas, leading to the development of diagnostic guidelines and treatment frameworks. Furthermore, collaborations between governments, healthcare organizations, and gaming companies are exploring innovative ways to minimize risks. These include introducing features like gaming time-tracking tools, promoting healthy gaming habits, and embedding responsible gaming guidelines into game design.

Tailored interventions are another promising area of focus. Integrating gaming into therapeutic contexts has shown potential to enhance its benefits while minimizing harm. For instance, serious games designed for mental health treatment can improve cognitive flexibility, alleviate symptoms of anxiety and depression, and support rehabilitation programs. By leveraging the positive potential of gaming, policymakers can encourage its use as a tool for education, social engagement, and mental health support while simultaneously addressing the challenges posed by excessive gaming (Long et al., 2022).

These combined efforts at the national and global levels underscore the critical need for a balanced approach that maximizes the benefits of gaming while mitigating its risks. Continued research, public awareness, and cross-sector collaboration are essential to developing sustainable solutions that protect individuals and promote healthier gaming practices.

Prevention and Early Identification

Preventing gaming disorders and facilitating early identification require a comprehensive strategy involving education, community engagement, and policy development. Raising public awareness about the symptoms of gaming addiction, such as excessive preoccupation, withdrawal symptoms, and neglect of responsibilities—empowers individuals and families to recognize problematic behavior early (Werner et al., 2024). Schools, healthcare providers, and community organizations can play a vital role in promoting balanced lifestyle choices, encouraging healthy habits like physical activity, face-to-face social interactions, and diverse recreational activities that do not rely solely on gaming.

Regulatory measures are another critical component of prevention. Governments and industry stakeholders can collaborate to implement policies that reduce gaming-related harm, such as limiting exposure to addictive game mechanics, setting time restrictions for minors, as was mentioned before the China case, or ensuring transparency in game design to prevent exploitative practices. Educational campaigns targeting parents and caregivers can equip them with tools to set boundaries and encourage responsible gaming habits.

Early identification also involves training healthcare professionals to recognize the signs of gaming disorders and differentiate them from intensive but non-pathological gaming. This enables timely intervention, which may include counseling, behavioral therapies, or family support programs. Furthermore, incorporating gaming disorder awareness into school curricula and youth programs can help young people understand the risks and develop self-regulation skills.

By combining education, proactive healthcare measures, and thoughtful regulation, societies can address the challenges posed by gaming disorders while maintaining the positive contributions of gaming to entertainment, education, and social connection (Wirkus et al., 2024).

Intervention and Treatment Approaches

Gaming can enhance cognitive skills, improve social interaction, and provide stress relief, especially when used in moderation. However, excessive gaming or addiction can lead to negative outcomes such as social isolation, poor physical health, and impaired mental well-being. The paper also discusses strategies to mitigate these risks, including setting time limits, promoting healthy gaming habits, and integrating educational or therapeutic elements into games to maximize their positive effects (Sharma & Weinstein, 2024).

Cognitive behavioral therapy (CBT) is the most extensively researched psychological intervention for gaming disorder, supported by multiple randomized controlled trials (RCTs). These studies demonstrate CBT's effectiveness in reducing gaming disorder severity and symptoms. Other promising psychological approaches for gaming disorder management include mindfulness-based interventions, relapse prevention strategies, abstinence protocols, and family therapy. While the quality of evidence supporting such interventions is considered moderate, the effectiveness of these methods suggests that a multifaceted approach tailored to individual needs could improve outcomes for individuals struggling with gaming disorder. The specific techniques and therapeutic strategies employed within each approach vary, highlighting the further research to refine these methods and establish best practices for their implementation (Sharma & Weinstein, 2024).

Pharmacological interventions for gaming disorder have shown some promise, especially when it co-occurs with conditions such as anxiety and major depressive disorder (MDD) (Sharma & Weinstein, 2024). Bupropion and escitalopram are two medications that have demonstrated, in conjunction with other treatments, potential benefits in reducing gaming disorder symptoms. For instance, studies have shown that bupropion, when combined with educational interventions, may effectively decrease cravings and reduce gaming time spent when compared to placebo. However, the quality of evidence for pharmacological approaches is rated as low to moderate, highlighting the need for larger, well-designed RCTs to confirm such preliminary findings and to further investigate optimal treatment protocols. Furthermore, the observed benefits of medication are noted to often be temporary, and the treatment's efficacy appears to be dependent on the presence of co-occurring mental health conditions. This suggests a crucial need for further research examining the role of pharmacological interventions within a multimodal treatment framework for gaming disorder.

Beyond established treatments, several emerging approaches hold significant promise in addressing gaming disorders, similar to those being developed for other addiction disorders. These include non-invasive brain stimulation (tDCS) and repetitive transcranial magnetic stimulation (rTMS), as well as electroacupuncture (EA). Studies suggest that tDCS, targeting the dorsolateral prefrontal cortex, may reduce gaming time and improve related symptoms, although research is limited by small sample sizes and a lack of randomized controlled designs. Similarly, rTMS, when applied to the left dorsolateral prefrontal cortex, has shown potential in reducing addictive symptoms and enhancing executive function in a single case study. EA has also demonstrated efficacy in decreasing gaming disorder symptoms and modulating brain activity, potentially offering benefits over traditional CBT in terms of impulsivity control and neural protection. While these emerging modalities show potential, further large-scale, well-designed RCTs are critical to confirm their efficacy.

Functional magnetic resonance (fMRI) studies have provided valuable insights into the neural mechanisms underlying gaming disorder and how different treatments impact brain activity. These studies often examine cue reactivity and craving, analyzing brain responses before and after interventions such as craving behavioral intervention (CBI) or acupuncture. Research using fMRI has shown that CBI can reduce cue-induced craving and modulate activity in brain regions associated with reward processing, attention, and emotional regulation. Similarly, fMRI studies of acupuncture have indicated altered functional connectivity within brain networks involved in reward and habit formation following treatment. Importantly, while these fMRI studies offer valuable insights into the neural mechanisms of gaming disorder and treatment response, many lack rigorous design of randomized controlled trials, limiting the generalizability of their findings and emphasizing the need for further research to strengthen the evidence base.

Conclusion

Gaming disorder represents a complex and evolving public health challenge, influenced by the widespread growth of gaming and its integration into daily life. While gaming offers numerous benefits, including cognitive enhancement and social connectivity, its potential for harm cannot be ignored. The inclusion of gaming disorder in diagnostic manuals such as the ICD-11 and its consideration in the DSM-5-TR reflect the growing global recognition of its impact on mental health. However, significant challenges remain in achieving consensus on diagnostic criteria, addressing heterogeneity in prevalence data, and distinguishing between intensive gaming and pathological behavior. Effective responses to gaming disorder require a multifaceted approach that includes public awareness, tailored interventions, and robust research.

Prevention strategies must emphasize early identification through education and training for families, schools, and healthcare providers. Regulatory measures, such as those implemented in countries like China and South Korea, demonstrate the potential of targeted policies to mitigate gaming-related harms. Additionally, the development of specialized treatment programs and the integration of gaming into therapeutic contexts can provide balanced solutions that minimize risks while leveraging the positive aspects of gaming.

Continued collaboration among governments, healthcare professionals and systems, researchers, and the gaming industry is essential to address the challenges posed by gaming disorders. There are emerging treatments coming and being promising, including non-invasive brain stimulation and pharmacological approaches, which offer promise but require further validation through rigorous studies. By fostering a nuanced understanding of gaming's benefits and risks, public health efforts can better support individuals and families, promote responsible gaming habits, and establish sustainable frameworks to prevent and treat gaming disorders effectively.

References

American Psychiatric Association. (2022). *Diagnostic and statistical manual of mental disorders* (5th ed., Text rev.). American Psychiatric Association.

Dąbrowska, J., Almpanopoulou, A., Brem, A., Chesbrough, H., Cucino, V., Di Minin, A., Giones, F., Hakala, H., Marullo, C., Mention, A.-L., Mortara, L., Nørskov, S., Nylund, P. A., Oddo, C. M., Radziwon, A., & Ritala, P. (2022). Digital transformation, for better or worse: A critical multi-level research agenda. *R&D Management, 52*, 930–954. https://doi.org/10.1111/radm.12531

Kattula, D., Philip, S., Shekhar, A., & Basu, A. (2021). A game of policies and contexts: Restricting gaming by minors. *Lancet Psychiatry, 8*(12), 1035–1036. https://doi.org/10.1016/S2215-0366(21)00427-2. PMID: 34801124.

Long, J., Bhad, R., Potenza, M. N., Orsolini, L., Phan, V., Kanabar, M., & Achab, S. (2022). Public health approaches and policy changes after the inclusion of gaming disorder in ICD-11: Global needs. *BJPsych International, 19*(3), 63–66. https://doi.org/10.1192/bji.2021.57. PMID: 36287819; PMCID: PMC9540648.

Marques, L. M., Uchida, P. M., Aguiar, F. O., Kadri, G., Santos, R. I. M., & Barbosa, S. P. (2023). Escaping through virtual gaming-what is the association with emotional, social, and mental health? A systematic review. *Frontiers in Psychiatry, 14*, 1257685. https://doi.org/10.3389/fpsyt.2023.1257685. PMID: 38025467; PMCID: PMC10663235.

Melodia, F., Canale, N., & Griffiths, M. D. (2022). The role of avoidance coping and escape motives in problematic online gaming: A systematic literature review. *International Journal of Mental Health and Addiction, 20*, 996–1022. https://doi.org/10.1007/s11469-020-00422-w

Reynaldo, C., Christian, R., Hosea, H., & Gunawan, A. A. S. (2021). Using video games to improve capabilities in decision making and cognitive skill: A literature review. *Procedia Computer Science, 179*, 211–221., ISSN 1877-0509. https://doi.org/10.1016/j.procs.2020.12.027

Sachan, T., Chhabra, D., & Abraham, B. (2025). Social capital in online gaming communities: A systematic review examining the role of virtual identities. *Cyberpsychology, Behavior and Social Networking, 28*, 147–161. https://doi.org/10.1089/cyber.2024.0375. Epub ahead of print. PMID: 39772759.

Satapathy, P., et al. (2024). Burden of gaming disorder among adolescents: A systematic review and meta-analysis. *Public Health in Practice, 9*, 100565., ISSN 2666-5352. https://doi.org/10.1016/j.puhip.2024.100565

Sharma, R., & Weinstein, A. M. (2024). Recent treatment and novel imaging studies evaluating treatment of internet gaming disorder: A narrative review. *Frontiers in Psychiatry, 15*, 1408560. https://doi.org/10.3389/fpsyt.2024.1408560. PMID: 38938461; PMCID: PMC11210348.

Stevens, M. W., Dorstyn, D., Delfabbro, P. H., & King, D. L. (2021). Global prevalence of gaming disorder: A systematic review and meta-analysis. *Australian and New Zealand Journal of Psychiatry, 55*(6), 553–568. https://doi.org/10.1177/0004867420962851

Townsend, C., Humpston, C., Rogers, J., et al. (2022). The effectiveness of gaming interventions for depression and anxiety in young people: Systematic review and meta-analysis. *British Journal of Psychiatry Open, 8*(1), e25. https://doi.org/10.1192/bjo.2021.1078. PMID: 34991767; PMCID: PMC8811791.

Werner, M., Kapetanovic, S., Nielsen, M., Gurdal, S., Andersson, M. J., Panican, A., & Claesdotter-Knutsson, E. (2024). When the relationship is at stake: Parents' perception of the relationship with a child with problematic gaming and their perceived need for support. *Healthcare (Basel), 12*(8), 851. https://doi.org/10.3390/healthcare12080851. PMID: 38667613; PMCID: PMC11049846.

WHO. (2022). *International classification of diseases eleventh revision (ICD-11)*. World Health Organization. License: CC BY-ND 3.0 IGO.

Wirkus, T., Czernecka, R., Bühringer, G., & Kräplin, A. (2024). Individual risk factors and prediction of gambling disorder in online sports bettors – The longitudinal RIGAB study. *Frontiers in Psychiatry, 15*, 1320592. https://doi.org/10.3389/fpsyt.2024.1320592. PMID: 38476615; PMCID: PMC10929711.

Correction to: Social Interaction and Connectivity in Gaming Communities

Pedro Makoto Uchida, Lara Poor, Gabriel Kadri, Raphael Santos, Gabriel Rêgo, and Lucas Murrins Marques

Correction to:
L. Murrins Marques et al. (eds.), *Social and Affective Neuroscience of Gaming*, https://doi.org/10.1007/978-3-032-00325-6_12

In Chapter 12, the author's name was incorrectly updated as "Laura Poor", but the correct name of the author is "Lara Poor". The correct name should be as follows:

Pedro Makoto Uchida, Lara Poor, Gabriel Kadri, Raphael Santos, Gabriel Rêgo, and Lucas Murrins Marques

---

The updated version of this chapter can be found at
https://doi.org/10.1007/978-3-032-00325-6_12

L. Murrins Marques et al. (eds.), *Social and Affective Neuroscience of Gaming*,
https://doi.org/10.1007/978-3-032-00325-6_19

Index

L. Murrins Marques et al. (eds.), *Social and Affective Neuroscience of Gaming*, https://doi.org/10.1007/978-3-032-00325-6

MIX
Papier aus verantwortungsvollen Quellen
Paper from responsible sources
FSC® C105338

If you have any concerns about our products,
you can contact us on
ProductSafety@springernature.com

In case Publisher is established outside the EU,
the EU authorized representative is:
Springer Nature Customer Service Center GmbH
Europaplatz 3, 69115 Heidelberg, Germany

Printed by Libri Plureos GmbH
in Hamburg, Germany